POLYMER RHEOLOGY

POLYMER RHEOLOGY

R. S. LENK

*Senior Lecturer in Plastics Technology,
Polytechnic of the South Bank,
London, UK*

APPLIED SCIENCE PUBLISHERS LTD

LONDON

APPLIED SCIENCE PUBLISHERS LTD
RIPPLE ROAD, BARKING, ESSEX, ENGLAND

ISBN: 0 85334 765 4

WITH 17 TABLES AND 214 ILLUSTRATIONS

© APPLIED SCIENCE PUBLISHERS LTD 1978

Printed in Great Britain by Galliard (Printers) Ltd, Great Yarmouth *πα ντα ρηει*

(*Everything Flows*)

HERACLITOS

Foreword

Everything flows, so rheology is a universal science. Even if we set aside claims of such width, there can be no doubt of its importance in polymers. It joins with chemistry in the polymerisation step but polymer engineering is supreme in all the succeeding steps. This is the area concerned with the fabrication of the polymer into articles or components, with their design to meet the needs in service, and with the long and short term performance of the article or component. This is a typical area of professional engineering activity, but one as yet without its proper complement of professional engineers.

An understanding of polymer rheology is the key to effective design and material plus process selection, to efficient fabrication, and to satisfactory service, yet few engineers make adequate use of what is known and understood in polymer rheology. Its importance in the flow processes of fabrication is obvious. Less obvious, but equally important, are the rheological phenomena which determine the in-service performance. There is a gap between the polymer rheologist and the polymer engineer which is damaging to both parties and which contributes to a less than satisfactory use of polymers in our society. It is important that this gap be filled and this book makes an attempt to do so. It presents an outline of what is known in a concise and logical fashion. It does this starting from first principles and with the minimum use of complex mathematics. Nevertheless, the approach is quite rigorous and the book should rightly find its way onto any rheologist's bookshelf. There must be some complexity in describing the effects of the interaction of viscosity and elasticity so subtly present in polymers, but the treatment adopted should not be more difficult than most engineers can normally handle.

The book does not attempt to spoonfeed. Any designer or processor of polymers on reading it will gain a much better understanding of the materials he is handling, but he will rightly be left to make his own translation into the engineering aspects of design and processing. The bridge is there, firmly and well constructed, but the engineer must walk across. There is no assumption that he wishes to be carried over. I therefore commend it especially to all engineers in the polymer field who wish to have a solid base for their day to day activities.

A. A. L. CHALLIS
Director
Polymer Engineering Directorate
Science Research Council
London, UK

Preface

Since 'Plastics Rheology' was published in 1968† important further developments have created a need for a new book to serve the practitioner. The present volume is intended to fill this need whilst simultaneously promoting a fundamental understanding of the behaviour of polymer materials.

The material which was retained from the earlier English and the subsequent expanded and revised German‡ books has been completely reorganised and augmented. There were nine chapters in the first and eleven in the second; this volume has twenty-two. We begin with an introduction which discusses the philosophy of rheology and the special nature of plastics in the spectrum of materials and end with a chapter which attempts to reinforce a conclusion that all properties are ultimately dependent on structural parameters on a supramolecular and molecular level.

Like Caesar's Gaul, 'Polymer Rheology' is divided into three parts, but unlike the historical model the divisions are not such as to isolate one area from the rest. If the rheology of melt processing represents one part, and the mechanical properties of manufactured components a second, then the third part is provided by the consideration in depth of the phenomena which affect flow and deformation, the interpretation of these phenomena on a structural basis and a description of the methods which are used to gain an understanding of the materials; thus the practitioner is assisted in his rational approach to processing and design problems. Portions of the third part are intercalated where the need arises, without disturbing the natural sequence of subject matter. Serving as a theoretical cement this 'third part' acts as a leaven in what might otherwise be a diet of strictly *ad hoc* problems.

In preparing this book I had the pleasure and privilege of receiving contributions from a number of distinguished colleagues who very kindly agreed to cover the subjects appropriate to their special expertise, namely

† 'Plastics Rheology', R. S. Lenk, Applied Science Publishers, London, 1968.
‡ 'Rheologie der Kunststoffe', R. S. Lenk, Carl Hanser Verlag, Munich, 1972.

Neil Cogswell (Chapter 12), Ian Barrie (Chapter 13), Juergen Pohrt (Chapters 17, 18 and 19) and Frigyes Thamm (Chapter 21).

It is good to see Euro-cooperation which brings research workers in industry and academe together in a joint enterprise. To the guest writers, their companies and institutions, thanks.

As regards the inevitable problem of mathematics: derivations are given only when they serve to illustrate a method, are needed to stimulate a quantitative approach, or when an essential relationship is to be established which the enquiring mind would prefer to emerge from basic principles rather than out of limbo. With the one possible exception of Chapter 5, the treatment does not transcend the standard techniques with which every scientist and technologist is familiar.

The principal problem was that of reconciling simplicity of presentation and clarity of the emerging concepts with the practicalities of process and component design based upon the scientific analysis of the parameters which determine polymer behaviour. The book must be judged by the extent to which this problem has been solved.

R. S. LENK
Polytechnic of the South Bank
London, UK

Contents

Introduction

Rheology has been defined as 'a branch of physics which concerns itself with the mechanism of deformable bodies'. Deformation is a phenomenon which is of necessity associated with volume elements. The pioneers of modern rheology concerned themselves principally with the *bulk mechanical* manifestations of the deformational effects of stresses applied to bodies in the liquid state. However, it has become increasingly apparent that rheology cannot be encompassed within such arbitrary confines for the following reasons:

1. The definition of the 'liquid state' is rather arbitrary. A liquid, ideally, is a body which deforms irreversibly as a result of flow. But we know that bodies such as metals—which are indisputably solids—do flow and so deform irreversibly if a force of sufficient magnitude is allowed to act upon them for a sufficient length of time. Flow in solids is known as 'creep'. Structural engineers are well aware of the problems which creep presents. They attempt to cope by designing their structures in such a way as to limit the freedom of relative displacement of the material constituents. They endeavour to 'rigidify' a structure by locking the volume elements so that the design load is insufficient to cause a significant change in the existing spatial arrangement.

The fact that under certain conditions such a change *is* possible implies that at ordinary temperatures there still exists an irreversible flow potential even in such materials as metals. Since the energy required to produce such flow is irrecoverable it must be dissipated as heat. It is true, of course, that an overwhelming proportion of the energy used to deform a piece of metal is instantaneously recovered upon removal of the stress, but it is equally true that some small (and sometimes significant) amount of energy is lost in creep or irreversible flow. The actual percentage of this energy loss depends upon the case with which the constituent parts of the stressed body can be made to alter their positions in space relative to one another, to slip over one another, to flow.

The resistance of a material to irreversible positional change of its constituent volume elements and the concomitant conversion of mechanical energy into heat is denoted by a parameter known as the 'viscosity'. The term viscosity immediately conjures up the concept of a liquid rather than a solid. In order that metals could continue to be regarded as solids, the viscosity of solids is often referred to as the 'internal' or 'frictional' viscosity, as if it were a special kind of viscosity. This devious expertise is both misleading and unnecessary. There is no need for a *qualitative* dividing line between solids and liquids, although the transition from one to the other usually involves a *quantitative* viscosity change which may cover many decades within possibly very close limits of changes in environmental conditions.

The same reasoning can be applied to the gas/liquid transition, since a gas is merely a fluid of particularly low viscosity when compared with a liquid. It is therefore logical to regard the gaseous, liquid and solid states as special aspects of one generalised continuous and universal *fluid state*, with the primary transitions occurring fairly sharply in some materials and less sharply in others. The unifying principle of the universal fluid state lies precisely in the fact that the principal attribute of the liquid state, namely viscosity, exists (obviously) in the gaseous state and (less obviously, but nevertheless demonstrably) in the solid state.

2.   By inverting the argument, it is also easily seen that materials which are indisputably liquids do not necessarily dissipate *all* the deformational energy. Some of the energy is recoverable and since this is so the liquid has some of the principal attribute of the solid state, namely *elasticity*. Of course, in a typical liquid the magnitude of the viscous response mechanism to an applied stress may be overwhelmingly greater than any manifestation of reversible (recoverable, elastic) deformation.

This can be demonstrated when a high speed ciné film of the impact of a drop of water on a glass plate is examined frame by frame. It is then seen that the drop actually bounces like a ball and returns stored energy (*a*) by rebounding, and (*b*) by recovering its spherical shape after impact instead of maintaining the squat deformed shape which the contact with the glass surface has momentarily imparted to it. We are not immediately concerned with the nature of the internal or interfacial forces which manifest themselves by forcing an elastic response. Suffice it to say that liquids have not only viscosity but possess some of the main attributes of solid state behaviour, in the same way as solids are not only elastic but possess *some* of the main attributes of the liquid state.

The best way to describe real materials is to regard them as *viscoelastic*.

Some people use two terms (viscoelastic and elasticoviscous) in order to emphasise the predominance of the viscous or elastic response respectively, but the writer feels that this distinction is somewhat pedantic.

3. Rheology should not be restricted to bulk mechanical deformation. Since volume elements are involved in *all* deformations, whatever the force field, there is no reason why the mathematical equations developed to describe mechanical deformations should not equally apply to deformations induced by an electrical, magnetic or any other force field. Volume elements will be different in each case, to be sure; in a plastic under tension or in melt flow, viscosity manifests itself by the spatial rearrangement of polymer chain segments which constitute quite large volume elements. An electric field acting on polar plastics will cause energy dissipation as frictional heat when the dipoles do work against their environment in their endeavour to conform to the polarity of the applied field. The volume element is smaller, but the same argument applies.

Again, excitation by light rays (visible or otherwise) will cause deformations, some of which will be irreversible. In the infra-red the volume elements involved are specific atomic bond conformations and the energy dissipation is implicit in the frictional heat generated when these bonds are partially constrained in their vibrational and rotational evolutions. The volume elements are perfectly real, although they require a highly specific mode of excitation before they can manifest themselves.

Similar arguments can be applied to nuclear magnetic and electron-spin resonance where the volume elements are subatomic but no less real for that. Hence rheology is not just 'a branch of physics'. It is far more than that. It is the key to the understanding of the behaviour of materials when subjected to *any* kind of force field. Indeed, far from being 'a branch' of any science, rheology emerges as *the* central science with such branches as chemistry, physics, engineering or biology.

Having considered the scope of rheology in general, let us now look at the rheology of polymers in particular. What has made polymer rheology particularly fascinating is the fact that plastics exhibit a spectrum of deformational responses to stresses which concerns itself in particular with the border region between the 'solid' and 'liquid' states and uses this platform as a springboard for the further penetration of the more typical regions of these two states. In addition, industry has discovered the fundamental importance of studying plastics by methods which had to be specially developed, since the methods which were acceptable for traditional materials proved to be totally inadequate if a full understanding of the formulation, tailoring, compounding, processing, design and functional performance of plastics is to be gained. What is more, this realisation has caused the polymer chemist, physicist, technologist and engineer to concern themselves with rheology. It has enabled them to enhance their own efficiency and with it the research, development, production and applications performance of their industry.

Rheology has made important contributions to advances in food technology, medicine, paint and printing-ink technology, building and structural engineering, adhesives, cosmetics, oilwell drilling operations, and elsewhere. But it cannot be denied that it is due to its tremendous scope in the polymer field that rheology has now been recognised as a major scientific discipline affording (i) deep fundamental insights, and (ii) immediate practical rewards when the new understanding gained is applied to process and product design.

1

The Characterisation of Viscous Flow. Viscosity, Shear Rate and Shear Stress

When considering flow characterisation in what are commonly regarded as liquids it becomes necessary to examine a number of variables. The overall characterisation is attained when all the relationships—including the interdependent cross-relationships—of an all-embracing rheological equation have been evaluated. The most general rheological equation is

$$\eta = F(\dot{\gamma}, T, t, P, c, \ldots)$$

where η = coefficient of viscosity, $\dot{\gamma}$ = shear rate (itself a function of the shear stress), T = temperature, t = time, P = pressure (itself a function of volume), c = concentration, and where the multiple dots which follow include, for example, molecular parameters such as molecular weight (MW), molecular weight distribution (MWD), compositional variables such as crystallinity, the presence of additives (plasticisers, fillers, slip agents, mould release agents, pigments and dyes, stabilisers, products of decomposition and other impurities) and factors which relate to the processing history (orientation, residual stresses, etc.). Clearly such an equation is unrealistic and we shall therefore consider each of the principal variables in turn, assuming that the others remain constant.

This chapter deals with the shear rate and time dependence of viscosity. Subsequent chapters will deal with other factors that influence viscosity. The simplest relationship between viscosity and shear rate is implicit in the definition of viscosity itself. The classical (Newtonian) definition is

$$\eta = \frac{\tau}{\dot{\gamma}}$$

implying that viscosity is a kind of modulus by analogy with Hooke's law

$$E = \frac{\sigma}{\varepsilon}$$

1

where E is Young's modulus (in tension), σ is the (tensile) stress and ε is the strain, even though η and E differ dimensionally. In a liquid which obeys Newton's law, τ and $\dot{\gamma}$ are linearly related.

Flow characterisation involves the measurement of τ at various arbitrary values of $\dot{\gamma}$ (or vice versa) within the region of laminar flow. In a Newtonian liquid the entire experimentally observable range of the $\tau/\dot{\gamma}$ relationship is linear and passes through the origin of a direct plot, so that a single point on the $\tau/\dot{\gamma}$ plot completely describes the behaviour. The viscosity is the slope of the straight line. In a double logarithmic plot a linear function is also obtained, but its slope is unity.

The shear stress τ has the dimensions of force per unit area. The shear rate is defined by

$$\dot{\gamma} = \frac{\mathrm{d}v_i}{\mathrm{d}j}$$

where v = velocity, i denotes a length along the flow direction and j denotes a length alongside i on one of the two principal planes parallel to the i-axis. It is seen that $\dot{\gamma}$ has the dimensions of reciprocal time.

It is worth noting that, in a Hookean solid, strain is defined as the ratio of two lengths and is therefore dimensionless whether it is expressed as the conventional strain, $\varepsilon = l/l_0$, or the 'engineering strain', $\varepsilon = \Delta l/l_0$, where l = final length, l_0 = original length, and Δl = elongation and $l = l_0 + \Delta l$.

Hence, the tensile (Young's) modulus of a Hookean solid has the dimensions of stress, whilst the viscosity of a Newtonian liquid has the dimensions of stress per unit time.

Whilst one can measure deformation in a solid, one cannot normally do this in a liquid; but one *can* determine the deformation *rate* (the shear rate) caused by an applied shear stress or vice versa. In practice, Newtonian behaviour is confined to low molecular weight liquids. Polymer melts obey Newton's law only at shear rates close to zero and polymer solutions only at concentrations close to zero. Before considering the rheology of polymer melts let us describe the recognised phenotypes of rheological behaviour based upon $\tau/\dot{\gamma}$ curves (Fig. 1.1).

These phenotypes are named Newtonian, pseudoplastic, dilatant, plastic, Bingham and Ostwald and we shall discuss them in detail presently. Suffice it to say at this stage that in pseudoplastic and dilatant liquids the viscosity is no longer constant. In the former it decreases and in the latter it increases with increasing shear rate; that is to say, the shear stress increases with increasing shear rate less than proportionately in a pseudoplastic and

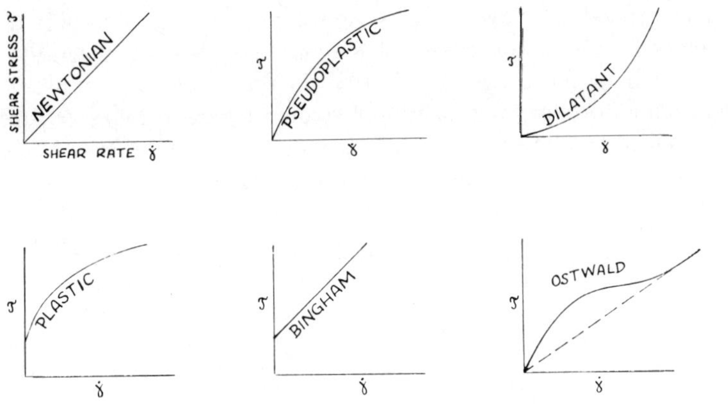

Fig. 1.1. Flow phenotypes.

more than proportionately in a dilatant. These two flow phenotypes can be described by an equation

$$\tau = \eta_N \dot{\gamma}^n$$

(the 'Power Law'), where η_N is the zero shear (Newtonian) viscosity. The exponent n is greater than unity for a dilatant and less than unity for a pseudoplastic. A Newtonian is then seen to be a special case, with $n = 1$. A Bingham body would be described by the equation

$$\tau - \tau_y = \eta \dot{\gamma}$$

where τ_y is the 'yield stress' or 'yield value'. Below τ_y the material will not flow at all, hence $\dot{\gamma} = 0$ and $\eta = \infty$. However, as soon as τ exceeds τ_y the material—to all intents and purposes a solid at $\tau < \tau_y$—suddenly behaves like a liquid with a viscosity which remains constant with increasing shear rate.

Materials which exhibit this type of behaviour include drilling muds, sewage sludge, toothpaste, greases and fats, as well as the clay slurries originally observed by Bingham. The 'plastic' differs from a Bingham body in that it has a curvilinear transition instead of a sharp knee, whilst the existence of the Ostwald type will presently be seen to be exceedingly doubtful. Lenk[1] has shown that all flow phenotypes (except Ostwald) form part of a general response pattern which may be summarised in a general flow curve. This curve can be derived intuitively from the structural changes which may be assumed to occur in laminar flow with increasing shear rate.

The generalised flow curve is shown in Fig. 1.2, alongside a fully developed stress–strain curve for a typical tough solid in tension after conversion of the conventional stress in the latter to true stress. This conversion (force per *original* cross-sectional area to force per *actual* cross-sectional area) can be effected if a continuous record of the changes in cross-sectional area of the specimen under test is kept.

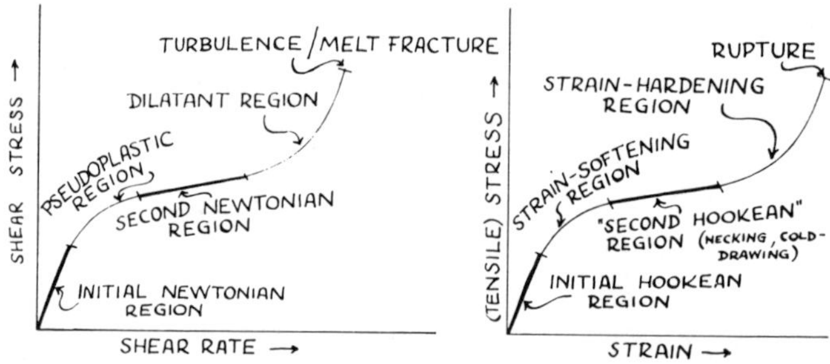

FIG. 1.2. *Left:* Generalised flow curve; *Right:* Typical fully developed stress–strain curve as found in tough plastics under appropriate conditions. The conventional stress has been converted to the true stress.

It is seen that the shapes of the two curves are absolutely identical and that the regions into which they divide have analogous physical significance.

It is now necessary to see how the different established flow types fit into the concept of a generalised flow curve.

The *Newtonian type* represents the behaviour of any liquid which is subjected to a shear rate between zero and some finite value beyond which the deviations from linearity cannot be experimentally demonstrated. This does not mean that the deviations from linearity do not exist, it merely means that turbulent flow conditions occur before deviation from Newtonian behaviour has a chance to manifest itself. Indeed, it is an essential postulate of the generalised flow theory that deviation from Newtonian behaviour would become manifest if non-turbulent flow conditions could be generated at shear rates which are high enough.

The *pseudoplastic type* represents the behaviour of any liquid which is subjected to a shear rate from zero upwards and which shows a convex curvature with respect to the shear rate axis.

It is well known that at extremely low shear rates the slope of the curve is constant and that there exists some very low but finite and sometimes experimentally determinable threshold shear rate beyond which deviation from linearity commences. The slope of the linear portion of the curve is known as the 'limiting viscosity', the 'zero shear viscosity' or the 'Newtonian viscosity'. It is best to call it a first, or initial, Newtonian viscosity. Beyond the low shear rate region characterised by the initial Newtonian regime the material is shear-softened, a phenomenon which has its precise counterpart in the solid state where it is known as strain softening.

Continuing into the pseudoplastic portion proper it has frequently been seen that an upper threshold can be reached beyond which no further reduction in viscosity occurs. The curve then enters a second linear region of proportionality, the slope of which is the second Newtonian viscosity. This has sometimes been called the viscosity at infinite shear rate, but it will presently become obvious that such a term is incorrect.

Pseudoplastic systems are common in polymer melt rheology, and the existence of first and second Newtonian regions has long been recognised. It need hardly be said that the second Newtonian region can only be experimentally demonstrated if non-turbulent flow conditions can be maintained beyond the upper threshold of pseudoplastic behaviour.

The *dilatant type* is less common among plastics under ordinary conditions, but it can be found in heavily filled (crowded) systems and in some PVC pastes. The dilatant type represents the entire generalised flow curve in which the initial Newtonian and pseudoplastic regions have, however, degenerated to a vanishingly small portion of the curve as a whole. The initial Newtonian and pseudoplastic regions therefore tend to coincide with the origin and are not experimentally demonstrable. In some cases of dilatancy a linear region may be distinguished before the curvature (concave with respect to the shear rate axis) appears. In other cases no distinctly linear portion can be seen at the low shear rate end. It is reasonable to argue that a linear low shear rate region in a dilatant represents the second Newtonian viscosity, and that in the *absence* of a linear low shear rate region the second Newtonian region has also become vanishingly small, just like the first Newtonian and pseudoplastic regions which preceded it.

It is by no means certain that a '*plastic*' *type* does in fact exist. But if it does, then the generalised flow curve can readily account for it on the assumption that the initial Newtonian viscosity coincides with the shear stress coordinate. This is equivalent to saying that the liquid is characterised by an initial Newtonian viscosity of infinity, since the low shear portion of the curve coincides with the shear stress coordinate up to a point where a certain 'yield stress' is reached. Beyond this point the same arguments hold which were used in the pseudoplastic type.

The *Bingham type* is not common among plastics, but it has been observed in a number of crowded disperse systems, notably those involving clay and other mineral slurries containing a high volume fraction of disperse phase. The Bingham flow curve is readily derived from the generalised flow curve if it is assumed, as in the plastic type described before, that the initial Newtonian viscosity is infinite, and by postulating additionally that the pseudoplastic region has degenerated to vanishing point so that the curvature characteristic for the pseudoplastic type disappears. Again, we have a 'yield stress' which characterises the departure from the initial Newtonian viscosity which in this case, as in the degeneracy of the intervening pseudoplastic region, follows the initial Newtonian region directly and which denotes the start of the second Newtonian region.

Ostwald[2] has reported a flow type which was stated to be pseudoplastic, but in which it is further claimed that a true Newtonian region followed the pseudoplastic region and that the linear course of that part of the curve extrapolated to the origin. It is believed that extrapolation to the origin is fortuitous and insignificant and that the linearity is itself spurious. If there *is* a second Newtonian region in the Ostwald curve it is degenerate and is represented by the point of inflexion of the plot. The region which Ostwald claims to be Newtonian at high shear rates is in fact the dilatant region which starts with the curvature towards the shear rate axis and which then flattens out into a quasi-linear course which is asymptotic to the dilatancy parabola.

Before the 'generalised flow curve' can be regarded as a 'generalised flow theory' it is necessary (i) to construct an acceptable model which can explain the changes in flow regime (i.e. transitions from one part of the curve to another) on a molecular basis, and (ii) to produce evidence of the existence at unusually high shear rates of a second Newtonian and perhaps even a dilatant region in materials which behave as pseudoplastics under 'normal' shear rates.

(i) It is suggested that a polymer melt at rest exists in a 'random ground state'. This implies that Brownian movement and random entanglement of chain sequences have been attained after a long rest period during which any prior strain-induced memory effects have become obliterated.

If vanishingly small shear rates are applied to a melt in this condition, then the random ground state will remain undisturbed and the slope of the flow curve will remain constant so long as the shear rate increase fails to affect the overall molecular configuration. The random ground state thus coincides with the region of initial Newtonian linearity.

The random ground state structure is broken down by the application of a shear field. It is accompanied by the build-up of a shear-oriented structure and a *decrease* in the entropy of the system. The process begins at a threshold shear rate at which the random ground state is beginning to be disturbed and the molecules tend to arrange themselves increasingly in such a way as to present minimum resistance to flow.

In the case of polymer melts this will favour alignment along the flow lines created by the shear field, a process which commences at very low shear rates; with simple liquids very high shear rates may be necessary to secure an end-on alignment of the small molecules and since at such high shear rates turbulence is almost invariably encountered, regions beyond the first Newtonian are not experimentally accessible.

In high molecular weight materials, on the other hand, the least disturbance will tend to cause alignment so that the ground state structure is virtually confined to the rest condition. The moment at which the random ground state structure begins to break down and a shear-oriented structure begins to build up in its stead is the point of transition from the initial Newtonian to the pseudoplastic condition. The discontinuity in flow regime is clearly seen in the log–log plot provided by the power law. In this presentation the power index changes abruptly from a value of unity to one which is less than unity.

If random ground state structure is progressively broken down and a shear-oriented structure is created in its stead which offers reduced resistance to the force field, it is clear that this process must terminate when it is complete. If this process characterises the pseudoplastic region, then the moment at which all ground state structure has been sheared out and the maximum possible alignment along the flow lines has been attained is characteristic for the inception of the second Newtonian region. On a molecular level this means that further increases in shear rate will only bring about proportional increases in shear stress because the molecular chains, having aligned themselves to the maximum degree, have no means of further conforming to the shear field. We have thus reached a limiting viscosity which is often (but incorrectly!) referred to as the viscosity at infinite shear rate. Linearity in the curve is reattained, but the slope can obviously not be extrapolated to the origin. This is the second Newtonian region. In the log–log plot which represents the power law, the transition from the pseudoplastic to the second Newtonian regime is characterised by a return to unity of the power index. Although the log–log slopes of the first and second Newtonian regions are both the same (unity), the second Newtonian region differs from the first in that the line cannot pass through the origin but must give a positive intercept with the shear rate coordinate on extrapolation.

When very high shear rates are operating on a polymer melt the chain segments of the polymer molecules, aligned along the stream lines to the greatest extent possible, will begin to exert a frictional force on their neighbours. This will modify the flow pattern which will, however, not necessarily become turbulent. Secondary flow patterns may be created which will alter the dimensions and the shape of the flow units. The new shapes are pictured as micro whorls or flow nuclei which will tend to flow in 'balled-up' fashion and so present more resistance to flow than the aligned segments did in the second Newtonian region which preceded the new state of the system. When the balled-up flow nuclei attain large dimensions (as they may well do in polymer melts), flow may become altogether impossible. The flow curve will then become asymptotic to the shear stress axis. On the other hand, the balled-up flow nuclei may attain a limiting diameter. If flow is then still possible, the curve will tend to become asymptotic to the straightening extrapolated curvature of the dilatancy parabola. In the log–log plot of the power law the transition from the second Newtonian region to the dilatant regime is characterised by a discontinuous change in the slope of the power index from unity to some value in excess of unity.

(ii) The existence of a first and second Newtonian region has long been recognised. The reader is referred to the work of Wright and Crouse[3] who used the two Newtonian regions for flow reference. Galt and Maxwell[4] studied the flow of polyethylene melts by direct visual and photographic observation, using a particle tracer technique as Clegg[5] had done earlier. They clearly demonstrated the existence of shear discontinuities at the wall of the capillary rheometer which they used. The zones where the shear discontinuities occurred could extend to as much as one-sixth of a channel radius towards the axis.   More importantly, the boundary between the 'wall zone' and the cylindrical zone which it surrounds showed an abrupt transition in the velocity profile and hence in the viscosity profile. Moreover, the transition became more pronounced and extended more deeply towards the axis the more the severity of the shear conditions was stepped up. Galt and Maxwell attributed the annular region at the wall to the 'highly elastic behaviour of polymer melts that do not flow continuously. The highly elastic melt elements are created by long chain branching which causes the generation of molecular clusters'. In other words, Galt and Maxwell detected dilatancy in polyethylene melts which have always been regarded as exemplary models of pseudoplastic behaviour.

Moreover, they detected this precisely where it is expected to occur (if at all), namely in the wall region of the die channel where the shear conditions are most severe. The fact that they were actually able to observe this is not only due to the tracer technique but also to the experimental geometry of the system which constrained the highly pressurised flow while at the same time preventing it from becoming turbulent.

Finally, van der Vegt and Smit[6] studied polymer melts in capillary rheometers, specifically polypropylene, polyethylene, *cis*-1,4-poly-butadiene, and *cis*-1,4-polyisoprene. They were thus dealing with typical plastics as well as unvulcanised elastomers. The melts were subjected to varying shearing conditions. Van der Vegt and Smit observed that at high shear rate conditions the apparent melt viscosity increased sharply with shear rate and that this viscosity increase may be so great as to completely inhibit flow. The same effect was also observed by others[7] on PVC and polystyrene. Van der Vegt and Smit suggested that the observed viscosity increase is due to shear-induced orientation crystallisation in the melt, and the most striking feature of this concept is that this manifestation of crystallinity is possible in polymers which are normally considered amorphous as well as in those which possess substantial crystallinity at submelt temperatures. The assumption of melt crystallinity was confirmed by the X-ray diffraction patterns of the contents of the rheometer capillaries after quenching. It was clearly seen that a highly oriented crystalline structure is present during flow at high shear rates, while at lower shear rates the crystallisation is random and isotropic and at even lower rates it is altogether absent. Since crystallisation in the melt is comparable to orientation crystallisation in highly stretched rubber vulcanisates we incidentally obtained a further significant feature of correspondence between the 'liquid' and 'solid' states. The effect of shear-rate-induced orientation crystallisation in the melt cannot be connected with the pressures necessary to produce these high shear rates since these pressures, as van der Vegt and Smit point out, are quite moderate—of the order of $100\,\mathrm{kg\,cm^{-2}}$ at $165\,°C$ for polypropylene. Far greater pressures are required to cause genuine pressure-induced viscosity increases of the Bridgman type.

Galt and Maxwell's 'molecular clusters' and van der Vegt and Smit's 'crystallites' in the melt are obviously identical, and they represent, in fact, the balled-up flow nuclei, the existence of which was postulated as an essential requirement for a satisfactory molecular interpretation of the generalised flow curve.

The power law is undoubtedly simple and convenient. It has proved to be satisfactory for the characterisation of the flow behaviour of polymer melts and solutions. However, there arises a peculiar and somewhat disturbing problem; shear stress and shear rate have fixed dimensions—yet a change in dimensions seems to be implied in the power law equation for all values of n other than unity, unless it is assumed that n itself has compensating dimensions. This particular nettle was grasped by Scott Blair.[8]

We consider a viscous liquid with random ground state structure and assume that the number of bonds that contribute to this structure (S) is linearly related to $\log \tau$. This simple assumption is only inapplicable at infinitely low shear stress when $dS/d\tau$ becomes $-\infty$ at $\tau = 0$. Once the process of structural breakdown has started following the traverse of the first Newtonian region, the rate of further breakdown would decrease exponentially because of the progressive reduction of the number of remaining bonds which maintain the diminishing ground state structure, so that

$$\tau = k \exp(-aS)$$

where a and k are constants. Thus $\ln \tau = k' - aS$, or $-dS/d\tau = a/\tau$. If the shear rate $\dot{\gamma}$ is kept constant or if it is increased, then any potential structure producing volume elements are prevented from remaining in contact sufficiently long to enable the restoring forces to effect some restructuring and to produce an increase in S. Equally, it is simplest to assume that the number of 'structural' bonds which remain intact despite the existence of the shear field is a linear function of $\log \dot{\gamma}$, so that we can write, analogously, $-dS/d\dot{\gamma} = b/\dot{\gamma}$, b being another constant. Thus S—the number of structural bonds per unit volume—is proportional to both $\log \tau$ and $\log \dot{\gamma}$. Dividing the two previous equations

$$d\dot{\gamma}/d\tau = a\dot{\gamma}/b\tau \qquad \text{or} \qquad d\dot{\gamma}/\dot{\gamma} = a\,d\tau/b\tau$$

and integrating yields

$$\ln \dot{\gamma} = (a/b)\ln \tau + c' \qquad \text{or} \qquad \tau = c\dot{\gamma}^{b/a}$$

where the ratio b/a is obviously identical with n of the power law. (Note: In a Newtonian $a \equiv b$, in a pseudoplastic $a > b$, in a dilatant $a < b$.)

Scott Blair drew attention to the consideration of the dimensional problem by Prentice[9] who in turn referred to the work of Nedonchelle and Schütz.[10] It was suggested that any given system has an innate relationship which connects the proportionality constant k and the power law exponent n, since the equation can be reduced to dimensionless form containing three dimensionally stable constants. These constants define mutually independent properties which the material possesses under the given conditions.

As has been seen, Scott Blair showed that $n = b/a$. If the power law is written

$$\dot{\gamma}/\dot{\gamma}_0 = (\tau/\tau_0)^{1/n} = (\tau/\tau_0)^{a/b}$$

then

$$(\dot{\gamma}/\dot{\gamma}_0)^b = (\tau/\tau_0)^a$$

and on taking logarithms one obtains

$$a \ln \tau - b \ln \dot{\gamma} = a \ln \tau_0 - b \ln \dot{\gamma}_0$$

Comparing this with the earlier logarithmic form which, after multiplying with b becomes $a \ln \tau - b \ln \dot{\gamma} = -bc$, it is seen that the two equations are identical *provided* $(-bc)$ represents the difference $a \ln \tau_0 - b \ln \dot{\gamma}_0$; i.e. *provided* these two magnitudes represent the two independent integration constants of Scott Blair which in turn refer to the breakdown and build-up of 'structure'. If τ_0 is determined at some value $\dot{\gamma}_0$ which is low enough to be still in the first Newtonian region, and if we know one of the integration constants, then we need only follow Scott Blair's assumption that a, b and c change dimensions in the same way as the power exponent n to make the power law dimensionally acceptable. Following the serious aspersions against the power law on the grounds of inconstancy, this law has now been rehabilitated with a little goodwill on the part of a slightly sympathetically biased tribunal.

Attention is drawn to a treatment which exemplifies a fairly simple linear viscoelastic approach. Tobolsky and Chapoy[11] developed an equation which links viscosity and shear rate on the basis of linear viscoelasticity in which a tensile as well as a shear viscosity are implicit (this will receive further attention later on). The equation was applied to thermoplastics well above the glass transition temperature, such as pseudoplastic polymer melts under normal processing conditions.

It can be shown that the time-dependent tensile stress in a viscoelastic Maxwell–Wiechert body is given by

$$\sigma(t) = \dot{\gamma} \sum_i \tau_i G_i [1 - \exp(t/\tau_i)]$$

which, according to Hopkins, can also be written

$$\sigma(t) = \dot{\gamma} \int_0^t G_R(t)\, dt$$

where $G_R(t)$ is the relaxation modulus in tension. Applying this to a polymer melt, one reaches the experimental limit when the specimen breaks or yields; this occurs after a time t_B when the modulus has reached its maximum value. The corresponding deformation is given by

$$\gamma_B = \dot{\gamma} t_B$$

If the experiment is carried out at a low elongation rate, t_B is very much greater than the longest relaxation time which could apply to even the most tardily relaxing volume element. The specimen yields and achieves steady-state flow, the time required to do so being t_B. The stress under the steady-state conditions is given by

$$\sigma_{SS} = \dot{\gamma} \sum_i \tau_i G_I = \dot{\gamma} \int_0^{t=\infty} G_R(t)\,dt$$

and we obtain the steady-state viscosity η_{SS} (in tension) on dividing by $\dot{\gamma}$:

$$\eta_{SS} = \frac{\sigma_{SS}}{\dot{\gamma}} = \sum_i \tau_i G_i = \int_0^\infty G_R(t)\,dt$$

However, if the experiment is carried out at higher extension rates, then the sample will either break, or the stress–strain curve becomes non-linear and reaches an experimental maximum at t_B. This experimental maximum will be *below* the theoretical maximum (Fig. 1.3).

This is explained by the fact that t_B is now too short to enable the slowest of the volume elements to relax fully. The (tensile) viscosity $\eta(t_B)$ then becomes

$$\eta(t_B) = \frac{\sigma(t)}{\dot{\gamma}} = \sum_i \tau_i G_i(1 - \exp(-t_B/\tau_i)) = \int_0^{t=\gamma_B/\dot{\gamma}} G_R(t)\,dt$$

for a given extension rate $\dot{\gamma}$.

Tobolsky and Chapoy considered that a parallel situation exists when the melt viscosity is determined on rotational or extrusion rheometers, although the strain is in shear rather than in tension. The analogy between $\eta(t_B)$ and $\eta(\dot{\gamma})$ is implicit in the relationship

$$\eta(\dot{\gamma}) = \int_0^{t_B} G_R(t)\,dt = \int_0^{\gamma_B/\dot{\gamma}} G_R(t)\,dt$$

where $G_R(t)$ is a *shear* modulus, whereas in the preceding equations it was a *tensile* modulus relating to a *tensile* viscosity. The last equation establishes the connection between shear viscosity, shear modulus and the deformation limit in polymer melts. If $G_R(t)$ is known at any given temperature, it is also possible to use the time/temperature superposition principle to obtain $G_R(t)$ at the temperature at which $\eta(\dot\gamma)$ has been determined. Alternatively,

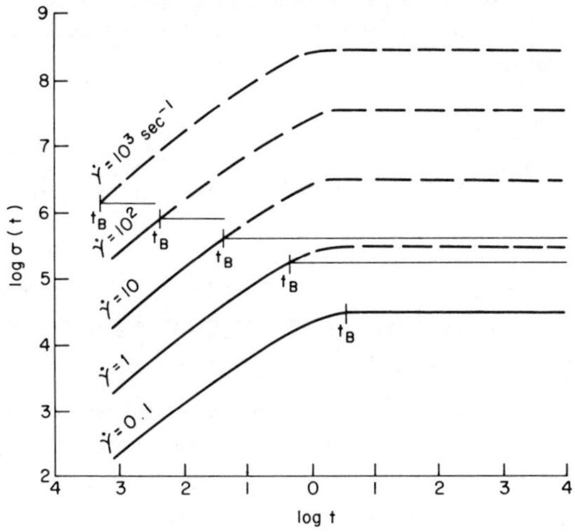

FIG. 1.3.   The time function of shear stress at various shear rates.

$\dot\gamma_B$ may be calculated from experimental values of $G_R(t)$ and $\eta(\dot\gamma)$ by means of a computer.

To test this theory, Tobolsky and Chapoy used a series of commercial polystyrenes at temperatures between 200 and 240 °C. They found excellent agreement between experimental and calculated values of $\eta(\dot\gamma)$ at various shear rates. The results are illustrated in Fig. 1.3. It is seen that the maximum value for the shear stress at various shear rates is reached, in each case, within the time t_B. The fully developed (theoretical) curve only appears at *very low* shear rates. This curve, however, may be used as a master curve which can be shifted to other positions in terms of shear rate, provided a part (even one point) of the appropriate curve is known. The theoretical curve can then be drawn (broken lines). When comparing the theoretical and experimental curves we particularly note the sharp discontinuity, in each case above the master curve, at t_B. The experimental shear stress maximum is read off as the ordinate at the discontinuity, and the corresponding value for t_B follows. If full relaxation of even the most sluggish volume elements were possible at higher shear rates, then one would obtain a correspondingly higher shear stress maximum after a somewhat longer time t_B. Hence it follows that the difference between the theoretical maximum (at the discontinuity) is proportional to the number of volume elements which are unable to relax fully under the given conditions because of their long relaxation times.

Numerous rheological equations have been proposed which are based upon theories of viscoelasticity and using continuum mechanics. These have also been applied to non-Newtonian liquids. None of them is *universally* applicable and most of them are difficult to apply. The equations are presented and discussed by Han[12] who concludes that

> 'What is most needed today from the practical standpoint is either to modify existing theories or to develop new ones which are general enough to explain the observed behaviour of polymeric systems undergoing large deformations, but still simple enough to provide a basis for engineering analysis of complicated flow problems arising in polymer processing operations. . . .'

It would appear that until this is done it would be best—and certainly convenient—to stick to the power law which has served the polymer technologist well and which is easily applied to many process design operations, as will be seen in later chapters.

REFERENCES

1. R. S. LENK, *J. Appl. Poly. Sci.*, **11**, 1033 (1967).
2. M. REINER, *Deformation, Strain and Flow*, p. 233, H. K. Lewis, London (1960).
3. W. A. WRIGHT and W. W. CROUSE, *ASLE Trans.*, **8**, 181 (1965).
4. J. GALT and B. MAXWELL, *Mod. Plastics*, **42**, 115 (1964).
5. P. L. CLEGG, in *Rheology of Elastomers*, p. 174, P. Mason and N. Wookey (editors), Proc. Welwyn Garden City Conference, Pergamon Press, Oxford (1957).
6. A. K. VAN DER VEGT and P. P. A. SMIT, 'Crystallisation phenomena in flowing polymers', paper read at the Conference on Advances in Polymer Science and Technology, London, 1966.
7. E. ATKINSON, discussion following the paper quoted in reference 6.
8. G. W. SCOTT BLAIR, *Rheol. Acta*, **4**, 53 (1968).
9. J. H. PRENTICE, *Nature*, **217**(5124), 157 (1968).
10. Y. NEDONCHELLE and R. A. SCHÜTZ, *CR Acad. Sci.*, **16**, 265C (1967).
11. A. V. TOBOLSKY and L. L. CHAPOY, *Polymer Letters*, **6**, 493 (1965).
12. C. D. HAN, *Rheology in Polymer Processing*, p. 33, Academic Press, New York and London (1976).

2

The Time Dependence of Viscous Flow. Thixotropy and Rheopexy

We have considered viscosity as a function of shear rate, which, in turn, is a function of shear stress. In the course of this consideration it has become clear that both viscous flow and changes in the flow regime (if any) are governed by rate constants. It is therefore logical to consider the effect of time on viscous flow as a next step.

Time effects assume major importance when the random ground state structure of a liquid changes *gradually* rather than instantaneously when a shear field is applied. In this context a 'gradual' change is defined as one in which the transition from the initial state (and the rate at which it occurs) can be experimentally monitored. This is a strictly pragmatic view. From a fundamental theoretical standpoint *all* processes are time dependent and those which might be thought of as being instantaneous just happen to have high rate constants, so that the available techniques are insufficiently sensitive to observe and measure the changes. On this basis, typical polymer melts and solutions are time-independent, so long as they are flowing in homogeneous fashion. This introduces a further qualification which makes it necessary to examine what can cause the flow to become inhomogeneous.

Essentially, flow inhomogeneity is caused by the presence of two or more phases which mutually interact such as to produce local perturbations in the streamlines. These may be due to interfacial forces, hydrogen bonding, other molecular interactions or a tendency on the part of the disperse phase to crystallise and so to increase its volume fraction at the expense of the continuous phase. The continuous phase provides both free volume and lubrication for the volume elements of the disperse phase. It is therefore inevitable that flow inhomogeneities are bound to appear and cause increasing disturbance as the system becomes more and more crowded. In plastics this would arise in melts which contain large amounts of fillers such as china clay or glass fibre, in plastisols such as PVC pastes, and in solutions or dispersions of sufficiently high concentration. When considering the rates of breakdown and rebuilding of rheological structure, one observes

15

the effect of raising and lowering the shear rate. The rate of change in the shear rate (the shear acceleration $\ddot{\gamma}$) now becomes another experimental variable. Time dependency is found to occur in pseudoplastic and in dilatant materials. It shows itself in that the upcurves and downcurves of a $\tau/\dot{\gamma}$ plot are non-coincident and form a hysteresis loop. This will only be observed if the structural changes are not too rapid and if the rate constants differ sufficiently in magnitude.

In a pseudoplastic showing time dependency, the hysteresis loop consists of two parts: an upcurve which is convex with respect to the shear rate axis, and a linear downcurve terminating at the origin when fully developed. The various degrees to which hysteresis loops may be developed will be seen in a later diagram. Whatever the detailed shape of the hysteresis loop, it will be characteristic for the material, provided shearing started when the liquid was sufficiently rested to be in its random ground state, the shear acceleration remains constant, and the arbitrarily selected top shear rate remains the same. This type of time dependent behaviour is known as *thixotropy*. Thixotropy is defined as the isothermal and reversible time-dependent loss of viscosity as a result of the application of shear. Thixotropic materials must inevitably be pseudoplastic, but a pseudoplastic need not necessarily be thixotropic.

In many thixotropic materials the rebuilding process is extremely slow. In these the downcurve is seen to be a perfect straight line; furthermore, an immediate re-run will fail to reproduce the original upcurve. Instead the repeat upcurve will coincide with the previously obtained downcurve, as if the material were a Newtonian. The reason for this is obvious: none of the structure which has been sheared out when running the first upcurve has been rebuilt, since rebuilding happens to be a very slow process. Given sufficient rest, however, the liquid will regain its original ground state structure, whereupon the originally obtained hysteresis loop can be reproduced.

In dilatant liquids structural *build-up* occurs with increasing shear rate. If the downcurve shows any divergence from the upcurve a time dependent effect is evidently present. The resulting hysteresis loop, however, will be in the reverse direction to that obtained with thixotropic materials. This is known as 'negative thixotropy' or *rheopexy*. Its occurrence is not common and it is of no significance in polymer processing where dilatant materials are avoided in any case because they are rheologically unsuitable. The explanation of the phenomenon is not difficult: at any given shear rate there is more structure present on the downcurve than on the upcurve, although structure, on an absolute scale, may be decreasing progressively on the downcurve. In some cases, however, the build-up will actually continue for part of the downcurve as well and go through a maximum, followed by fast breakdown—in the extreme, even all the way down to almost zero shear rate and the shear stress acquired in the course of the preceding shear history will only dissipate after a long rest period. It may be said that an elastic memory effect has been superimposed on the rheopectic viscous flow

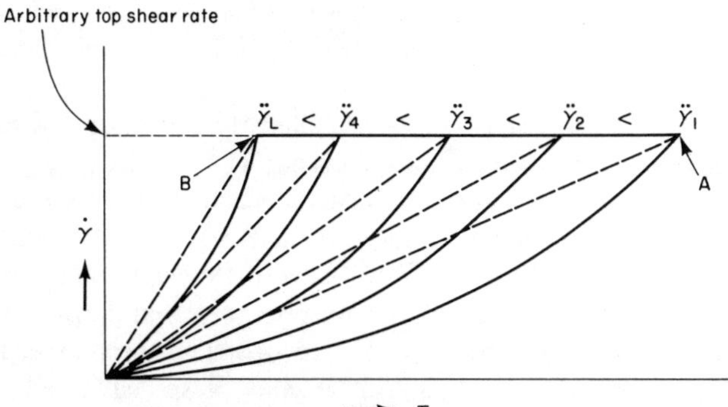

FIG. 2.1. Hysteresis loops showing thixotropy at various shear accelerations. (Upcurves continuous, downcurves broken.)

pattern. Every rheopectic material is dilatant, but not every dilatant material is rheopectic.

To return to thixotropy. If structural breakdown is a process with a rate constant which is appreciably greater than that of the reverse process, it follows that an infinite number of hysteresis loops can be obtained when raising the shear rate to an arbitrarily selected top shear rate. The loop obtained depends on the shear acceleration. If the shear acceleration $\ddot{\gamma}$ is high, little structural breakdown can occur in the short time available—the shear stress and hence the viscosity will be high (see A in Fig. 2.1). At lower $\ddot{\gamma}$ progressively greater reduction in structure will produce lower shear stress and hence lower viscosity, until a limiting $\ddot{\gamma}_L$ is reached (see B in Fig. 2.1) in which all the structure that *can* be sheared out *has* been sheared out, since the time available for this to happen is now sufficient.

When plotting viscosity vs. time at different shear rates $\dot{\gamma}$ one obtains curves as shown in Fig. 2.2. If viscosity is plotted vs. *logarithmic* time (see Fig. 2.3) then it is seen that the viscosity drop is linear until the equilibrium condition is reached. If this is done at a series of arbitrarily selected top shear rates, then a family of parallel straight lines is obtained. The

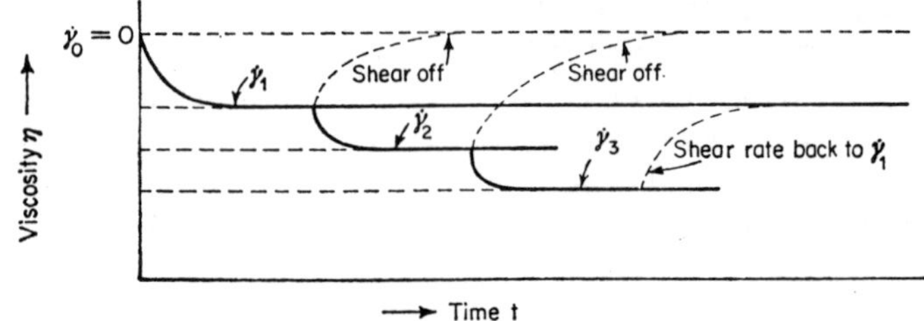

FIG. 2.2.   Viscosity as a function of time ($\dot{\gamma}_3 > \dot{\gamma}_2 > \dot{\gamma}_1 > \dot{\gamma}_0$).

downward slope ($\mathrm{d}\eta/\mathrm{d}\ln t$) has been called the coefficient of thixotropic breakdown with time (Green and Weltmann[1,2]).

It should be pointed out that whilst the shear stress in a pseudoplastic increases less than proportionately with increasing shear rate, the shear stress can never actually *decrease*. If it nevertheless *appears* to do so, this is due to an easily explained artefact. It simply shows that the maximum shear rate at which meaningful results can be recorded has been exceeded. Voids may have appeared in the liquid, it may have 'cavitated', 'split' or 'fractured', in short, laminar flow has ceased. In a cone–plate viscometer this is seen when material begins to climb out of the gap (Fig. 2.4). Normal forces (the Weissenberg effect) are in evidence. It is clear that an apparent *reduction* in shear stress with increasing shear rate in a pseudoplastic is principally due to a reduction in the volume which is being sheared. The normal force perpendicular to the shear plane is visualised when a cone with vertical channels and with its axis normal to a plate is rotated with a viscous liquid placed between cone and plate. Liquid will climb into those channels (Fig. 2.5). The arrows in Fig. 2.5 indicate that a force exists which acts on the liquid, normal to the shear plane. This force can be measured without

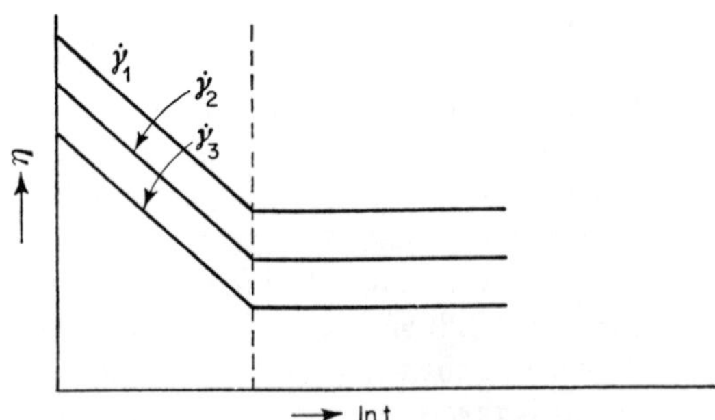

FIG. 2.3.   Viscosity as a function of time (semi-log plot).

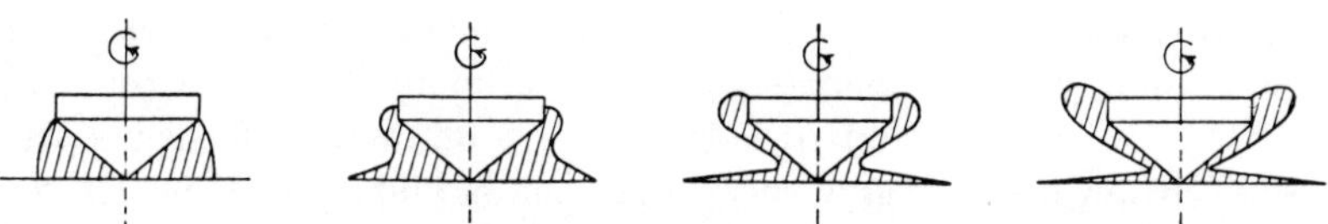

FIG. 2.4.   Cavitation in a cone–plate rotational viscometer.

drilling channels into the cone (and thus losing material from the gap) by placing pressure transducers in contact with the liquid at the cone face and by measuring the pressure exerted on the cone by the liquid.

The simultaneous study of normal force and shear phenomena is the study of viscoelasticity which has been mentioned earlier. The theoretical treatment of viscoelasticity is complex, but the practical implications in polymer processing are considerable. Moreover, the elastic—that is to say, time-dependent—effects play a most important part in die swell, extrusion defects and melt fracture which will be dealt with in later chapters. Suffice it to say, at this stage, that elastic effects become more and more important as shear stresses increase. Looking at calendering, extrusion and injection moulding, the shear stress increases by about a decade from one process to the next in the order just given. It is no coincidence that the first is an 'open', the second a 'semi–closed' and the third an almost 'totally closed' system. In calendering, the end of workability will be signalled by cavitation or splitting (just as in rotational viscometry) in extrusion by melt fracture (just as in extrusion rheometry). In injection moulding, turbulent flow certainly occurs, but does not necessarily impede production, indeed it may be even desirable in the interest of isotropy, if directional strength uniformity is required in a moulding.

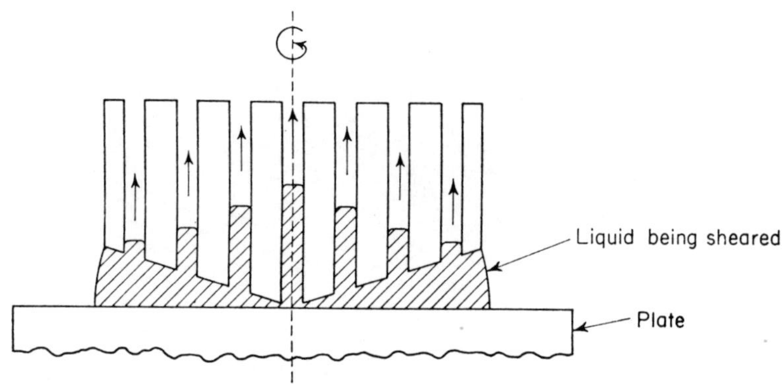

FIG. 2.5.   Experiment demonstrating the normal force (Weissenberg) effect.

It is obvious that the study of time dependency requires continuous observation of *the same* material. This is accomplished when rotational viscometers are used, but the open nature of these instruments set a relatively low upper limit of shear which can be employed. However, they are useful instruments for the observation of low shear behaviour generally. More recently rotational viscometers have been designed which are closed and which can therefore be pressurised and used at higher shears (see later section on pressure dependence of viscous flow). Extrusion type instruments are generally *unsuitable* for the study of time dependence owing to the continuous passage of fresh melt material through the die channel.

REFERENCES

1. H. GREEN and R. WELTMANN, *Ind. Eng. Chem.*, **18,** 167 (1946).
2. H. GREEN, and R. WELTMANN, *Industrial Rheology and Rheological Structures*, Chapman & Hall, London (1949).

3

The Temperature Dependence of Viscous Flow.
Free Volume

We now consider the viscosity–temperature functions of non-Newtonian (specifically power law) liquids such as polymer melts.

It has already been pointed out that viscosity is also a function of shear rate $\dot{\gamma}$ or of shear stress τ, the latter two being interdependent. The viscosity/temperature function can be studied either at constant shear stress or at constant shear rate. In general, the viscosity change with temperature at constant shear rate and at constant shear stress is *not* equal and this is proved as follows.

The total differential of viscosity as a function of shear stress and temperature can be written as the sum of two partial differentials

$$d\eta = \left(\frac{\partial \eta}{\partial T}\right)_{\tau} dT + \left(\frac{\partial \eta}{\partial \tau}\right)_{T} d\tau$$

If $\dot{\gamma}$ is constant this becomes

$$d\eta = \left(\frac{\partial \eta}{\partial T}\right)_{\tau} dT + \left(\frac{\partial \eta}{\partial \tau}\right)_{T}\left(\frac{\partial \tau}{\partial T}\right)_{\dot{\gamma}} dT \quad \text{or} \quad \left(\frac{\partial \eta}{\partial T}\right)_{\dot{\gamma}} = \left(\frac{\partial \eta}{\partial T}\right)_{\tau} + \left(\frac{\partial \eta}{\partial \tau}\right)_{T}\left(\frac{\partial \tau}{\partial T}\right)_{\dot{\gamma}}$$

Rearranging

$$1 = \frac{(\partial \eta/\partial T)_{\tau}}{(\partial \eta/\partial T)_{\dot{\gamma}}} + \frac{(\partial \eta/\partial \tau)_{T}(\partial \tau/\partial T)_{\dot{\gamma}}}{(\partial \eta/\partial T)_{\dot{\gamma}}}$$

or

$$\frac{(\partial \eta/\partial T)_{\tau}}{(\partial \eta/\partial T)_{\dot{\gamma}}} = 1 - \left(\frac{\partial \eta}{\partial \tau}\right)_{T}\left(\frac{\partial \tau}{\partial T}\right)_{\dot{\gamma}}\left(\frac{\partial T}{\partial \eta}\right)_{\dot{\gamma}} = 1 - \left(\frac{\partial \eta}{\partial \tau}\right)_{T}\left(\frac{\partial \tau}{\partial \eta}\right)_{\dot{\gamma}}$$

Differentiation of $\tau = \eta\dot{\gamma}$ with respect to η at constant $\dot{\gamma}$ gives

$$(\partial \tau/\partial \eta)_{\dot{\gamma}} = \gamma$$

so that

$$\frac{(\partial \eta/\partial T)_{\tau}}{(\partial \eta/\partial T)_{\dot{\gamma}}} = 1 - \dot{\gamma}\left(\frac{\partial \eta}{\partial \tau}\right)_{T}, \quad \text{q.e.d.} \tag{1}$$

It is seen that there are two special cases when the two temperature derivatives of viscosity are equal after all, namely when the second term on the right of eqn. (1) is zero. This is the case either when $\partial \eta / \partial \tau$ is zero—at constant viscosity (the Newtonian case)—or when $\dot{\gamma}$ is zero—i.e. at rest, when all liquids are Newtonians. It is worth noting that the differential coefficient on the right may be either positive or negative, depending on how the viscosity changes with increasing shear rate. It is seen that in pseudoplastics the right-hand side is greater than unity, whilst in dilatants it is less than unity.

Since viscous flow is a rate process it is possible to express the temperature dependence of viscous flow also in terms of an Arrhenius-type equation involving the activation energies at constant shear rate or at constant shear stress, as the case may be:

$$\eta = A \exp(-E\dot{\gamma}/RT) \quad \text{and} \quad \eta = A \exp(-E\tau/RT) \qquad (2)$$

Taking the partial derivatives we obtain

$$\left(\frac{\partial \eta}{\partial T}\right)_{\dot{\gamma}} = -\eta \frac{E_{\dot{\gamma}}}{RT^2} \quad \text{and} \quad \left(\frac{\partial \eta}{\partial T}\right)_{\tau} = -\eta \frac{E_{\tau}}{RT^2} \qquad (3)$$

The ratio of the left-hand sides is the same as that of eqn. (1), whilst the ratio of the right-hand sides is seen to be the ratio of the respective activation energies, so that

$$\frac{E_{\tau}}{E_{\dot{\gamma}}} = 1 - \dot{\gamma}\left(\frac{\partial \eta}{\partial \tau}\right)_{T} \qquad (4)$$

In power law liquids the ratio of the activation energies is in fact identical with the power index n. This can be proved as follows:

$$\eta_{\text{app}} = \frac{\tau}{\dot{\gamma}} \quad \text{and} \quad \tau = K\dot{\gamma}^n \quad \text{or} \quad \dot{\gamma} = \left(\frac{\tau}{K}\right)^{1/n}$$

Therefore

$$\eta_{\text{app}} = \frac{\tau}{(\tau/K)^{1/n}} = \tau^{(n-1)/n} K^{1/n}$$

where $K = $ constant at fixed temperature.

Differentiating η_{app} with respect to τ

$$\left(\frac{\partial \eta}{\partial \tau}\right)_{T} = \frac{n-1}{n}\tau^{-1/n}K^{1/n} = \frac{n-1}{n}\left(\frac{K}{\tau}\right)^{1/n} = \frac{n-1}{n}\left(\frac{1}{\dot{\gamma}}\right)$$

Substituting in eqn. (4)

$$1 - \dot{\gamma}\left(\frac{\partial \eta}{\partial \tau}\right)_T = 1 - \dot{\gamma}\left(\frac{n-1}{n}\frac{1}{\dot{\gamma}}\right) = \frac{1}{n}$$

Therefore

$$\frac{E_\tau}{E_{\dot{\gamma}}} = \frac{1}{n} \qquad \text{or} \qquad \frac{E_{\dot{\gamma}}}{E_\tau} = n \qquad (5)$$

We now examine the viscosity changes which a temperature change ΔT will bring about in a non-Newtonian liquid.

First consider the liquid which is being sheared in the annular space of a coaxial cylindrical viscometer, the inner cylinder being driven at constant speed. As the temperature is raised, so the viscosity will decrease and less torque will be exerted. The shear rate is constant and the viscosity change is given by

$$d\eta = \left(\frac{\partial \eta}{\partial T}\right)_{\dot{\gamma}} dT$$

which, because of eqns. (3), can also be written

$$d\eta = -\eta\left(\frac{E_{\dot{\gamma}}}{RT^2}\right) dT \qquad (6)$$

Dividing by η, integrating and taking the limits we get

$$\frac{d\eta}{\eta} = \ln\eta = -\frac{E_{\dot{\gamma}}}{RT^2} \qquad \text{and hence} \qquad \ln\frac{\eta_2}{\eta_1} = \frac{E_{\dot{\gamma}}\,\Delta T}{RT_1T_2} \qquad (7)$$

(The negative sign indicates that the viscosity *decreases* with increasing temperature.)

Supposing, however, that the drive is a *constant torque* device. In that case τ would be constant and the shear rate would increase with increasing temperature. By analogous reasoning—*mutatis mutandis*—one obtains

$$\ln\frac{\eta_2}{\eta_1} = -\frac{E_\tau\,\Delta T}{RT_1T_2} \qquad (8)$$

Before giving examples of applying these equations it is necessary to show that the power law $\tau = K(\dot{\gamma})^n$ can be written in different forms.

Since

$$\frac{\tau}{\dot{\gamma}} = \eta_{\text{app}} \qquad \text{therefore} \qquad \eta_{\text{app}} = K(\dot{\gamma})^{n-1}.$$

Let an arbitrary reference state be defined by a zero superscript such that $\eta = \eta^0$ when $\dot\gamma = \dot\gamma^0$. $\dot\gamma^0$ is conveniently chosen to be $1\,\mathrm{s}^{-1}$. Now

$$\eta^0_{\mathrm{app}} = K(\dot\gamma^0)^{n-1}$$

and

$$\frac{\eta}{\eta^0_{\mathrm{app}}} = \left(\frac{\dot\gamma}{\dot\gamma^0}\right)^{n-1}$$

Since the ratio of the apparent viscosities may be taken to be virtually the same as the ratio of the true viscosities, and since $\dot\gamma^0$ is numerically equal to unity, it is now possible to write

$$\eta = \eta^0 \dot\gamma^{n-1} \tag{9}$$

By exactly analogous reasoning

$$\dot\gamma = K'\tau^{1/n}$$

If an arbitrary reference state is now defined as that where $\eta = \eta^0$ when $\tau = \tau^0 = 1\,\mathrm{dyn\,cm}^{-2}$, then it can also be shown that

$$\eta = \eta^0 \tau^{(n-1)/n} \tag{10}$$

The two standard reference viscosities η^0 are not, however, identical in eqns. (9) and (10).

According to eqn. (7) one may equally write

$$\ln\frac{\eta^0_2}{\eta^0_1} = \frac{E_{\dot\gamma}}{R}\frac{\Delta T}{T_1 T_2} \tag{11}$$

Example

At a temperature of 174 °C and at a shear rate of $1\,\mathrm{s}^{-1}$ the viscosity of a polyethylene melt was found to be 31 500 p. Assuming the power law to be applicable over the ranges of shear rate and temperature involved, estimate the viscosity of the melt at 230 °C and at $100\,\mathrm{s}^{-1}$, using the activation energy data shown in Table 3.1 which had been obtained by Philippoff and Gaskins:[1]

Solution: The given viscosity represents η^0_1 at the arbitrary reference state $\dot\gamma^0_1$, where the subscript 1 refers to the temperature T_1 which is 174 °C.

The value of $E_{\dot\gamma}$ at $\dot\gamma = 1$ is read off from Philippoff and Gaskins' table—10·3 kcal mol^{-1}—and eqn. (11) then yields η^0_2 which represents the standard reference viscosity at $1\,\mathrm{s}^{-1}$ at temperature T_2, i.e. at $230\,°\mathrm{C}$. One obtains $\eta_2^0 = 8720\,\mathrm{p}$. In order to obtain the viscosity η_2 at the higher shear rate $\dot\gamma_2 = 100\,\mathrm{s}^{-1}$ we use the modified power law

$$\eta_2 = \eta_2^0 \dot\gamma_2^{\bar{n}-1}$$

where $\bar{n}$ represents the average value of the melt flow index between the shear rates of 1 and $100\,\mathrm{s}^{-1}$.

TABLE 3.1

$\dot\gamma$ (s^{-1})	$E_{\dot\gamma}$ $(kcal\,mol^{-1})$	τ $(dynes\,cm^{-2})$	E_τ $(kcal\,mol^{-1})$
0	12·8	0	12·8
10^{-1}	11·4	10^4	15·0
10^0	10·3	10^5	17·8
10^1	8·5	10^6	19·0
10^2	7·2		
10^3	6·1		

From the table data the value of $E_{\dot\gamma}$ at $1\,\mathrm{s}^{-1}$ is 10·3, as has already been seen. With a viscosity of the order of $10^4\,\mathrm{p}$ for η_2^0 the shear stress at the standard reference shear rate $\dot\gamma_2^0 = 1$ is seen to be of the order of magnitude 10^4, so that the activation energy E_τ corresponding to $\tau = 10^4$ is 15·0 and

$$n_1 = \frac{E_{\dot\gamma}}{E_\tau} = \frac{10\cdot3}{15\cdot0} = 0\cdot68$$

In order to find n_2, the power index at the shear rate $\dot\gamma_2 = 100\,\mathrm{s}^{-1}$, the activation energies must be known. Now $E_{\dot\gamma}$ opposite a shear rate of $100\,\mathrm{s}^{-1}$ is 7·2, but before reading off the correct value for E_τ one must know τ.

If the melt were Newtonian the viscosity would be constant at an order of magnitude of $10^4\,\mathrm{p}$; at a shear rate of 10^2 the shear stress generated would be 10^6. But we have a pseudoplastic and may reasonably assume that the viscosity at $10^2\,\mathrm{s}^{-1}$ will be significantly less than at $1\,\mathrm{s}^{-1}$, perhaps by an order of magnitude. We therefore assume—provisionally, and subject to later verification—that the viscosity will be of the order of $10^3\,\mathrm{p}$ and that the shear stress is therefore of the order of $10^3 \times 10^2 = 10^5$. E_τ for that value of τ is now seen to be 17·8, so that

$$n_2 = \frac{E_{\dot\gamma}}{E_\tau} = \frac{7\cdot2}{17\cdot8} = 0\cdot40$$

$\bar{n}$, the mean of n_1 and n_2, is therefore 0·54 and we now have

$$\eta_2 = 8720 \times (100)^{0·54-1} = 1100\,\text{p}$$

This also proves that the assumption that η_2 is of the order of 10^3 (and with it the assumption that the corresponding shear stress of 10^5 produced a value of 17·8 for E_τ under those conditions) was correct.

[Had we guessed wrongly, say η_2 to be of the order 10^4 so that τ would have been of the order of 10^6 and $E_\tau = 19·0$, then n_2 would have been 0·38, $\bar{n} = 0·53$ and $\eta_2 = 1000\,\text{p}$. On checking back, however, the correct order of magnitudes is now established and the iterated procedure will give the correct result for η_2.]

It has now been shown that general flow phenomena—and viscosity in particular—are determined by the structure of the fluid (in the widest sense) and by the constraints exercised upon it by the environment. Continuing the process of examining these constraints and their environmental agents, however, it is already abundantly clear that shear rate, shear stress, time and temperature are mutually interdependent. The same will be seen to be true for volume and pressure as well. Thus, viscosity decreases with increasing temperature because 'free volume' increases; when free volume increases the flow units become less restricted and with increasing temperature they also become more highly energised, less highly organised, and their relaxation times decrease. This leads to classical volume/temperature/entropy/time relationships. Pressure is obviously related to volume and an increase in pressure will have exactly the opposite effect to that observed when increasing temperature.

We shall deal with pressure effects presently, but the effect of temperature on viscosity will be considered first in terms of free volume.

The apparent volume V of a polymer mass at temperature T consists of two portions—that part which is contributed by the molecules themselves (the occupied volume V_o), and the free volume V_f. If there exists a temperature T_0 at which the free volume is zero, then no significant volume contraction can occur below that temperature. Furthermore, if the coefficient of thermal expansion $(\partial V/\partial T)_P$ is known, then the free volume will be

$$V_f = (T - T_0)(\partial V/\partial T)_P$$

and

$$V_o/V_f = \frac{1}{\alpha(T - T_0)}$$

where α is a proportionality constant.

Ordinarily T_0 is taken to be the absolute zero, but in polymers it is very much higher. Miller[2] has shown that published melt viscosity data for polystyrene and PIB conform to a modified Arrhenius equation

$$\eta = A \exp \frac{B}{\alpha(T - T_0)}$$

whilst the more usual Arrhenius relationship

$$\eta = A \exp(=E/RT)$$

is restricted to narrow temperature ranges and does not hold at all well around the glass transition point T_g.

The free volume concept was developed by Doolittle[3] on the basis of the viscosity behaviour of *n*-alkanes:

$$\eta = A \exp(BV_0/V_f)$$

Cohen and Turnbull[4] showed that Miller's equation was equivalent to Doolittle's modified Arrhenius equation, that V_f is zero at T_0 and that V_f increases linearly with temperature above T_0.

Williams[5] used the glass transition T_g as a reference point and found that the specific volume of the liquid above T_g was

$$V = V_g + (T - T_0)(\partial V/\partial T)_P$$

where V_g is the specific volume at T_g.

Substituting for V in Doolittle's equation and using viscosity/temperature data obtained on a series of polystyrene fractions by Flory and Fox, the values of V_0, A and B were computed. Miller pointed out that the substitution for V_0/V_f (see above) into the Doolittle equation produces

$$\eta = A \exp \frac{B}{\alpha(T - T_0)}$$

Moreover if Flory and Fox's data for a polystyrene fraction of molecular weight 1675 ($T_g = 40\,°C$) are plotted as $\log \eta$ vs. $1/(T - T_0)$ using trial values for T_0, then a family of curves is obtained as shown in Fig. 3.1. The temperatures for curves (a) through (e) were $-273°$, $0°$, $5°$, $10°$ and $40\,°C$ (T_g) respectively. Linearity is seen to occur when T_0 is taken to be $5\,°C$ and that temperature is identified as the reference temperature for relaxation processes in the liquid state. The temperature T_0, then, is the temperature at

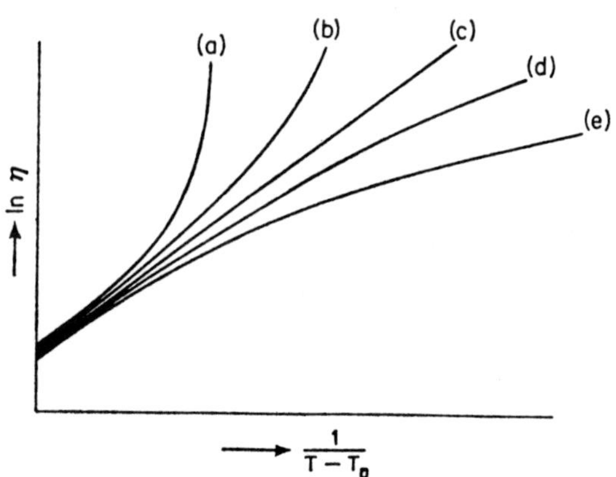

FIG. 3.1.   Modified Arrhenius plots.

which V_f is zero. The intercept of the straight line yields A, whilst B/α is identical with $-E/R$, where E is the true activation energy of the process of flow and R is the gas constant.

The values for A and E calculated by Williams agreed well with those obtained by Miller's curve fitting method. Williams also showed that the activation energy changed only slightly with molecular weight: at MW of 1675, E was 2·9, at 134 000 it was 3·1 kcal/mol^{-1}.

The *apparent* activation energy at any one temperature T, i.e. the slope of the curve at that point, is obtained by differentiating the modified Arrhenius equation

$$\frac{\mathrm{d}\ln\eta}{\mathrm{d}(1/T)} = \frac{E_{\mathrm{app}}}{R}\left(\frac{T}{T-T_0}\right)^2$$

At temperatures which are very much higher than T_0, $[T/(T-T_0)]^2$ approximates to unity and $1/(T-T_0)$ approximates to zero. At such temperatures the apparent activation energy is close to the value for the true activation energy which is a limiting value. With decreasing temperature the true activation energy increases exponentially and attains extremely high values at T_g.

The work of Miller on Flory and Fox's viscosity/temperature data, together with Doolittle and Williams' free volume approach shows that there is a unique zero point (T_0, V_0) below the glass transition which can be related to the glass transition (T_g, V_g). That zero point is a true reference point for kinetic relaxation processes occurring in the liquid above T_g; it is not directly accessible through viscosity measurements because of the kinetic barrier in the glass, but it can nevertheless be determined as shown. Free volume data thus provide a further insight into the melt flow phenomena of polymers.

Miller has obtained similarly satisfactory results with PIB.

REFERENCES

1. W. PHILIPPOFF and F. H. GASKINS, *J. Poly. Sci.*, **21,** 205 (1956).
2. A. A. MILLER, *J. Poly. Sci.*, **A1**(6), 1857 (1963).
3. A. K. DOOLITTLE, *J. Appl. Phys.*, **22,** 1031 and 1471 (1951) and **23,** 418 (1952).
4. M. H. COHEN and D. TURNBULL, *J. Chem. Phys.*, **31,** 1164 (1959).
5. M. L. WILLIAMS, *J. Appl. Phys.*, **29,** 1395 (1958).

4

The Influence of Pressure on the Viscosity of Polymer Melts. Viscosity and Molecular Weight

Polymer melts are compressible liquids of bulk modulus $\sim 10^9 \, \text{N} \, \text{m}^{-2}$. Most polymer melt processes operate at pressures of 10^6 to $10^7 \, \text{N} \, \text{m}^{-2}$ (10 to 100 atm) with a bulk compression of less than 1%. However, in some— notably injection moulding—pressures of the order of $10^8 \, \text{N} \, \text{m}^{-2}$ (1000 atm) may be employed. Bulk compression then is becoming appreciable and the consequent reduction in free volume reduces the mobility of the molecules. This manifests itself as a viscosity increase. Large mouldings may require very high pressures indeed. In such cases allowance should be made for the reduced melt fluidity in the processing specifications.

The study of pressure dependence of viscous flow is clearly of major importance in the injection moulding of large objects, but it may also be significant in screw extrusion. Furthermore, such data are of fundamental interest for the study of molecular mobility and may be correlated with thermodynamic data obtained by entirely independent methods.

Cogswell[1] described the design and use of an instrument which was suitable for studying the flow behaviour of polymer melts under pressure. This was a pressurised Couette-Hatschek cup/bob rheometer in which the melt was kept under ram pressure in an outer rotating cylinder. The inner (fixed) cylinder has a torsion member with a mirror which deflects a light beam and indicates the shear stress as a function of the torque exerted on the inner cylinder. Using a cylinder length/gap ratio of 25:1, end corrections were small, but still needed to be applied. Secondary flows were neglected as a first approximation.

The viscosity change with temperature at constant pressure $(\partial\eta/\partial T)_P$ and the viscosity change with pressure at constant temperature $(\partial\eta/\partial P)_T$ were determined. It was found that the viscosity gradients were linear functions when $\log\eta$ was plotted vs. pressure and temperature respectively. This suggested that the effect of pressure on viscosity could be related to the well understood temperature effect through a coefficient $-(\Delta T/\Delta P)_\eta$. Since this is approximately constant, it is implicit that a pressure increase ΔP will have

31

the same effect as a temperature decrease $-\Delta T$. It was found that the coefficient was independent of molecular weight and varied only slightly from one polymer to another.

The coefficient has the same form as the familiar thermodynamic functions $(\partial T/\partial P)_S$ and $(\partial T/\partial P)_V$. The first of these may be measured as the rate of temperature rise during adiabatic compression and agrees with the calculated value from the equation

$$(\partial T/\partial P)_S = \frac{-(\partial S/\partial P)_T}{(\partial S/\partial T)_P}$$

$(\partial T/\partial P)_V$ is calculated from the equation

$$(\partial P/\partial T)_V = (\partial S/\partial P)_T(-K/V)$$

where S = entropy, K = bulk modulus, and V = the specific volume.

The entropy/pressure and entropy/temperature functions at constant temperature and constant volume respectively are fundamental thermodynamic functions which may be obtained from specific heat, specific volume and temperature measurements.

Cogswell noted that polypropylene and high density polyethylene showed anomalous flow behaviour at temperatures within 50 °C of the crystalline melting points. With increasing shear rate the usual pseudoplastic behaviour was found to give way to a region of unusually high viscosity, the melt showed 'pulsating flow' and in extreme cases even cessation of flow. In the light of the generalised flow theory and the work of van Vegt and Smit[2] this is scarcely surprising. Equally, it is in no way odd that this should also occur at lesser shear rates when the *pressure* is high. Cogswell refers to a 'log jam' hypothesis of Lamb which coincides completely with the various explanations for transitions to a dilatant flow regime, since pressure-induced crystallisation is really the same phenomenon as shear-induced crystallisation.

The superposition of the pressure and temperature dependence of viscosity is exceedingly convenient when approaching melt flow engineering problems. Consider first viscosity as a function of temperature and then as a function of pressure. Under isoviscous conditions it then becomes possible to obtain a conversion factor $-(\Delta T/\Delta P)_\eta$, i.e. to determine the drop in temperature which is equivalent to the imposed pressure increase in producing the same melt viscosity.

The conversion factor $(-\Delta T/\Delta P)_\eta$, the change of temperature with pressure at constant entropy and the change of temperature with pressure at constant volume are shown in Table 4.1 for a number of polymer melts.

Consider, for example, LDPE at 220 °C and at a pressure of $10^8 \, \mathrm{N \, m^{-2}}$ (1000 atm). The conversion factor given in the first number column of Table 4.1 is $5 \cdot 3 \times 10^{-7}$. The temperature drop necessary to produce iso-viscous conditions at normal atmospheric pressure is therefore $5 \cdot 3 \times 10^{-7} \times 10^8 = 53$ °C. In other words, the flow behaviour of the melt at 220 °C and $10^8 \, \mathrm{N \, m^{-2}}$ pressure is equal to that at 167 °C and at

TABLE 4.1

Polymer	$-(\Delta T/\Delta P)_\eta$ $\times 10^7$	$(\partial T/\partial P)_S$ $\times 10^7$	$(\partial T/\partial P)_V$ $\times 10^7$
PVC	3·1	1·1	16
Nylon 66	3·2	1·2	11
PMMA	3·3	1·2	13
PSt	4·0	1·5	13
HDPE	4·2	1·5	13
Acetal copolymer	5·1	1·4	14
LDPE	5·3	1·6	16
Silicon polymer	6·7	1·9	9
PP	8·6	2·2	19

atmospheric pressure. In extrusion processes where pressures are perhaps a decade or more smaller than in injection moulding, the effect will also be a decade or so smaller and will not, therefore, amount to more than a few degrees effective temperature decrease—but even that could be significant in polymers with a high viscosity/temperature gradient.

We now consider the relationship between viscosity and molecular weight. As the molecular weight of a polymer increases, so the melt viscosity must also increase, because the activation energy of the process increases with any increase in the mass of the flow unit. However, as the polymer chains increase in length the chains will begin to entangle and cease to move independently. Chain segments rather than entire chains will represent the flow unit and these segments will depend on cooperation with other segments in their vicinity. The 'other segments' may be part of the same chain or they may belong to chains with which the first-considered chain has become entangled. The segmental nature of the flow units is proved by the fact that the activation energy of flow remains practically constant once a certain range of molecular weight level is attained. Further increases in molecular weight then cause only linear (and no longer exponential) increases in viscosity. That is to say, the proportionality constant A in the

Arrhenius equation becomes a function of viscosity and hence of molecular weight. The empirical relationship $A = \text{const} \times M^{3 \cdot 4}$ applies to many polymers, both linear and branched, in concentrated solutions as well as in the melt. The flow unit in linear polyethylene is about 50 carbon atoms long and so has a molecular weight of about 700. Molecular weights of repeating units will, however, also depend upon the presence or absence of heteroatoms and cyclic structures in the main chain or as substituents and pendant side groups, especially when the latter are bulky. Thus, Schmieder *et al.*[3] have shown that the unit of relaxation of polychloroprene could be as few as 17 carbon atoms even in a high molecular weight polymer.

Naturally, dilution—including plasticisation—will have the effect of increasing the relative independence of flow units as a result of the increase in free volume and the consequent loosening of the bulk structure. This, together with equivalent changes in mobility as a function of temperature has been intensively studied and dynamic mechanical and electrical methods have provided powerful tools for this purpose. The molecular interpretation of relaxation processes in terms of structural rearrangement has been intensively studied from the time of Boltzmann onwards; the whole literature on diffusion and energy barriers is relevant. The reader is referred to a paper by Müller[4] with 24 references, and to Treloar's monograph.[5]

The concept of a 'critical' molecular weight which must be exceeded before entanglements appear, stems from a consideration of the viscosity of polymers within the first Newtonian region. It is assumed that there is an abrupt change in viscosity at that molecular weight (M_c), such that below M_c the viscosity is proportional to the molecular weight itself, whilst above M_c it is proportional to the 3·4th power of the molecular weight. This was first proposed by Fox *et al.*[6] on an empirical basis which Bueche[7,8] has attempted to justify. However, Cross[10] has pointed out that a *discontinuous* from unity slope, to slope 3·4 has never been experimentally observed when plotting $\log \eta$ vs. $\log \text{MW}$. On the contrary, the transition is obviously *gradual*. However, it is possible to conceive of a molecular weight M_c—though in no sense a 'critical' molecular weight—at which the viscosity contributions of entangled and unentangled chains is exactly equal, so that, whilst in general

$$\eta = c_1 M + c_2 M^{3 \cdot 4}$$

at M_c we have

$$c_1 M_c = c_2 M_c^{3 \cdot 4}$$

Thus it is not necessary to exclude the possibility of entanglement at a molecular weight below M_c nor that of non-entanglement at a molecular weight above M_c. Instead one need only regard the region of transition from the lower to the higher slope in the log–log plot as one where the *probability* of entanglement increases from zero to unity, and that halfway through the transition region there is just a 50/50 chance of entanglement.

However, it has also been reported that much higher slopes than 3·4 have been observed. Thus, Porter and Johnson[9] found 7 in polyethyleneglycol, which cannot be accounted for by Bueche's theory. Cross[10] has worked out a theory which is as convincing as it is simple: the probability of entanglement is proportional to the product of the chain lengths (and hence the molecular weights) of the entangling chains. If chains thus primarily entangled in turn entangle, four chains are involved and higher degrees of entanglement are also possible, though less probable. If we call these entanglements 'primary', 'secondary', 'tertiary', etc., then each of these will make a contribution to the overall viscous drag force F. In addition, there may also be a few unentangled chains which make a viscosity contribution.

If F is given by

$$F = k_0 M + k_1 M^2 + k_2 M^4 + k_3 M^6 + \cdots$$

then the viscosity is given by $\eta = FN$, where N is the number of molecules per unit volume and is therefore inversely proportional to the molecular weight, so that

$$\eta = c_0 + c_1 M + c_2 M^3 + c_3 M^5 + \cdots$$

In polymers, M is a large number, so that c_0 may be neglected and Cross finally obtains

$$\eta = c_1 M + c_2 M^3 + c_3 M^5 + \cdots$$

The three coefficients c_1, c_2 and c_3 may be obtained after viscosity determinations of molecular weight fractions have been carried out, those fractions being as closely monodisperse as possible.

Cross had no difficulty in finding support for his theory from data reported in the technical literature. Thus, using data obtained by Flory[11] on poly(decamethylene adipate), the $\log \eta$–$\log M$ plot according to Cross' equation produced a continuous curve. If tangents of slope 1 and of slope 3·4 are drawn to this curve, then these tangents intersect at the so-called 'critical' molecular weight M_c. But it has been seen that

$$c_1 M_c = c_2 M^{3·4} = \eta/2$$

Reading off the viscosity at M_c it is indeed seen that the value from the tangent intersection is just 0·3 log-unit less than that from the curve based on the experimental data and Cross' equation; that is to say, the observed viscosity is just twice the viscosity which would be expected if entanglement were abrupt rather than gradual. The plot shown (Fig. 4.1) also shows that at higher molecular weights even steeper slopes are encountered and that

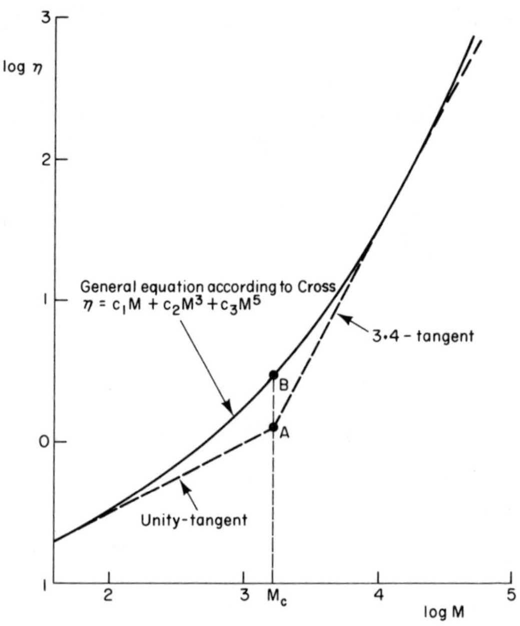

FIG. 4.1.   Viscosity–molecular weight relationship.

higher orders of entanglements are thus becoming significant. Indeed, the fortuitous circumstance that a 3·4-slope is so often closely approximated is an indication that this merely represents a statistical mean value of entanglements of higher order (4, 6 or more chains).

The determination of molecular weight *distribution* is carried out by absolute methods such as fractionation, light scattering or osmometry. These methods are lengthy and experimentally difficult. Relative methods, on the other hand, such as gel permeation chromatography, ultracentrifugation and intrinsic viscosity determinations require calibration by absolute methods. All these determinations are carried out in solution. Attempts to obtain molecular weight distributions (MWD) from measurements on polymer melts are complex and are also based upon theories whose validity is not fully proved. Locati *et al.*[12] have put forward a simple relative method which rests on two parameters—weight average molecular weight $\bar{M}_w$ and an 'index of polydispersity' Q. It involves the measurement of two rheological properties, one in the melt and one in solution and it is limited to polymers which are known to follow a normal MWD.

According to Wesslau[13]

$$w(M)\,dM = \frac{1}{B\sqrt{\pi}M}\exp\left[-\left(\frac{1}{B}\ln\frac{M}{M_o}\right)^2\right]dM$$

where $w(M)\,dM$ is the molecular weight fraction between M and $M+dM$, M_o is the average value and B the breadth of the distribution. Q is given by $\bar{M}_w/\bar{M}_N$, and M_o and B are related to $\bar{M}_w$ and B by the following equations:

$$M_o = \bar{M}_w\exp\left(-B^2/4\right) \qquad \text{and} \qquad B^2 = 2\ln Q$$

According to earlier work by Locati and his colleagues[14] the zero shear viscosity η_0 of polymers with a normal MWD may be written:

$$\log\eta_0 = \log k + \alpha\log\bar{M}_w + \beta\log Q \tag{1}$$

where the viscosity is obtained in the melt and where $\log k$ and α reduce to the values found for monodisperse systems. In general, the specific values of $\log k$, α and β can be obtained experimentally.

Having derived the zero-shear viscosity from $\bar{M}_w$ and Q, the same method is applied to the intrinsic (solution) viscosity $[\eta]$

$$[\eta] = k'M^a$$

where $0\cdot5 < a < 1\cdot0$ in monodisperse systems.
Extended to polydisperse systems

$$[\eta] = k'\bar{M}_v^a$$

where $\bar{M}_v$, the viscosity average molecular weight, is

$$\bar{M}_v = \left[\frac{\int_0^\infty w(M)M^a\,dM}{\int_0^\infty w(M)\,dM}\right]^{1/a}$$

By introducing Wesslau's equation and integrating

$$\bar{M}_v = \bar{M}_w Q^{(a-1)/2}$$

so that

$$\log[\eta] = \log k' + \gamma \log \bar{M}_{\mathrm{w}} + \delta \log Q \tag{2}$$

where $\gamma = a$, and $\delta = -(a/2)(1-a)$.

Equations (1) and (2) can be used to evaluate $\bar{M}_{\mathrm{w}}$ and Q. For this one first has to calibrate the coefficients for the given polymer species. It should be noted that δ cannot be positive and is a function of a only; therefore the determination of δ only requires measurements on monodisperse systems of polymers. For a polymer of the same species $\bar{M}_{\mathrm{w}}$ and Q are obtained by introducing the values of η_0 and of $[\eta]$ which are measured at the same temperature, and solving eqns. (1) and (2).

It is seen that this method enables one to determine $\bar{M}_{\mathrm{w}}$ and Q for polymers with normal logarithmic MWD, such as, for example, linear polyethylene. It requires the determination of the zero-shear viscosity in the melt and the intrinsic viscosity in solution and these measurements can be carried out with high precision. Both the zero-shear melt viscosity and the intrinsic viscosity have a logarithmic dependence on $\bar{M}_{\mathrm{w}}$ and Q. The

TABLE 4.2

Sample	$\bar{M}_{\mathrm{w}} \times 10^{-3}$		Q	
	Calculated	Literature	Calculated	Literature
F1	21·5	19·2	1·28	—
F2	56·9	53·8	1·47	—
F3	97·5	97·3	1·72	—
F4	624·0	724·0	2·34	—
F5	7·2	6·04	1·16	1·22
F6	19·1	17·4	1·13	1·24
F7	48·7	44·7	1·88	1·25
F8	77·7	75·1	1·76	1·29
F9	173·0	182·0	1·47	1·29
F10	430·0	44·0	3·16	—

coefficients can be determined by multiple regression analysis on a series of samples for which $\bar{M}_{\mathrm{w}}$ and Q are known. Errors in the zero-shear viscosity are not serious, but errors in the intrinsic viscosity do contribute heavily to errors in the calculated parameters.

Table 4.2 gives examples of values of $\bar{M}_{\mathrm{w}}$ and Q obtained by calculation and compared with published literature values. The intrinsic viscosity was determined in tetralin at 130°C, the melt viscosity at 190°C. The polymer was linear polyethylene.

REFERENCES

1. F. N. Cogswell, *Plastics & Polymers*, 39 (Feb. 1973).
2. A. K. van der Vegt and P. P. A. Smit, 'Crystallisation phenomena in flowing polymers', paper read at the Conference on Advances in Polymer Science and Technology, London, 1966.
3. W. C. Schmieder, W. C. Carter *et al.*, *J. Am. Chem. Soc.*, **67,** 959 (1945).
4. F. H. Müller, *Kolloid-Z.*, **134**(2/3), 207 (1953).
5. L. R. G. Treloar, *The Physics of Rubber Elasticity*, Oxford University Press (1958).
6. T. G. Fox and S. Loshack, *Rheology*, R. F. Eirich (editor), **1,** p. 431, Academic Press, New York and London (1956).
7. F. Bueche, *J. Chem. Phys.*, **20,** 1959 (1952) and **25,** 599 (1956).
8. F. Bueche, *Physical Properties of Polymers*, pp. 67–83, Interscience, London and New York (1962).
9. R. S. Porter and J. F. Johnson, *Trans. Soc. Rheol.*, **6,** 107 (1962).
10. M. M. Cross, paper presented at the autumn meeting of the British Society of Rheology, Glasgow, 1969, and subsequently published in *Polymer*.
11. P. J. Flory, *J. Am. Chem. Soc.*, **62,** 1057 (1940).
12. G. Locati, L. Gargani and A. De Chirico, *Rheol. Acta*, **13,** 278 (1974).
13. W. Wesslau, *Makromol. Chem.*, **20,** 111 (1956).
14. G. Locati, L. Gargani and A. De Chirico, *Polymer Letters*, **11**(2), 95 (1973).

5

Vectors and Tensors. Fundamental Equations

Vectors are entities which possess direction as well as magnitude. Thus, a force applied to one point of a body is a *vector* since it is characterised by both magnitude and direction. If the force is applied not merely at one point but at many points simultaneously with interacting effects, as for example in a continuous body subjected to tension, then we are dealing with a vector of a higher order which we call a *tensor*. Conversely, we could regard vectors as degenerate tensors.

If we chose any arbitrary system of reference coordinates in three dimensions such that the point of application of the force coincides with the origin, then we can represent a vector by a straight line headed by an arrow indicative of its direction and by a length indicative of its magnitude. Velocity is a vector since it obviously has both magnitude and direction. If we represent the velocity vector in a plane by $\mathbf{v}$ (Fig. 5.1) we can clearly resolve it algebraically by referring to its components $\mathbf{v}_x$ and $\mathbf{v}_y$. In other words, we can realise a vector by applying component forces $\mathbf{v}_x$ and $\mathbf{v}_y$ to a point in such a way that a resultant force $\mathbf{v}$ is obtained. This was known to prehistoric man who used bows and slings; he would apply two component forces to the missile and would propel the missile: (1) with a velocity which was a function of the magnitude of these forces, and (2) in a direction which was a function of the direction of the two component forces. These functions are additive, so that one can write:

$$\mathbf{v} = \mathbf{v}_x + \mathbf{v}_y$$

but the rules of addition are different from those which apply to directionless magnitudes or *scalars*.

We are not concerned here with the *derivation* of the rules which apply to the manipulation of vectors. Suffice it to say that all algebraic operations (addition and subtraction, multiplication and division, integration and differentiation) can be carried out in an appropriate manner the justification of which can be found in standard mathematical textbooks.

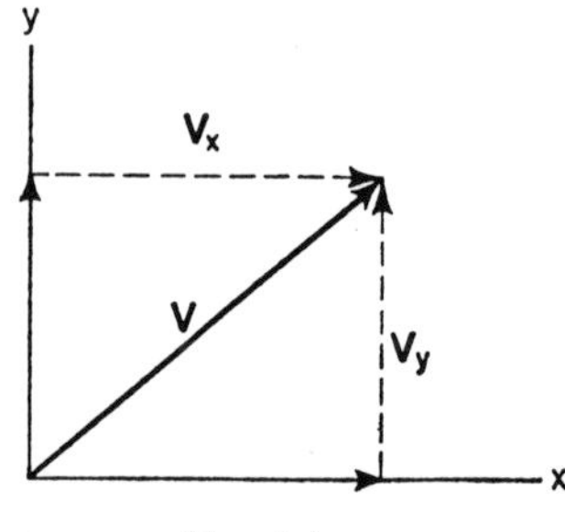

FIG. 5.1.

It is, however, necessary to state that the scalar and directional attributes of a vector can be separated for reasons of algebraic convenience and that this is done by introducing the concept of the *unit vectors* **i, j, k,** such that:

$$\mathbf{v}_x = \mathbf{i}v_x$$
$$\mathbf{v}_y = \mathbf{j}v_y$$
$$\mathbf{v}_z = \mathbf{k}v_z$$

where **i, j, k** always have a magnitude of unity and a direction coincident with that of the appropriate reference coordinate and where v_x, v_y, v_z are scalars.

It will be apparent that with different algebraic operational methods it is essential to use a method of notation which differs from that which is used in the manipulation of scalars. There are additional compelling reasons for such a notation, namely:

1. The number of components which have to be considered and which make up our operating equations may be large and the expressions, although generally of a simple cyclic nature, become cumbersome. Vector notation provides a shorthand way of expressing them clearly and enables one to carry out the necessary manipulations in a simple manner. The final expression can then be expanded using one's memory or tables.

2. The final equations are rigorous and make no simplifying assumptions. If assumptions become necessary in order to obtain a numerical solution from experimental data which may be incomplete because of difficulties in determining certain components, then these assumptions and corresponding provisos in the assessment of the result can be made. This leaves the door open for refining the result by more rigorous treatment in the future as experimental techniques become more sophisticated.

3. In addition to being concise, equations in vector notation are independent of arbitrary coordinate systems. The expansion of the final equation will take the form appropriate to the coordinate system which is defined by the geometry of the force field, but it is not necessary to pay attention to the latter during the generalised algebraic manipulation until the final equation is obtained.

VECTOR AND TENSOR NOTATION

It is necessary to define the notation which will be used in the following. It is not the *only* notation in common use, but it is a reasonably simple one.

(i) Vectors and tensors are characterised in print by bold type. In writing it is convenient to denote them by a squiggle under the symbol.

(ii) Single subscripts denote a component. Thus, a vector **v** has a component v_x with unit vector **i** and magnitude v_x. The velocity vector in a rectangular coordinate system x, y, z, will therefore be completely defined by the unit vectors **i, j, k** and the component scalars v_x, v_y, v_z.

(iii) If more than one subscript is present, then we are dealing with the components of a tensor, characterised by the following convention: The second subscript defines the direction of the force and the first subscript defines the plane which is perpendicular to the stated axis. For example, in a system of rectangular coordinates, the stress tensor τ will have the components shown in Fig. 5.2.

It is seen that the stress tensor has nine components. Those with equal subscripts represent normal (tensile) forces, whilst those with unequal subscripts represent tangential (shear) forces. It is therefore clear that the shear stress tensor can be represented by an array, the full development of which is:

$$\begin{matrix} \tau_{xx} & \tau_{xy} & \tau_{xz} \\ \tau_{yx} & \tau_{yy} & \tau_{yz} \\ \tau_{zx} & \tau_{zy} & \tau_{zz} \end{matrix}$$

(iv) The *nabla function* or 'vector operator' $\mathbf{V}$ is a shorthand method of writing the expression:

$$\mathbf{i}\frac{\partial}{\partial x} + \mathbf{j}\frac{\partial}{\partial y} + \mathbf{k}\frac{\partial}{\partial z}$$

for rectangular coordinates.

It never stands alone, since the partial differentials must state the parameter which is subjected to the operation.

Thus, when we write $\nabla\phi$, we mean that parameter ϕ is subjected to the operation:

$$\left(\mathbf{i}\frac{\partial}{\partial x} + \mathbf{j}\frac{\partial}{\partial y} + \mathbf{k}\frac{\partial}{\partial z}\right)\phi = \mathbf{i}\frac{\partial\phi}{\partial x} + \mathbf{j}\frac{\partial\phi}{\partial y} + \mathbf{k}\frac{\partial\phi}{\partial z}$$

for rectangular coordinates.

$\nabla\phi$ is also called the 'gradient of ϕ' or 'grad ϕ', where ϕ is a scalar.

(v) Vector multiplication, like vector addition, has its special algebraic rules. These need not be derived here; but it must be stated that there exist two kinds of products, a '*scalar product*' or '*dot product*' and a *vector product* or '*cross product*'.

(vi) The *scalar or dot product* is written $\mathbf{p}.\mathbf{q}$. If it involves the nabla function (which is itself a vector), then $\nabla.\mathbf{a}$ is called the '*divergence of a*' or '*div a*', so that

$$\mathrm{div}\,\mathbf{a} = \nabla.\mathbf{a} = \left(\mathbf{i}\frac{\partial}{\partial x} + \mathbf{j}\frac{\partial}{\partial y} + \mathbf{k}\frac{\partial}{\partial z}\right)(\mathbf{i}a_x + \mathbf{j}a_y + \mathbf{k}a_z)$$

div $\mathbf{a}$ therefore involves the scalar entities of the components a_x, a_y and a_z of the vector $\mathbf{a}$.

It must be added that by the rules of vector manipulation the products of the unit vectors $\mathbf{i}$, $\mathbf{j}$ and $\mathbf{k}$ are as follows:

$$\left.\begin{array}{ll} \mathbf{i}.\mathbf{i} = 1 & \mathbf{i}.\mathbf{j} = 0 \\ \mathbf{j}.\mathbf{j} = 1 & \mathbf{i}.\mathbf{k} = 0 \\ \mathbf{k}.\mathbf{k} = 1 & \mathbf{j}.\mathbf{k} = 0 \\ & \mathbf{j}.\mathbf{i} = 0 \\ & \mathbf{k}.\mathbf{i} = 0 \\ & \mathbf{k}.\mathbf{j} = 0 \end{array}\right\} \quad or: \quad \begin{array}{ll} \mathbf{i}.\mathbf{j} = 1 & \text{when } \mathbf{i} = \mathbf{j} \\ \mathbf{i}.\mathbf{j} = 0 & \text{when } \mathbf{i} \neq \mathbf{j} \end{array}$$

so that

$$\mathrm{div}\,\mathbf{a} = \nabla.\mathbf{a} = \frac{\partial a_x}{\partial x} + \frac{\partial a_y}{\partial y} + \frac{\partial a_z}{\partial z}$$

The quantities on the right-hand side of this expression are the scalar entities of the components of the vector $\mathbf{a}$ and their sum is therefore a scalar.

In general, the dot product of two vectors is a scalar and this is the reason

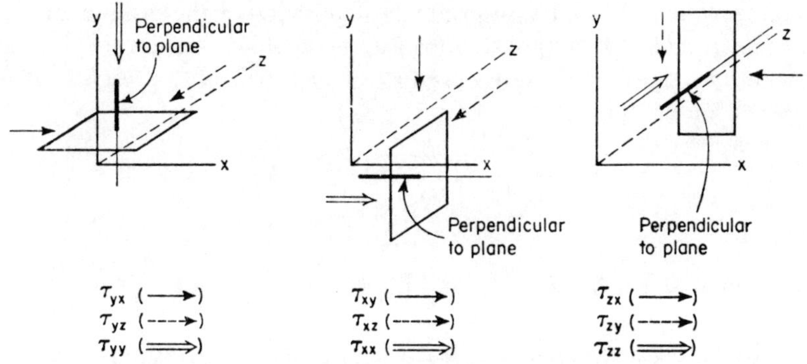

FIG. 5.2. The components of the stress tensor τ.

why the dot product is also called (paradoxically at first sight, but nevertheless logically) the scalar product.

(viii) The '*vector product*' or '*cross product*' of two vectors is written $\mathbf{p} \wedge \mathbf{q}$. If it involves the nabla function (which is itself a vector), then $\nabla \wedge \mathbf{b}$ is called the 'curl of b' or '*curl b*', where

$$\operatorname{curl} b = \nabla \wedge \mathbf{b} = \left(\mathbf{i}\frac{\partial}{\partial x} + \mathbf{j}\frac{\partial}{\partial y} + \mathbf{k}\frac{\partial}{\partial z} \right) \wedge (ib_x + jb_y + kb_z)$$

According to the rules of cross product multiplication, the order of writing the multiplicands (unlike in ordinary algebra) is *not* interchangeable, but

$$\mathbf{p} \wedge \mathbf{q} = -\mathbf{q} \wedge \mathbf{p}$$

Furthermore, the cross products of unit vectors are as follows:

$$\mathbf{i} \wedge \mathbf{i} = 0 \qquad \mathbf{i} \wedge \mathbf{j} = -\mathbf{j} \wedge \mathbf{i} = \mathbf{k}$$
$$\mathbf{j} \wedge \mathbf{j} = 0 \qquad \mathbf{j} \wedge \mathbf{k} = -\mathbf{k} \wedge \mathbf{j} = \mathbf{i}$$
$$\mathbf{k} \wedge \mathbf{k} = 0 \qquad \mathbf{k} \wedge \mathbf{i} = -\mathbf{i} \wedge \mathbf{k} = \mathbf{j}$$

Now, if the right-hand side of the curl $\mathbf{b}$ equation is multiplied out we obtain:

$$\operatorname{curl} \mathbf{b} = \nabla \wedge \mathbf{b} = \mathbf{i} \wedge \mathbf{i}\,\frac{\partial b_x}{\partial x} + \mathbf{i} \wedge \mathbf{j}\,\frac{\partial b_y}{\partial x} + \mathbf{i} \wedge \mathbf{k}\,\frac{\partial b_z}{\partial x}$$

$$+ \mathbf{j} \wedge \mathbf{i}\,\frac{\partial b_x}{\partial y} + \mathbf{j} \wedge \mathbf{j}\,\frac{\partial b_y}{\partial y} + \mathbf{j} \wedge \mathbf{k}\,\frac{\partial b_z}{\partial z}$$

$$+ \mathbf{k} \wedge \mathbf{i}\,\frac{\partial b_x}{\partial z} + \mathbf{k} \wedge \mathbf{j}\,\frac{\partial b_y}{\partial z} + \mathbf{k} \wedge \mathbf{k}\,\frac{\partial b_z}{\partial z}$$

Making use of the stated cross products of unit vectors, the terms involving $\mathbf{i} \wedge \mathbf{i}$, $\mathbf{j} \wedge \mathbf{j}$ and $\mathbf{k} \wedge \mathbf{k}$ disappear. Substituting the third unit vector (with the appropriate sign) for the cross product of each pair of unit vectors as it appears, we obtain:

$$\operatorname{curl} \mathbf{b} = \mathbf{\nabla} \wedge \mathbf{b} = \mathbf{k}\frac{\partial b_y}{\partial x} - \mathbf{j}\frac{\partial b_z}{\partial x} - \mathbf{k}\frac{\partial b_x}{\partial y} + \mathbf{i}\frac{\partial b_z}{\partial y} + \mathbf{j}\frac{\partial b_x}{\partial z} - \mathbf{i}\frac{\partial b_y}{\partial z}$$

or

$$\operatorname{curl} \mathbf{b} = \mathbf{\nabla} \wedge \mathbf{b} = \mathbf{i}\left(\frac{\partial b_z}{\partial y} - \frac{\partial b_y}{\partial z}\right) + \mathbf{j}\left(\frac{\partial b_x}{\partial z} - \frac{\partial b_z}{\partial x}\right) + \mathbf{k}\left(\frac{\partial b_y}{\partial x} - \frac{\partial b_x}{\partial y}\right)$$

A strict cyclic arrangement is clearly in evidence. It will be noted that the right-hand side of the above equation is identical with the expansion of the determinant:

$$\begin{vmatrix} \mathbf{i} & \mathbf{j} & \mathbf{k} \\ \dfrac{\partial}{\partial x} & \dfrac{\partial}{\partial y} & \dfrac{\partial}{\partial z} \\ b_x & b_y & b_z \end{vmatrix}$$

It will also be noted that whilst the dot product of two vectors is a scalar, the cross product of two vectors is a vector the components of which are the three right-hand side terms of the equation for the curl.

(viii) Another useful shorthand expression is that of the *multiple-dot product* which is used to indicate that the summation is to be applied to more than one kind of component. This does not arise with vectors (tensors of the first order) which have only components of one kind, but it *does* arise with tensors of the second order because they have components of two kinds, namely those where the subscripts are identical, and those where the subscripts are non-identical.

If

$$\mathbf{\nabla} . \boldsymbol{\theta} = \frac{\partial}{\partial x}\theta_x + \frac{\partial}{\partial y}\theta_y + \frac{\partial}{\partial z}\theta_z$$

then

$$\mathbf{v}(\mathbf{\nabla}\boldsymbol{\theta}) = v_x\frac{\partial \theta_x}{\partial x} + v_y\frac{\partial \theta_y}{\partial y} + v_z\frac{\partial \theta_z}{\partial z}$$

where $\mathbf{v}$ is a vector (say, velocity).

If, on the other hand, we treat the second order tensor τ in an analogous fashion, then we have two types of components, namely the normal force (tensile) components of the kind τ_{ii} and the tangential force (shear) components of the kind τ_{ij}, where i and j (in rectangular coordinates) can have any of the connotations x, y and z in turn.

In such a case, in order to emphasise that there are two kinds of forces involved which have to be summed, it is convenient to write, *not* $\boldsymbol{\tau}\,.\,(\boldsymbol{\nabla}\mathbf{v})$, but $\boldsymbol{\tau}:(\boldsymbol{\nabla}\mathbf{v})$.

Using the same rules as before, the expansion of $\boldsymbol{\tau}:(\boldsymbol{\nabla}\,.\,\mathbf{v})$ becomes (for rectangular coordinates) in general:

$$\boldsymbol{\tau}:(\boldsymbol{\nabla}\mathbf{v}) = \sum_{ii} \tau_{ii}(\boldsymbol{\nabla}\mathbf{v}) + \sum_{\{{i\,j}\atop{j\,i}\}} \tau_{\{{i\,j}\atop{j\,i}\}}(\boldsymbol{\nabla}\mathbf{v})$$

or in particular:

$$\boldsymbol{\tau}:(\boldsymbol{\nabla}\,.\,\mathbf{v}) = \left\{\tau_{xx}\frac{\partial v_x}{\partial x} + \tau_{yy}\frac{\partial v_y}{\partial y} + \tau_{zz}\frac{\partial v_z}{\partial z}\right\} + \left\{(\tau_{xy} + \tau_{yx})\left(\frac{\partial v_x}{\partial y} + \frac{\partial v_y}{\partial x}\right)\right.$$
$$\left. + (\tau_{xz} + \tau_{zx})\left(\frac{\partial v_x}{\partial z} + \frac{\partial v_z}{\partial x}\right) + (\tau_{yz} + \tau_{zy})\left(\frac{\partial v_y}{\partial z} + \frac{\partial v_z}{\partial y}\right)\right\}$$

Whilst the expansion can be easily and purely mechanically written down without being an undue burden on the memory, it is obviously extremely cumbersome and quite unnecessary if the elegant and perfectly unambiguous left-hand expression can be used instead.

Another fact emerges from the expansion however: if normal force components are so small as to be negligible compared to the shear force components, then the first three terms can be ignored. Furthermore, if $\tau_{ij} \equiv \tau_{ji}$ (as is the case in symmetrical tensors) then the expansion can be simplified to:

$$\frac{\boldsymbol{\tau}:(\boldsymbol{\nabla}\,.\,\mathbf{v})}{2} = \left\{\tau_{xy}\left(\frac{\partial v_x}{\partial y} + \frac{\partial v_y}{\partial x}\right) + \tau_{yz}\left(\frac{\partial v_y}{\partial z} + \frac{\partial v_z}{\partial y}\right) + \tau_{zx}\left(\frac{\partial v_z}{\partial x} + \frac{\partial v_x}{\partial z}\right)\right\}$$

A double dot product is used in conjunction with the strain rate tensor $\boldsymbol{\Delta}$ with which we shall be dealing in due course.

OTHER COORDINATE SYSTEMS

So far it has been assumed that we are only dealing with rectangular coordinates. But it is obvious that other coordinate systems are equally capable of defining the position of points in space.

Of the coordinate systems other than the rectangular those of the greatest practical importance are the cylindrical and the spherical systems.

Assuming that a velocity vector **v** acts on a point in space, then this vector can be resolved into component vectors in the following ways (Fig. 5.3).

In order to convert the expansions of grad **v**, div **v** and curl **v** from the rectangular to the cylindrical or spherical coordinate system, it is only necessary to substitute the correct function of r and θ for x and y in the

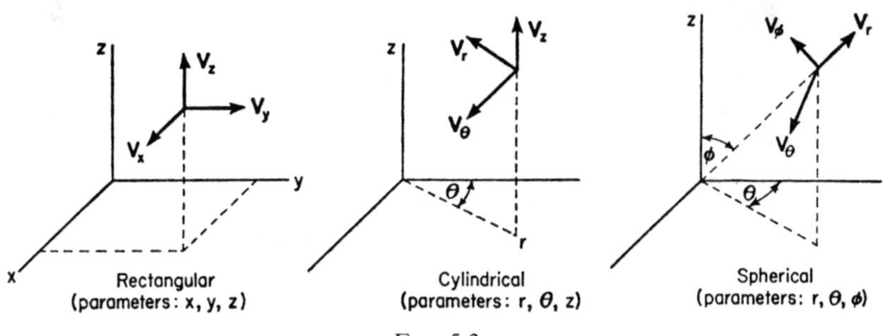

Fig. 5.3.

cylindrical system, and the correct functions of r, θ and ϕ for x, y and z in the spherical system. In extruder flow problems the cylindrical geometry is as important as the rectangular geometry.

The way in which a vector can be split into its components in rectangular, cylindrical and spherical coordinates is perhaps most usefully illustrated in the case of the *heat flux vector*: Heat itself is a scalar, but the conductive transfer of heat from point A to point B involves a gradient along the line AB. This means that conductive heat flux is directional and therefore a vector.

The scalar entities of the components of the conductive heat flux vector **q** are as shown in Table 5.1, where k is the thermal conductivity of the material in question.

In extrusion flow problems it is, of course, necessary to express the stress tensor in the terms in which the data are stated. The velocity in the axial direction of an extruder (the z-axis in a cylindrical geometry) is generally available. There is also flow in the radial direction; this is possibly accessible with considerable experimental sophistication and may not be negligibly small. Flow in the third direction should also be included in a rigorous flow equation, but this is not experimentally determinable so far and is, in any case, thought to be negligibly small.

Resolving the stress tensor τ into its components of which the scalar

TABLE 5.1[a]

Rectangular	Cylindrical	Spherical
$q_x = -k\dfrac{\partial T}{\partial x}$	$q_r = -k\dfrac{\partial T}{\partial r}$	$q_r = -k\dfrac{\partial T}{\partial r}$
$q_y = -k\dfrac{\partial T}{\partial y}$	$q_\theta = -k\dfrac{1}{r}\dfrac{\partial T}{\partial \theta}$	$q_\theta = -k\dfrac{1}{r}\dfrac{\partial T}{\partial \theta}$
$q_z = -k\dfrac{\partial T}{\partial z}$	$q_z = -k\dfrac{\partial T}{\partial z}$	$q_\phi = -k\dfrac{1}{r\sin\theta}\dfrac{\partial T}{\partial \phi}$

[a] Reprinted with permission from R. B. Bird, W. E. Steward and E. N. Lightfoot, *Transport Phenomena*, John Wiley (1960).

entities are τ_{ii} and τ_{ij} (and of which, as has been shown, there are nine) is not sufficient. In order to solve flow problems, the stress tensor τ must ultimately be expressed in terms of the scalars of the velocity components which are relevant to the prevailing geometry. Tables for conversion of the stress tensor components have been worked out and are reproduced in Table 5.2 in slightly modified form.[1] In these expansions η_N is the Newtonian viscosity of the liquid and K the compressibility.

The equations in Table 5.2 look formidably complex, but it should be noted that they can be generalised (by the shorthand notation described earlier on) as follows:

(a)
$$\tau_{ii} = \eta_N\left[2\frac{\partial v_i}{\partial i} - \frac{2}{3}\operatorname{div}\mathbf{v}\right] + K\operatorname{div}\mathbf{v}$$

(b)
$$\tau_{ij} = \tau_{ji} = \eta_N\left[\frac{\partial v_i}{\partial j} + \frac{\partial v_j}{\partial i}\right]$$

where i and j can have any value of

$$\left\{\begin{array}{c}x, y, z \\ r, \theta, z \\ r, \theta, \phi\end{array}\right\} \quad \text{in a} \quad \left\{\begin{array}{c}\text{rectangular} \\ \text{cylindrical} \\ \text{spherical}\end{array}\right\}$$

coordinate system, as the case may be.

They are further simplified in practice as follows: If the liquid in flow is assumed to be incompressible, $K = 0$ then, despite the fact that large pressure drops occur along an extruder channel, the velocity components v_i

TABLE 5.2
COMPONENTS OF THE STRESS TENSOR τ [a]

Rectangular coordinates

$$(a) \quad \begin{cases} \tau_{xx} = \eta_N\left[2\dfrac{\partial v_x}{\partial x} - \dfrac{2}{3}(\mathbf{\nabla}.\mathbf{v})\right] + K(\mathbf{\nabla}.\mathbf{v}) \\[2ex] \tau_{yy} = \eta_N\left[2\dfrac{\partial v_y}{\partial y} - \dfrac{2}{3}(\mathbf{\nabla}.\mathbf{v})\right] + K(\mathbf{\nabla}.\mathbf{v}) \\[2ex] \tau_{zz} = \eta_N\left[2\dfrac{\partial v_z}{\partial z} - \dfrac{2}{3}(\mathbf{\nabla}.\mathbf{v})\right] + K(\mathbf{\nabla}.\mathbf{v}) \end{cases}$$

where

$$(\mathbf{\nabla}.\mathbf{v}) = \operatorname{div}\mathbf{v} = \frac{\partial v_x}{\partial x} + \frac{\partial v_y}{\partial y} + \frac{\partial v_z}{\partial z}$$

$$(b) \quad \begin{cases} \tau_{xy} = \tau_{yx} = \eta_N\left[\dfrac{\partial v_x}{\partial y} + \dfrac{\partial v_y}{\partial x}\right] \\[2ex] \tau_{yz} = \tau_{zy} = \eta_N\left[\dfrac{\partial v_y}{\partial z} + \dfrac{\partial v_z}{\partial y}\right] \\[2ex] \tau_{zx} = \tau_{xz} = \eta_N\left[\dfrac{\partial v_z}{\partial x} + \dfrac{\partial v_x}{\partial z}\right] \end{cases}$$

Cylindrical coordinates

$$(a) \quad \begin{cases} \tau_{rr} = \eta_N\left[2\dfrac{v_r}{r} - \dfrac{2}{3}\operatorname{div}\mathbf{v}\right] + K\operatorname{div}\mathbf{v} \\[2ex] \tau_{\theta\theta} = \eta_N\left[2\left(\dfrac{1}{r}\dfrac{\partial v_\theta}{\partial\theta} + \dfrac{v_r}{r}\right) - \dfrac{2}{3}\operatorname{div}\mathbf{v}\right] = K\operatorname{div}\mathbf{v} \\[2ex] \tau_{zz} = \eta_N\left[2\dfrac{\partial v_z}{\partial z} - \dfrac{2}{3}\operatorname{div}\mathbf{v}\right] + K\operatorname{div}\mathbf{v} \end{cases}$$

where

$$\operatorname{div}\mathbf{v} = \frac{1}{r}\frac{\partial}{\partial r}rv_r + \frac{1}{r}\frac{\partial v_\theta}{\partial\theta} + \frac{\partial v_z}{\partial z}$$

TABLE 5.2—*continued*

$$(b) \quad \begin{cases} \tau_{r\theta} = \tau_{\theta r} = \eta_N \left[r \frac{\partial}{\partial r}\left(\frac{v_\theta}{r}\right) + \frac{1}{r}\frac{\partial v_r}{\partial \theta} \right] \\[3mm] \tau_{\theta z} = \tau_{z\theta} = \eta_N \left[\frac{\partial v_\theta}{\partial z} + \frac{1}{r}\frac{\partial v_z}{\partial \theta} \right] \\[3mm] \tau_{zr} = \tau_{rz} = \eta_N \left[\frac{\partial v_z}{\partial r} + \frac{\partial v_r}{\partial z} \right] \end{cases}$$

Spherical coordinates

$$(a) \quad \begin{cases} \tau_{rr} = \eta_N \left[2\frac{\partial v_r}{\partial r} - \frac{2}{3}\,\mathrm{div}\,\mathbf{v} \right] + K\,\mathrm{div}\,\mathbf{v} \\[3mm] \tau_{\theta\theta} = \eta_N \left[2\left(\frac{1}{r}\frac{\partial v_\theta}{\partial \theta} + \frac{v_r}{r} \right) - \frac{2}{3}\,\mathrm{div}\,\mathbf{v} \right] + K\,\mathrm{div}\,\mathbf{v} \\[3mm] \tau_{\phi\phi} = \eta_N \left[2\left(\frac{1}{r\sin\theta}\frac{\partial v_\phi}{\partial \phi} + \frac{v_r}{r} + \frac{v_\theta \cot\theta}{r} \right) - \frac{2}{3}\,\mathrm{div}\,\mathbf{v} \right] + K\,\mathrm{div}\,\mathbf{v} \end{cases}$$

$$(b) \quad \begin{cases} \tau_{r\theta} = \tau_{\theta r} = \eta_N \left[r\frac{\partial}{\partial r}\left(\frac{v_\theta}{r}\right) + \frac{1}{r}\frac{\partial v_r}{\partial \theta} \right] \\[3mm] \tau_{\theta\phi} = \tau_{\phi\theta} = \eta_N \left[\frac{\sin\theta}{r}\frac{\partial}{\partial \theta}\left(\frac{v_\phi}{\sin\theta}\right) + \frac{1}{r\sin\theta}\frac{\partial v_\theta}{\partial \phi} \right] \\[3mm] \tau_{\phi r} = \tau_{r\phi} = \eta_N \left[\frac{1}{r\sin\theta}\frac{\partial v_r}{\partial \phi} + r\frac{\partial}{\partial r}\left(\frac{v_\phi}{r}\right) \right] \end{cases}$$

where

$$\mathrm{div}\,\mathbf{v} = \frac{1}{r^2}\frac{\partial}{\partial r}(r^2 v_r) + \frac{1}{r\sin\theta}\frac{\partial}{\partial \theta}(v_\theta \sin\theta) + \frac{1}{r\sin\theta}\frac{\partial v_\phi}{\partial \phi}$$

[a] Reprinted with permission from R. B. Bird, W. E. Steward and E. N. Lightfoot, *Transport Phenomena*, John Wiley (1960).

are constant in the *i*-direction. Therefore div **v** (see expansion, p. 44) and the first term in the bracket of equations (*a*) disappear. This means that the normal forces can be ignored. If the further approximation is made that in an extruder and die channel the velocity vector components

$$\frac{\partial}{\partial r}\left(\frac{v_\theta}{r}\right) \qquad \frac{\partial v_r}{\partial \theta} \qquad \frac{\partial v_\theta}{\partial z} \qquad \frac{\partial v_z}{\partial \theta} \qquad \text{and} \qquad \frac{\partial v_r}{\partial z}$$

are all zero, then the only velocity vector component remaining is $\partial v_z/\partial r$ and the stress tensor is represented by the magnitude of τ_{rz} alone and all other components disappear, so that: the *magnitude* of τ is given by $\eta_N(\partial v_z/\partial r)$, a result which it is extremely simple to apply and which we shall be using extensively in due course.

In aerodynamics where the fluid is a gas and not a liquid, compressibility is, of course, a most important factor even under isothermal conditions and div **v** cannot therefore be ignored. Fortunately we do not need to concern ourselves with the great complexities which arise as a consequence since we are exclusively interested in polymer melts.

THE EQUATIONS OF CONTINUITY, MOMENTUM AND ENERGY

These equations[2] formulate fundamental principles of physics. They are not in themselves sufficient to evaluate flow problems since they are independent of the rheological behaviour of the liquid, but together with the appropriate rheological equation they provide all the necessary data for solving flow problems.

A volume element of liquid in motion is defined by the scalar quantities of density (ρ), pressure (P), temperature (T) and by the velocity vector **v**. The stress τ which results from the shear forces is a tensor of the second order. The continuity equation considers these parameters as a function of time (t) and position as defined by the components in any generalised system of coordinates.

Imagine that the volume element is isolated from the bulk of the moving volume but continues to follow the motion of the latter without exerting any force upon it. The volume element can then exchange energy—but not matter—with the bulk volume across its boundaries. Since, according to the principle of the conservation of matter, the mass of a closed system remains constant, any local increase or decrease in density with time within the volume element must be exactly reflected by the rate of flow of material into or away from such loci, so that the overall density change within the volume element is zero. This is expressed by the *continuity equation*

$$\frac{\mathrm{d}\rho}{\mathrm{d}t} = -\rho(\nabla \cdot \mathbf{v}) = -\rho\,\mathrm{div}\,\mathbf{v}$$

On expanding this general equation for rectangular, cylindrical and spherical coordinates the following equations are obtained:

Rectangular coordinates (x, y, z):

$$\frac{\partial \rho}{\partial t} = -\left[\frac{\partial(\rho v_x)}{\partial x} + \frac{\partial(\rho v_x)}{\partial y} + \frac{\partial(\rho v_z)}{\partial z}\right]$$

Cylindrical coordinates (r, θ, z):

$$\frac{\partial \rho}{\partial t} = -\left[\frac{\partial(\rho_r v_r)}{r\,\partial r} + \frac{\partial(\rho v_\theta)}{r\,\partial \theta} + \frac{\partial(\rho v_z)}{\partial z}\right]$$

Spherical coordinates (r, θ, ϕ):

$$\frac{\partial \rho}{\partial t} = -\left[\frac{\partial(\rho_r^2 v_r)}{r^2\,\partial r} + \frac{\partial(\rho v_\theta \sin \theta)}{r \sin \theta\,\partial \theta} + \frac{\partial(\rho v_\phi)}{r \sin \theta\,\partial \phi}\right]$$

If the liquid can be assumed to be incompressible, then div $\mathbf{v}$ will be zero and no local changes can occur in the density of the volume element as a function of time. This is generally the case in polymer melts during processing, but obviously, if the enclosed material is gaseous this assumption does not hold and tremendous complications may be expected to arise, for example, in the extrusion of expanded polystyrene. The continuity equation was included here for the sake of completeness and because it demonstrates, once again, the elegant and convenient way in which a relationship can be generally expressed through the use of vector and tensor notation.

Newton's second law of motion requires[2] 'that the rate of increase of momentum of the fluid element be equal to the sum of all the forces acting upon it'.

The rate of increase of momentum is equal to $\rho\,d\mathbf{v}/dt$.

The body forces are represented by a function of:

(i)   the pressure differential in the net direction of flow; this is a negative quantity given by $-\nabla P$ or grad P,
(ii) the stress tensor; this is given by $\nabla . \tau$ or div τ,
(iii) the gravitational force; this is given by $\rho \mathbf{g}$, where the vector $\mathbf{g}$ represents 'the resultant of the body forces (per unit mass) acting on the fluid at any point'.[2]

$$\textbf{TABLE 5.3}^{a}$$

Rectangular coordinates

x-component:

$$\rho\left(\frac{\partial v_x}{\partial t} + v_x\frac{\partial v_x}{\partial x} + v_y\frac{\partial v_x}{\partial y} + v_z\frac{\partial v_x}{\partial z}\right) = -\frac{\partial P}{\partial x} + \left(\frac{\partial \tau_{xx}}{\partial x} + \frac{\partial \tau_{yx}}{\partial y} + \frac{\partial \tau_{zx}}{\partial z}\right) + \rho g_x$$

y-component:

$$\rho\left(\frac{\partial v_y}{\partial t} + v_x\frac{\partial v_y}{\partial x} + v_y\frac{\partial v_y}{\partial y} + v_z\frac{\partial v_y}{\partial z}\right) = -\frac{\partial P}{\partial y} + \left(\frac{\partial \tau_{xy}}{\partial x} + \frac{\partial \tau_{yy}}{\partial y} + \frac{\partial \tau_{zy}}{\partial z}\right) + \rho g_y$$

z-component:

$$\rho\left(\frac{\partial v_z}{\partial t} + v_x\frac{\partial v_z}{\partial x} + v_y\frac{\partial v_z}{\partial y} + v_z\frac{\partial v_z}{\partial z}\right) = -\frac{\partial P}{\partial z} + \left(\frac{\partial \tau_{xz}}{\partial x} + \frac{\partial \tau_{yz}}{\partial y} + \frac{\partial \tau_{zz}}{\partial z}\right) + \rho g_z$$

Cylindrical coordinates

r-component:

$$\rho\left(\frac{\partial v_r}{\partial t} + v_r\frac{\partial v_r}{\partial r} + \frac{v_\theta}{r}\frac{\partial v_r}{\partial \theta} - \frac{v_\theta-}{r} + v_z\frac{\partial v_r}{\partial z}\right)$$

$$= -\frac{\partial P}{\partial r} + \left[\frac{1}{r}\frac{\partial}{\partial r}(r\tau_{rr}) + \frac{1}{r}\frac{\partial \tau_{r\theta}}{\partial \theta} - \frac{\tau_{\theta\theta}}{r} + \frac{\partial \tau_{rz}}{\partial z}\right] + \rho g_r$$

θ-component:

$$\rho\left(\frac{\partial v_\theta}{\partial t} + v_r\frac{\partial v_\theta}{\partial r} + \frac{v_\theta}{r}\frac{\partial v_\theta}{\partial \theta} + \frac{v_r v_\theta}{r} + v_z\frac{\partial v_\theta}{\partial z}\right)$$

$$= -\frac{1}{r}\frac{\partial P}{\partial \theta} + \left[\frac{1}{r^2}\frac{\partial}{\partial r}(r^2\tau_{r\theta}) + \frac{1}{r}\frac{\partial \tau_{\theta\theta}}{\partial \theta} + \frac{\partial \tau_{\theta z}}{\partial z}\right] + \rho g_\theta$$

z-component:

$$\rho\left(\frac{\partial v_z}{\partial t} + v_r\frac{\partial v_z}{\partial r} + \frac{v_\theta}{r}\frac{\partial v_z}{\partial \theta} + v_z\frac{\partial v_z}{\partial z}\right)$$

$$= -\frac{\partial P}{\partial z} + \left[\frac{1}{r}\frac{\partial}{\partial r}(r\tau_{rz}) + \frac{1}{r}\frac{\partial \tau_{\theta z}}{\partial \theta} + \frac{\partial \tau_{zz}}{\partial z}\right] + \rho g_z$$

TABLE 5.3[a]—*continued*

Spherical coordinates

r-component:

$$\rho\left(\frac{\partial v_r}{\partial t} + v_r\frac{\partial v_r}{\partial r} + \frac{v_\theta}{r}\frac{\partial v_r}{\partial \theta} + \frac{v_\phi}{r\sin\theta}\frac{\partial v_r}{\partial \phi} - \frac{v_\theta^2 + v_\phi^2}{r}\right)$$

$$= -\frac{\partial P}{\partial r} + \left[\frac{1}{r^2}\frac{\partial}{\partial n}(r^2\tau_{rr}) + \frac{1}{r\sin\theta}\frac{\partial}{\partial \theta}(\tau_{r\theta}\sin\theta)\right.$$

$$\left. + \frac{1}{r\sin\theta}\frac{\partial \tau_{r\phi}}{\partial \phi} - \frac{\tau_{\theta\theta} + \tau_{\phi\phi}}{r}\right] + \rho g_r$$

θ-component:

$$\rho\left(\frac{\partial v_\theta}{\partial t} + v_r\frac{\partial v_\theta}{\partial r} + \frac{v_\theta}{r}\frac{\partial v_\theta}{\partial \theta} + \frac{v_\phi}{r\sin\theta}\frac{\partial v_\theta}{\partial \phi} + \frac{v_r v_\theta}{r} = \frac{v_\phi^2\cot\theta}{r}\right)$$

$$= -\frac{1}{r}\frac{\partial P}{\partial \theta} + \left[\frac{1}{r^2}\frac{\partial}{\partial r}(r^2\tau_{r\theta}) + \frac{1}{r\sin\theta}\frac{\partial}{\partial \theta}(\tau_{\theta\phi}\sin\theta)\right.$$

$$\left. + \frac{1}{r\sin\theta}\frac{\partial \tau_{\theta\phi}}{\partial \phi} + \frac{\tau_{r\theta}}{r} - \frac{\cot\theta}{r}\tau_{\phi\phi}\right] + \rho g_\theta$$

ϕ-component:

$$\rho\left(\frac{\partial v_\phi}{\partial t} + v_r\frac{\partial v_\phi}{\partial r} + \frac{v_\theta}{r}\frac{\partial v_\phi}{\partial \theta} + \frac{v_\phi}{r\sin\theta}\frac{\partial v_\phi}{\partial \phi} + \frac{v_\phi v_r}{r} + \frac{v_\theta v_\phi}{r}\cot\theta\right)$$

$$= -\frac{1}{r\sin\theta}\frac{\partial P}{\partial \phi} + \left[\frac{1}{r^2}\frac{\partial}{\partial r}(r^2\tau_{r\phi}) + \frac{1}{r}\frac{\partial \tau_{\theta\phi}}{\partial \theta}\right.$$

$$\left. + \frac{1}{r\sin\theta}\frac{\partial \tau_{\phi\phi}}{\partial \phi} + \frac{\tau_{r\phi}}{r} + \frac{2\cot\theta}{r}\tau_{\theta\phi}\right] + \rho g_\phi$$

---

[a] Reprinted with permission from R. B. Bird, W. E. Steward and E. N. Lightfoot, *Transport Phenomena*, John Wiley (1960).

The equation is therefore generally expressed as follows:

$$\rho\frac{d\mathbf{v}}{dt} = -\nabla P + \nabla\cdot\tau + \rho\mathbf{g} = -\operatorname{grad} P + \operatorname{div}\tau + \rho\mathbf{g}$$

and can be expanded as shown in Table 5.3.

Polymer Rheology

TABLE 5.4[a]

Rectangular coordinates

$$\rho C_v \left(\frac{\partial T}{\partial t} + v_x \frac{\partial T}{\partial x} + v_y \frac{\partial T}{\partial y} + v_z \frac{\partial T}{\partial z} \right) = - \left[\frac{\partial q_x}{\partial x} + \frac{\partial q_y}{\partial y} + \frac{\partial q_z}{\partial z} \right]$$

$$- T \left(\frac{\partial P}{\partial T} \right)_\rho \left(\frac{\partial v_x}{\partial x} + \frac{\partial v_y}{\partial y} + \frac{\partial v_z}{\partial z} \right) + \left\{ \tau_{xx} \frac{\partial v_x}{\partial x} + \tau_{yy} \frac{\partial v_y}{\partial y} + \tau_{zz} \frac{\partial v_z}{\partial z} \right\}$$

$$+ \left\{ \tau_{xy} \left(\frac{\partial v_x}{\partial y} + \frac{\partial v_y}{\partial x} \right) + \tau_{xz} \left(\frac{\partial v_x}{\partial z} + \frac{\partial v_z}{\partial x} \right) + \tau_{yz} \left(\frac{\partial v_y}{\partial z} + \frac{\partial v_z}{\partial y} \right) \right\}$$

Cylindrical coordinates

$$\rho C_v \left(\frac{\partial T}{\partial t} + v_r \frac{\partial T}{\partial r} + \frac{v_\theta}{r} \frac{\partial T}{\partial \theta} + v_z \frac{\partial T}{\partial_z} \right) = - \left[\frac{1}{r} \frac{\partial}{\partial r} (rq_r) + \frac{1}{r} \frac{\partial q\theta}{\partial \theta} + \frac{\partial q_z}{\partial z} \right]$$

$$- T \left(\frac{\partial P}{\partial T} \right)_\rho \left(\frac{1}{r} \frac{\partial}{\partial r} (rv_r) + \frac{1}{r} \frac{\partial v_\theta}{\partial \theta} + \frac{\partial v_z}{\partial z} \right) + \left\{ \tau_{rr} \frac{\partial v_r}{\partial r} + \tau_{\theta\theta} \frac{1}{r} \left(\frac{\partial v_\theta}{\partial \theta} + v_r \right) + \tau_{zz} \frac{\partial v_z}{\partial z} \right\}$$

$$+ \left\{ \tau_{r\theta} \left[r \frac{\partial}{\partial r} \left(\frac{v_\theta}{r} \right) + \frac{1}{r} \frac{\partial v_r}{\partial \theta} \right] + \tau_{rz} \left(\frac{\partial v_z}{\partial r} + \frac{\partial v_r}{\partial z} \right) + \tau_{\theta z} \left(\frac{1}{r} \frac{\partial v_z}{\partial \theta} + \frac{\partial v_\theta}{\partial z} \right) \right\}$$

Spherical coordinates

$$\rho C_v \left(\frac{\partial T}{\partial t} + v_r \frac{\partial T}{\partial r} + \frac{v_\theta}{r} \frac{\partial T}{\partial \theta} + \frac{v_\phi}{r \sin \theta} \frac{\partial T}{\partial \phi} \right)$$

$$= - \left[\frac{1}{r^3} \frac{\partial}{\partial r} (r^2 q_r) + \frac{1}{r \sin \theta} \frac{\partial}{\partial \theta} (q_\theta \sin \theta) + \frac{1}{r \sin \theta} \frac{\partial q_\phi}{\partial \phi} \right]$$

$$- T \left(\frac{\partial P}{\partial T} \right)_\rho \left(\frac{1}{r^2} \frac{\partial}{\partial r} (r^2 v_r) + \frac{1}{r \sin \theta} \frac{\partial}{\partial \theta} (v_\theta \sin \theta) + \frac{1}{r \sin \theta} \frac{\partial v_\phi}{\partial \phi} \right)$$

$$+ \left\{ \tau_{rr} \frac{\partial v_r}{\partial r} + \tau_{\theta\theta} \left(\frac{1}{r} \frac{\partial v_\theta}{\partial \theta} + \frac{V_r}{r} \right) + \tau_{\phi\phi} \left(\frac{1}{r \sin \theta} \frac{\partial v_\phi}{\partial \phi} + \frac{v_r}{r} + \frac{v_\theta \cot \phi}{r} \right) \right\}$$

$$+ \left\{ \tau_{r\theta} \left(\frac{\partial v_\theta}{\partial r} + \frac{1}{r} \frac{\partial v_\theta}{\partial \theta} - \frac{v_\theta}{r} \right) + \tau_{r\phi} \left(\frac{\partial v_\phi}{\partial r} + \frac{1}{r \sin \theta} \frac{\partial v_r}{\partial \phi} - \frac{v_\phi}{r} \right) \right.$$

$$\left. + \tau_{\theta\phi} \left(\frac{1}{r} \frac{\partial v_\phi}{\partial \theta} + \frac{1}{r \sin \theta} \frac{\partial v_\theta}{\partial \phi} - \frac{\cot \theta}{r} v_\phi \right) \right\}$$

In polymer melts $\rho\mathbf{g}$ is negligibly small compared to the other forces acting upon the liquid. We can also assume incompressibility, so that div τ disappears. Lastly, one can make the simplifying assumption that most of the velocity vector components in the cylindrical geometry of an extruder are zero. The result, then, is suitably simple and can be used to obtain solutions to flow problems.

The energy equation is based on the principle of conservation of energy as embodied in the first law of thermodynamics. In tensor notation it takes the form:

$$\rho C_v \frac{\mathrm{d}T}{v\,\mathrm{d}t} = -[\nabla \cdot \mathbf{q}] - T\left(\frac{\partial P}{\partial T}\right)_\rho [\nabla \cdot \mathbf{v}] + [\tau : \nabla \cdot \mathbf{v}]$$

where C_v is the specific heat at constant volume and $\mathbf{q}$, the conductive heat flux vector is given by Fourier's law of heat conduction:

$$\mathbf{q} = -k\nabla T$$

where k is the thermal conductivity of the material.

In compressible fluids such as gases, aerosols and smokes $[\nabla \cdot \mathbf{v}]$ or div $\mathbf{v}$ is clearly of the greatest importance, but in polymer melts it is zero if incompressibility is assumed, so that the energy equation again reduces to a very simply manageable expression. The expansion of the conductive heat flux vector has already been given.

The full rigorous expansion of the tensor form of the energy equation is given in Table 5.4 and again illustrates by its many terms and its consequent cumbersome nature what enormous advantages accrue from the use of generalised tensor notation.

An equation of state is required for problems in which the density is variable. This is manifestly important for gases, aerosols, smokes, and for materials in the temperature regions where discontinuous states coexist. But it is of no importance in the typical processing of polymer melts.

Equations which describe the temperature and pressure dependence of other fluid properties are required. These have already been considered in preceding chapters.

THE STRAIN RATE TENSOR

Bearing in mind the nature of the stress tensor τ, the strain rate $\dot{\gamma}$ must clearly also be a tensor function, since it has both magnitude and direction.

We use the symbol $\mathbf{\Delta}$ for the strain rate tensor and write the rheological equation

$$\tau = \eta_N \mathbf{\Delta} \qquad \text{and} \qquad \tau = \eta \mathbf{\Delta}$$

for a Newtonian and a power-law liquid respectively, incompressibility being assumed.

The strain rate tensor has components

$$\Delta_{ij} = \left(\frac{\partial v_i}{\partial j}\right) + \left(\frac{\partial v_j}{\partial i}\right)$$

Now, whilst the Newtonian viscosity η_N is by definition independent of the shear rate, the viscosity of a power-law liquid *is* dependent on the shear rate. Since viscosity is a scalar, it can be a function of the *scalar* invariants of the strain rate tensor only, that is to say, of those scalar quantities that remain unchanged upon rotation of the coordinate axes. There are three such invariants (e.g. in rectangular coordinates, Δ_{xy}, Δ_{xz} and Δ_{yz}) and two of these are zero in simple shear. In simple shear the only non-zero velocity component is v_x and the only non-zero derivative of v_x is $(\partial v/\partial y)$. If, in a power-law liquid $\tau = \eta \mathbf{\Delta}$ and $\eta = f(\mathbf{\Delta})$, then this double dependence of τ on $\mathbf{\Delta}$ is indicated by tensor notation as $\mathbf{\Delta}:\mathbf{\Delta}$, the double dot product.[†] The general form of the power law can be written:

$$\eta = \eta^0 \left[\frac{(\mathbf{\Delta}:\mathbf{\Delta})}{2}\right]^{(n-1)/2}$$

where the standard state viscosity (superscript o) is taken as the viscosity when the shear rate is $1\ \mathrm{s}^{-1}$. The dot product of two vectors is a scalar, and so is the double-dot product of two tensors. The full expansion of the double-dot product for the principal coordinate systems is given in Table 5.5; but noting that in simple shear these rigorous expansions have mostly zero terms, the expressions simplify considerably, so that for rectangular coordinates

$$\frac{\mathbf{\Delta}:\mathbf{\Delta}}{2} = \left(\frac{\partial v_x}{\partial y}\right)^2$$

The strain rate tensor $\mathbf{\Delta}$ is a three-dimensional entity and therefore provides a rigorous solution if its components are known. The shear rate $\dot{\gamma}$, on the other hand, represents the uni-dimensional degeneration of the

[†] $(\mathbf{\Delta}:\mathbf{\Delta}) = \sum_i \sum_j (\Delta_{ij})^2$

TABLE 5.5

EXPANSION OF THE FUNCTION $\frac{1}{2}(\boldsymbol{\Delta}:\boldsymbol{\Delta})$ IN RECTANGULAR, CYLINDRICAL AND SPHERICAL COORDINATES

Rectangular coordinates

$$\frac{1}{2}(\boldsymbol{\Delta}:\boldsymbol{\Delta}) = 2\left[\left(\frac{\partial v_x}{\partial x}\right)^2 + \left(\frac{\partial v_y}{\partial y}\right)^2 + \left(\frac{\partial v_z}{\partial z}\right)^2\right] + \left[\left(\frac{\partial v_y}{\partial x}\right) + \left(\frac{\partial v_x}{\partial y}\right)\right]^2$$

$$+ \left[\left(\frac{\partial v_z}{\partial y}\right) + \left(\frac{\partial v_y}{\partial z}\right)\right]^2 + \left[\left(\frac{\partial v_x}{\partial z}\right) + \left(\frac{\partial v_z}{\partial x}\right)\right]^2$$

Cylindrical coordinates

$$\frac{1}{2}(\boldsymbol{\Delta}:\boldsymbol{\Delta}) = 2\left[\left(\frac{\partial v_r}{\partial r}\right)^2 + \left(\frac{1}{r}\left(\frac{\partial v_\theta}{\partial \theta}\right) + \frac{v_r}{r}\right)^2 + \left(\frac{\partial v_z}{\partial z}\right)^2\right] + \left[r\frac{\partial}{\partial r}\left(\frac{v_\theta}{r}\right) + \frac{1}{r}\left(\frac{\partial v_r}{\partial \theta}\right)\right]^2$$

$$+ \left[\frac{1}{r}\left(\frac{\partial v_z}{\partial \theta}\right) + \left(\frac{\partial v_\theta}{\partial z}\right)\right]^2 + \left[\left(\frac{\partial v_r}{\partial z}\right) + \left(\frac{\partial v_z}{\partial r}\right)\right]^2$$

Spherical coordinates

$$\frac{1}{2}(\boldsymbol{\Delta}:\boldsymbol{\Delta}) = 2\left[\left(\frac{\partial v_r}{\partial r}\right)^2 + \left(\frac{1}{r}\left(\frac{\partial v_\theta}{\partial \theta}\right) + \frac{v_r}{r}\right)^2 + \left(\frac{1}{r\sin\theta}\left(\frac{\partial v_\phi}{\partial \phi}\right) + \frac{v_r}{r} + \frac{v_\theta\cot\theta}{r}\right)^2\right]$$

$$+ \left[r\frac{\partial}{\partial r}\left(\frac{v_\theta}{r}\right) + \frac{1}{r}\left(\frac{\partial v_r}{\partial \theta}\right)\right]^2 + \left[\frac{\sin\theta}{r}\frac{\partial}{\partial \theta}\left(\frac{v_\phi}{\sin\theta}\right) + \frac{1}{r\sin\theta}\left(\frac{\partial v_\theta}{\partial \phi}\right)\right]$$

$$+ \left[\frac{1}{r\sin\theta}\left(\frac{\partial v_r}{\partial \phi}\right) + r\frac{\partial}{\partial r}\left(\frac{v_\phi}{r}\right)\right]^2$$

strain rate tensor which is obtained when two of the three shear components of the strain rate tensor as well as all the normal force components are set to zero. If this is done, a rigorous solution to flow problems is sacrificed for the sake of convenience and simplicity at the cost of accuracy. If accuracy is to be regained, it is, however, necessary to devise means of determining the numerical value of those components which had previously been neglected, including the normal force components.

REFERENCES

1. *See also* J. M. McKelvey, *Polymer Processing*, John Wiley (1960).
2. J. M. McKelvey, *Polymer Processing*, Appendix A, John Wiley (1962).

6

The Analysis of Steady-State Flow in Rectangular and Cylindrical Channels

This analysis is based upon continuum mechanics and uses the fundamental equations which were considered in Chapter 5, as well as a suitable material-specific equation, a rheological equation. The flow analysis consists of finding expressions for the velocity and temperature profiles in the given flow geometries in which the boundary conditions are known. This analysis is generally complex unless certain simplifying assumptions are made.

We first consider flow between parallel plates. The geometry is rectangular (Cartesian) and the model is illustrated in Fig. 6.1. This is a model of a viscous liquid between two parallel plates. The top plate moves with a velocity V_x in the x-direction only and provides the driving force. The bottom plate is stationary. The distance between the two plates (H) represents the direction of the y-axis and is small compared with the other dimensions of the plates. The liquid is open to atmosphere and the hydrostatic pressure is therefore constant throughout, so that the density of the liquid as well as its viscosity can also be regarded as being constant. The body forces (such as gravity) which act upon the liquid may be taken to be negligible. The temperature at the plate surfaces is T_w.

The model represents a 'three-dimensional guards parade' in which each layer parallel to the two plates is composed of individual flow units which move with perfect precision in the x-direction, only. There is no sideways flow in the z-direction, nor any vertical (porpoising) flow in the y-direction. The forward velocity is constant in any given layer and it will be seen that its magnitude v_x is a function of its vertical position between the plates in terms of y relative to H.

Given these conditions, v_y and v_z are zero and hence all derivatives of v_y and v_z must also be zero. Since all flow units in any given layer move with the same velocity v_x, it follows that the derivatives of v_x relative to x and to z must also be zero. This leaves only one non-zero velocity derivative out of a

61

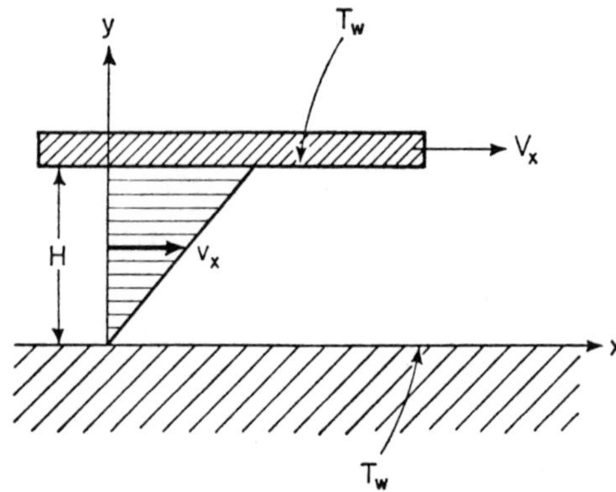

FIG. 6.1. Laminar flow between parallel plates.

possible nine, namely $(\partial v_x/\partial_y)$, since clearly there is a velocity gradient between adjoining vertical layers.

 The shearing stress which arises as a result of differential forward velocities in adjoining layers causes the development of frictional heat. Heat flow can only occur in the y-direction, since there is no temperature difference between flow units in the same layer (i.e. in the x- and z-directions. The only non-zero component of the conductive heat flux vector **q** is therefore q_y. The stress tensor τ arises from the velocity vector and also has only one non-zero component, τ_{yx}. The momentum and energy equations under those conditions reduce to

$$(\partial \tau_{yx}/\partial y) = 0 \qquad \text{(momentum)} \qquad (1)$$

and

$$\tau_{xy}(\partial v_x/\partial y) = (\partial q_y/\partial y) \qquad \text{(energy)} \qquad (2)$$

or in words: the rate of loss of mechanical energy as generated heat (per unit volume) at any point is equal to the rate at which heat is lost by conduction at that point.

 The y-component of the conductive heat flux vector is

$$q_y = -k(\partial T/\partial y)$$

so that substitution in eqn. (2) gives

$$\tau_{xy}\left(\frac{\partial v_x}{\partial y}\right) = -k\left(\frac{\partial^2 T}{\partial y^2}\right)$$

where k is the thermal conductivity of the liquid.

Now, τ is a symmetrical tensor, so that the subscripts of components are interchangeable and $\tau_{xy} = \tau_{yx}$, so that the last equation can be rewritten:

$$\tau_{yx}\left(\frac{\partial V_x}{\partial y}\right) = -k\left(\frac{\partial^2 T}{\partial y^2}\right) \tag{3}$$

The boundary conditions are:

$$\left\{\begin{array}{ll} v_x(y = 0) = 0 & T(y = 0) = T_w \\ \dot{v}_x(y = H) = V_x & T(y = H) = T_w \end{array}\right\} \tag{4}$$

The rheological equation which applies is the power law equation, and assuming Newtonian behaviour, i.e. $n = 1$,

$$\tau = \eta\dot{\gamma}$$

The only non-zero component of the stress tensor is τ_{yx}. The strain rate, in terms of the velocity vector, consists of six shear and three normal force components. The normal force components ($\partial v_i/\partial i$) are assumed to be zero (no elasticity). Since this is a case of simple shear, only one of the six shear components $\partial v_i/\partial j$ is non-zero. The only non-zero shear component is, of course, the one which arises from the movement relative to one another of the shear laminae in the planes parallel to the plates, as defined by the y-position coordinate $\partial v_x/\partial y$, so that the above equation can be written

$$\tau_{yx} = \eta\left(\frac{\partial v_x}{\partial y}\right) \tag{5}$$

We now have three independent eqns. (1), (3) and (5) obtained from the momentum, energy and rheological equation respectively. To solve them it is necessary to carry out several integration operations which bring forth integration constants. For the evaluation of these integration constants use is made of the known boundary conditions as stated in eqns. (4).

The series of operations involved are as follows:

(i) Integrate $\partial \tau_{yx}/\partial y = 0$ (eqn. (1)), which gives τ_{yx} directly:

$$\tau_{yx} = C_1 \tag{6}$$

(ii) Substitute for τ_{yx} in eqn. (5):

$$C_1 = \eta\frac{\partial v_x}{\partial y}$$

and integrate to give v_x:

$$v_x = \frac{C_1 y}{\eta} + C_2 \tag{7}$$

(iii) Substitute for τ_{yx} and v_x in eqn. (3) so that

$$\frac{\partial^2 T}{\partial y^2} = -\frac{C_1^2}{\eta k}$$

which, on integrating twice, gives

$$T = -\frac{C_1^2}{2\eta k} y^2 + C_3 y + C_4 \tag{8}$$

where C_3 and C_4 are the respective integration constants.

(iv) Evaluate the integration constants from the known boundary conditions as expressed in eqns. (4).

Since $C_1 = \eta(\partial v_x/\partial y)$, we can set $\partial y = H$, when ∂v_x becomes V_x and

$$C_1 = \eta \frac{V_x}{H}$$

C_2 follows directly from eqn. (7), using the boundary condition that $v_x = 0$ when $y = 0$, giving $C_2 = 0$.

Since $T = T_w$ when $y = 0$ it follows from eqn. (8) that

$$C_4 = T_w$$

Since $T = T_w$ also when $y = H$, substitution in eqn. (8) gives

$$T_w = -\frac{C_1^2}{2\eta k} H^2 + C_3 H + T_w$$

whence

$$C_3 = \frac{\eta V_x^2}{2kH}$$

(v) Introducing the expressions for the integration constants into eqns. (6), (7) and (8) one obtains:

$$\tau_{yx} = \eta \frac{V_x}{H} \tag{9}$$

$$v_x = y \frac{V_x}{H} \tag{10}$$

and

$$(T - T_w) = \frac{\eta}{2k} \left(\frac{V_x}{H}\right)^2 (H_y - y^2) \tag{11}$$

These final equations satisfy the equations of momentum, energy and the rheological equation and also the specific boundary conditions of the flow geometry. The procedure is the same for any flow behaviour and any geometry, provided the appropriate form of the equations and boundary conditions are employed (see later).

Some interesting facts emerge from the inspection of eqns. (9), (10) and (11). Equation (9) is perhaps not of *direct* importance in processing and can be left at this stage. According to eqn. (10) the 'dimensionless velocity' v_x/V_x, when plotted against the position variable y/H, should give a 45° linear function through the origin (slope of unity) and this is in fact observable under suitably arranged experimental conditions.

Equation (11) can be rewritten:

$$(T - T_w)\frac{2k}{\eta V_x^2} = \frac{y}{H}\left(1 - \frac{y}{H}\right)$$

which shows that the left-hand side is a function solely of the positional variable y/H, and that the function is parabolic with a maximum when $y/H = \frac{1}{2}$. It is, of course, obvious that the greatest temperature difference $(T - T_w)$ should be in the centre of the gap between the plates where the frictional shear heat cannot be dissipated as readily by conduction to the plates as elsewhere, but it does follow *a priori* that the function should be parabolic.

We now consider an incompressible liquid in laminar flow which obeys the power law and which has a power index $n = 1$ (Newtonian), but this time

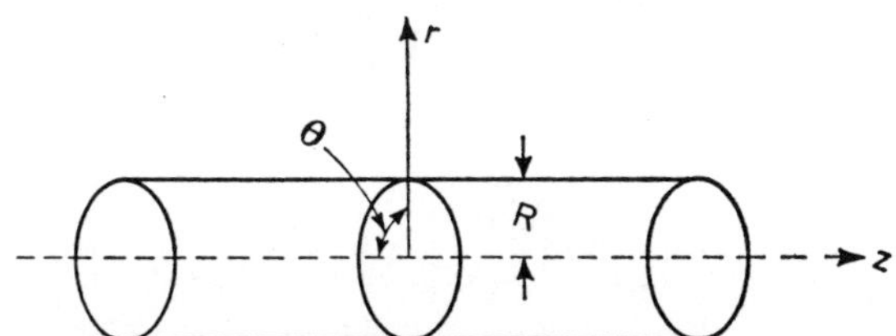

FIG. 6.2.   Laminar flow in a cylindrical channel.

we are using a long tube of circular cross-section of radius R instead of two parallel plates (see Fig. 6.2). The tube wall is at a constant temperature T_w. The objective of the flow characterisation (as before) is to calculate the velocity and temperature profiles across the tube far enough from the ends of the tube so as to be unaffected by possible entrance and exit disturbances.

The coordinates are cylindrical with axes r, θ and z. Consideration of the main flow characteristics (with simplifying assumptions analogous to those made before) lead to the realisations:

(i) All derivatives of temperature T, velocity components v_i and shear stress components τ_{ij} with respect to time t and the axes θ and z are zero;

(ii) The velocity components v_θ and v_r are themselves zero;

(iii) div v, because of incompressibility, is also zero. Using the tables given earlier on, the equations of momentum and energy and the rheological equation therefore reduce to:

$$\frac{\partial P}{\partial z} = \frac{1}{r}\frac{\partial}{\partial r}(r\tau_{rz}) \qquad \text{(Momentum)} \qquad (12)$$

$$\frac{1}{r}\frac{\partial}{\partial r}(rq_r) = \tau_{rz}\left(\frac{\partial v_z}{\partial r}\right) \qquad \text{(Energy)} \qquad (13)$$

$$\tau_{rz} = \eta\left(\frac{\partial v_z}{\partial r}\right) \qquad \text{(Rheological)} \qquad (14)$$

Introducing the only component of the heat flux vector which is non-zero (q_r), given by:

$$q_r = -k\frac{\partial T}{\partial r}$$

together with eqn. (14), into (12) and (13) respectively, we obtain:

$$\frac{\partial P}{\partial z} = \frac{\eta}{r}\frac{\partial}{\partial r}\left[r\left(\frac{\partial v_z}{\partial r}\right)\right] \qquad (15)$$

and

$$-\frac{k}{r}\frac{\partial}{\partial r}\left[r\left(\frac{\partial T}{\partial r}\right)\right] = \eta\left(\frac{\partial v_z}{\partial r}\right)^2 \qquad (16)$$

The boundary conditions are given by:

$$\left\{\begin{array}{llll} (a) & v_z(r = R) = 0 & (c) & \dfrac{\partial v_z}{\partial r}(r = 0) = 0 \\[3ex] (b) & T(r = R) = T_w & (d) & \dfrac{\partial T}{\partial r}(r = 0) = 0 \end{array}\right\}$$

Note that the left-hand side of the momentum eqn. (15) does not depend on r and that the equation can therefore be integrated directly, giving, on rearrangement

$$r\frac{\partial v_z}{\partial r} = \frac{r^2}{2\eta}\frac{\partial P}{\partial z} + C_1 \tag{17}$$

Because of boundary condition (c), the integration constant C_1 must be zero. The velocity profile is obtained from eqn. (17) upon a second integration which gives

$$v_z = \frac{r^2}{4\eta}\frac{\partial P}{\partial z} + C_2 \tag{18}$$

The integration constant C_2 follows from boundary condition (a), which when introduced into eqn. (18), shows that

$$C_2 = -\frac{R^2}{4\eta}\frac{\partial P}{\partial z}$$

so that eqn. (18) becomes:

$$v_z = -\frac{R^2}{4\eta}\left(\frac{\partial P}{\partial z}\right)\left[1 - \frac{r^2}{R^2}\right] \tag{19}$$

(The negative sign indicates that the liquid flows in the direction of decreasing pressure.)

The temperature profile is obtained by introducing $\partial v_z/\partial r$ as given by eqn. (17) into eqn. (16):

$$-\frac{k}{r}\frac{\partial}{\partial r}\left[r\left(\frac{\partial T}{\partial r}\right)\right] = \eta\left[\frac{r}{2\eta}\left(\frac{\partial P}{\partial z}\right)\right]^2$$

$$= \frac{r^2}{4\eta}\left(\frac{\partial P}{\partial z}\right)^2$$

or

$$-\frac{\partial}{\partial r}\left[r\left(\frac{\partial T}{\partial r}\right)\right] = \frac{r^3}{4\eta k}\left(\frac{\partial P}{\partial z}\right)^2$$

The first integration of this equation gives

$$-r\left(\frac{\partial T}{\partial r}\right) = \frac{r^4}{16\eta k}\left(\frac{\partial P}{\partial z}\right)^2 + C_3$$

Introducing boundary condition (*d*) it is seen that the integration constant C_3 is zero. The second integration gives

$$T = -\frac{r^4}{64\eta k}\left(\frac{\partial P}{\partial z}\right)^2 + C_4 \tag{20}$$

Boundary condition (*b*) is introduced into eqn. (20), showing that the integration constant C_4 is

$$C_4 = \frac{R^4}{64\eta k}\left(\frac{\partial P}{\partial z}\right)^2 + T_w$$

so that eqn. (20) becomes

$$T - T_w = \frac{1}{64\eta k}\left(\frac{\partial P}{\partial z}\right)^2 (R^4 - r^4)$$

or

$$T - T_w = \frac{R^4}{64\eta k}\left(\frac{\partial P}{\partial z}\right)\left[1 - \left(\frac{r}{R}\right)^4\right] \tag{21}$$

Again, some interesting facts emerge from a contemplation of the velocity eqn. (19) and the temperature difference eqn. (21).

Firstly, the velocity at the tube centre v_0 is obtained by setting $r = 0$ in eqn. (19), which then yields:

$$v_0 = -\frac{R^2}{4\eta}\left(\frac{\partial P}{\partial z}\right) \tag{22}$$

so that

$$v_0^2 = \frac{R^4}{16\eta^2}\left(\frac{\partial P}{\partial z}\right)^2 \tag{23}$$

Secondly, by setting $r = 0$ in eqn. (21), T in that equation becomes T_0 (the temperature at the centre of the tube) giving

$$T_0 - T_w = \frac{R^4}{64\eta k}\left(\frac{\partial P}{\partial z}\right)^2 \tag{24}$$

On comparing the right-hand sides of eqns. (23) and (24), it is seen that the latter differs from the former by a factor of $\eta/4k$, that is to say, that

$$T_0 - T_w = \frac{\eta v_0^2}{4k} \tag{25}$$

Thirdly, it follows directly from eqn. (22), that

$$\left(\frac{\partial P}{\partial z}\right) = -\frac{4\eta v_0}{R^2} \tag{26}$$

Lastly, eqns. (19) and (20) can be made dimensionless by dividing them by v_0 and $(T_0 - T_w)$ respectively, as given by eqns. (22) and (24), resulting in

$$\frac{v_z}{v_0} = 1 - \left(\frac{r}{R}\right)^2 \tag{27}$$

and

$$\frac{T - T_w}{T_0 - T_w} = 1 - \left(\frac{r}{R}\right)^4 \tag{28}$$

These equations are of considerable usefulness and surprising simplicity.

It is obvious that the velocity and temperature profiles are both parabolic between the limits of $+R$ and $-R$ and that the maximum for both v_z and $(T - T_w)$ is at the tube centre. But whilst the velocity profile is a square parabola, the temperature profile is a parabola of the fourth power, so that the rate of temperature increase towards the centre from the tube wall is much greater than the rate of velocity increase. In the next example the same assumptions are made as before, except that the rheological equation is the power law with $n \neq 1$. The momentum and energy equations are of the same form as before and give the following intermediates before integration:

$$\frac{\partial P}{\partial z} = \frac{1}{r}\frac{\partial}{\partial r}\left[r\eta\left(\frac{\partial v_z}{\partial r}\right)\right] \tag{29}$$

and

$$-\frac{k}{r}\frac{\partial}{\partial r}\left[r\left(\frac{\partial T}{\partial r}\right)\right] = \eta\left(\frac{\partial v_z}{\partial r}\right)^2 \tag{30}$$

corresponding to eqns. (15) and (16), but differing in that η is no longer the constant viscosity of a Newtonian, but the shear rate dependent viscosity of a power-law liquid where $n \neq 1$.

The shear rate dependent viscosity is given by the power law which is conveniently written in the form

$$\eta = \eta^0\left(\frac{\dot{\gamma}}{\dot{\gamma}^0}\right)^{n-1}$$

which for $\dot{\gamma}^0 = 1\,\mathrm{s}^{-1}$ reduces to

$$\eta = \eta^0 \dot{\gamma}^{n-1} \tag{31}$$

as shown earlier. η^0 is the standard viscosity at the standard reference shear rate of $\dot{\gamma}^0 = 1\,\mathrm{s}^{-1}$.

Introducing eqn. (31) into eqn. (29), we get

$$\frac{\partial P}{\partial z} = \frac{\eta^0}{r} \frac{\partial}{\partial r}(r\dot{\gamma}^n) \tag{32}$$

which, on integration with respect to r becomes:

$$r\dot{\gamma}^n = r\left(\frac{\partial v_z}{\partial r}\right)^n = \frac{r^2}{2\eta^0}\left(\frac{\partial P}{\partial z}\right) + C_1$$

Because the shear stress τ_{rz} at the centre of the tube is zero, therefore C_1 must be zero. Taking the nth root and putting $C_1 = 0$ we obtain

$$\dot{\gamma} = \frac{\partial v_z}{\partial r} = \left[\frac{r}{2\eta^0}\left(\frac{\partial P}{\partial z}\right)\right]^{1/n} \tag{33}$$

which on carrying out a second integration with respect to r gives

$$v_z = \frac{n}{n+1}r^{(n+1)/n}\left[\frac{1}{2\eta^0}\left(\frac{\partial P}{\partial z}\right)\right]^{1/n} + C_2 \tag{34}$$

The integration constant C_2 is obtained from the boundary condition:

$$v_z(r = R) = 0 \tag{35}$$

so that one finally obtains the equation:

$$v_z = -\frac{n}{n+1}R^{(n+1)/n}\left[\frac{1}{2\eta^0}\left(\frac{\partial P}{\partial z}\right)\right]^{1/n}\left[1 - \left(\frac{r}{R}\right)^{(n+1)/n}\right] \tag{36}$$

At the centre of the tube $v_z = v_0$ and $r = 0$, so that eqn. (36) becomes:

$$v_0 = -\frac{n}{n+1}R^{(n+1)/n}\left[\frac{1}{2\eta^0}\left(\frac{\partial P}{\partial z}\right)\right]^{1/n} \tag{37}$$

Equation (36) can be written in dimensionless form:

$$\frac{v_z}{v_0} = \left[1 - \left(\frac{r}{R}\right)^{(n+1)/n}\right] \tag{38}$$

Note that for $n=1$, eqn. (38) reduces to the Newtonian case, identical with eqn. (27).

To obtain the temperature profile we introduce eqns. (31) and (33) into the energy eqn. (30) and integrate with respect to r to obtain:

$$r\frac{\partial T}{\partial r} = -\frac{\eta^0}{k}\frac{nr}{3n+1}\left[\frac{1}{2\eta^0}\left(\frac{\partial P}{\partial z}\right)\right]^{(n+1)/n} r^{(2n+1)/n} + C_3 \tag{39}$$

Using the boundary condition that at the tube centre $(r = 0)$ $\partial T/\partial r$ is zero, substitution in eqn. (39) shows that C_3 must be zero. Integrating a second time with respect to r we obtain:

$$T = -\frac{\eta^0}{k}\left(\frac{n}{3n+1}\right)^2\left[\frac{1}{2\eta^0}\left(\frac{\partial P}{\partial z}\right)\right]^{(n+1)/n} r^{(3n+1)/n} + C_4 \tag{40}$$

The integration constant C_4 follows from the boundary condition that the temperature at radius $r = R$ is equal to T_w and eqn. (40) therefore becomes:

$$T - T_w = \frac{\eta^0}{k}\left(\frac{n}{3n+1}\right)^2 R^{(3n+1)/n}\left[\frac{1}{2\eta^0}\left(\frac{\partial P}{\partial z}\right)\right]^{(n+1)/n}\left[1 - \left(\frac{r}{R}\right)^{3(n+1)/n}\right] \tag{41}$$

The difference in the liquid temperature between the centre of the tube and the wall is obtained from the boundary condition that at $r = 0$ the temperature is T_0, so that, by putting $r = 0$, eqn. (41) becomes:

$$T_0 - T_w = \frac{\eta^0}{k}\left(\frac{n}{3n+1}\right)^2 R^{(3n+1)/n}\left[\frac{1}{2\eta^0}\left(\frac{\partial P}{\partial z}\right)\right]^{(n+1)/n} \tag{42}$$

and by dividing eqn. (41) by eqn. (42) the dimensionless form

$$\frac{T - T_w}{T_0 - T_w} = 1 - \left(\frac{r}{R}\right)^{(3n+1)/n} \tag{43}$$

is finally obtained—again a simple and elegant result.

Note that eqn. (43) becomes identical with (28) if $n = 1$, when it reduces to the Newtonian case.

The temperature difference across the die profile is obviously critically dependent on the magnitude of the power index n. It can be substantial under conditions where the shear stress is high. This is certainly the case under extrusion conditions, but even more so in the case of polymer melt passing through the runners and pinpoint gates of an injection moulding system. Indeed it is not at all uncommon that the temperature increase should be so large that thermal decomposition of the material occurs and mouldings show scorch marks. The remedy is obvious:

Either the viscosity is reduced by one of the following means:

(i) Selection of a lower viscosity (i.e. a lower molecular weight) grade of the same material. (This will be easier to mould but the mechanical strength of the moulding will be reduced.)

(ii) Increasing the melt temperature (paradoxically!).

(iii) Widening of gate diameter and/or runner diameters, *or* reducing the pressure drop. This always has a beneficial effect on extrudate quality because it reduces residual stresses, increases the dwell time in the die and thus minimises the risk of flow defects, but it naturally reduces the output rate. In moulding it is generally desirable to work at the lowest possible injection pressures in order to obtain the best quality moulding but in order to be able to fill the mould completely, the pressure must nevertheless be of a high order—about ten times as high as in extrusion. Naturally, it helps to fill the mould if the mould temperature is kept as high as possible, but whilst this assists in mould filling and minimises moulding stresses, it also increases cycle times and so reduces output.

It may be possible to increase the thermal conductivity of the melt by incorporating finely divided fillers with a specific conductivity greater than that of the polymer melt (e.g. metal powders), but this has not so far been regarded as a practical means of reducing temperature differences, because the changes in the properties of the extrudate or moulding (as the case may be) will obviously be rather extreme as a result of such inclusions.

In the calculations which have been carried out in this chapter it was assumed that the viscosity is constant across the die at any given shear stress. But this is patently incorrect since the temperature changes will cause corresponding viscosity changes. These may be ignored as a first approximation, but if the temperature difference across the die is more than (literally) one or two degrees centigrade the viscosity changes of polymer melts can be quite substantial and it is advisable to carry out a series of approximations.

The following is a simple numerical example of flow analysis:

A polymer melt with power index $n = 1$ has a viscosity of $10^4 p$ at $170\,°C$ and flows through a cylindrical die of circular cross-section (diameter $= 4\,mm$) at a rate of $10\,cm$ per s, the flow being laminar. Assuming that there are neither time dependent nor entrance effects, and assuming further that the simplifications which have led to eqns. (19) and (21) above are permissible:

(i) What is the pressure drop in the axial direction at the die centre in dynes $cm^{-2} cm^{-1}$?

(ii) What is the temperature at the die centre if the wall temperature is 170°C and if the thermal conductivity is $10^{-3} cal\ cm^{-1} s^{-1} (°C)^{-1}$?

(iii) How far from the centre of the die, to the nearest 0·1 mm, will the melt temperature be just 172°C?

Answer

(i) Set $r = 0$, then

$$v_z = v_0 = -\frac{R^2}{4\eta}\left(\frac{\partial P}{\partial z}\right)$$

$$\frac{\partial P}{\partial z} = -\frac{4v_0\eta}{R^2} = -\frac{4 \times 10 \times 10^4}{(0·2)^2} = -10^7\ dynes\ cm^{-2} cm^{-1}$$

(ii) According to eqn. (23)

$$T_0 - T_w = \frac{\eta v_0^2}{4k}$$

Since v_0 is the greatest velocity across the profile of the die, this is obviously the quoted extrusion rate (no draw being applied) of the material. Therefore, noting also that k must be converted to the mechanical equivalent of heat by multiplying with $4·2 \times 10^7$

$$T_0 = 170 + \frac{10^2 \times 10^4}{4 \times 10^{-3} \times 4·2 \times 10^7} = 170 + \frac{100}{16·8} = \underline{\underline{176°C}}$$

The temperature at the die centre will be 176°C, 6° greater than at the wall.

(iii) According to eqn. (28)

$$\frac{T - T_w}{T_0 - T_w} = 1 - \left(\frac{r}{R}\right)^4$$

Let $T - T_w = 2$, then

$$\frac{2}{6} = 1 - \left(\frac{r}{R}\right)^4 \qquad \text{or} \qquad r = R\sqrt[4]{\frac{2}{3}} = R \times 0·9 = \underline{\underline{0·18\ cm.}}$$

7

The Hagen–Poiseuille Equation and the Rabinowitsch Correction. The Pressure Drop in Tapered Channels

Consider the following balance in steady state flow through a circular (cylindrical) channel:

$$\tau \qquad \times \qquad 2\pi R L \qquad = \qquad \Delta P \qquad \times \qquad \pi R^2$$

Shear stress per unit area	Area of contact of liquid moving in a cylinder with circular cross-section (the area over which τ is exerted)	Pressure difference between entrance and exit of the channel	Area over which P is exerted

It is seen that

$$\tau = \frac{\Delta P R}{2L} = \eta_N \dot{\gamma}^n = \eta_N \left(\frac{\partial v}{\partial r}\right)^n \tag{1}$$

or, at any point in terms of r where $v_z = v_z(r)$,

$$\frac{\partial v_z}{\partial r} = \left(\frac{\Delta P}{2\eta_N L}\right)^{1/n} r^{1/n} \tag{2}$$

Integration gives

$$v_z = -\frac{n}{n+1}\left(\frac{\Delta P}{2\eta_N L}\right)^{1/n} r^{(n+1)/n} + C \tag{3}$$

The integration constant follows from the boundary condition that $v_z = 0$ when $r = R$, and substituting back into eqn. (3) we obtain

$$v_z = -\frac{n}{n+1}\left(\frac{\Delta P}{2\eta_N L}\right)^{1/n} R^{(n+1)/n}\left[1 - \left(\frac{r}{R}\right)^{(n+1)/n}\right] \tag{4}$$

At this point it should be noted that $v_z = v_{max}$ when $r = 0$ (at the centre) and putting $r = 0$ in eqn. (4) it is seen that

$$v_{max} = -\frac{n+1}{n}\left(\frac{\Delta P}{2\eta_N L}\right)^{1/n} R^{(n+1)/n} \tag{5}$$

and that the reduced velocity profile is

$$\frac{v_z}{v_{max}} = 1 - \left(\frac{r}{R}\right)^{(n+1)/n} \tag{6}$$

Imagine that the circular cross-sectional area is subdivided into a large number of concentric annular area elements, each with a mean radius r and breadth Δr, so that its area is $2\pi r\,\Delta r$. If each of these areas is multiplied with the appropriate flow velocity v_z that rules over that area element, and if all these products are summed from $r = 0$ (the centre) to $r = R$ (the wall), then the volume output Q would be approximated:

$$Q \cong \sum_{r=0}^{r=R} v_z 2\pi r\,\Delta r$$

In the limit

$$Q = 2\pi \int_0^R v_z r\,\mathrm{d}r$$

It should be noted that Q would follow immediately if the mean velocity $\bar{v}$ were known, since

$$Q = \pi R^2 \bar{v}$$

Let us set

$$a = \frac{n+1}{n}\left(\frac{\Delta P}{2\eta_N L}\right)^{1/n} \tag{7}$$

Substituting for v_z as per eqn. (4) and using eqn. (7), we get:

$$Q = 2\pi a \int_0^R (R^{(n+1)/n}r - r^{(n+1)/n}r)\,\mathrm{d}r$$

Integration gives

$$Q = 2\pi a \left(R^{(n+1)/n}\frac{r^2}{2} - \frac{r^{(3n+1)/n}}{\frac{3n+1}{n}} \right)\Bigg|_0^R$$

Taking the limits and simplifying we get, after resubstituting for a as per eqn. (7):

$$Q = \pi\frac{n}{3n+1}\left(\frac{\Delta P}{2\eta_N L}\right)^{1/n} R^{(3n+1)/n} \tag{8}$$

in a Newtonian $n = 1$, so that eqn. (8) becomes

$$Q = \frac{\pi \Delta P R^4}{8\eta_N L}$$

or

$$\eta_N = \frac{\pi \Delta P R^4}{8QL} \tag{9}$$

the Hagen–Poiseuille equation.

The mean velocity $\bar{v}$ is obtained by dividing Q by πR^2, using eqn. (8):

$$\bar{v} = \frac{n}{3n+1}\left(\frac{\Delta P}{2\eta_N L}\right)^{1/n} R^{(n+1)/n} \tag{10}$$

which for a Newtonian becomes

$$\bar{v}_N = \frac{\Delta P R^2}{8\eta_N L} \tag{11}$$

so that, in general

$$\bar{v} = \frac{n+1}{3n+1} v_{max}$$

and for a Newtonian

$$\bar{v}_N = \tfrac{1}{2} v_{max}$$

By definition

$$\eta_N = \frac{\tau}{\dot{\gamma}_N}$$

and

$$\tau = \frac{\Delta P R}{2L}$$

It follows from the Hagen–Poiseuille equation that

$$\frac{\pi \Delta P R^4}{8QL} = \frac{\Delta P R}{2L} \frac{1}{\dot{\gamma}_N}$$

whence

$$\dot{\gamma}_N = \frac{4Q}{\pi R^3} \tag{12}$$

Similarly, using the general eqn. (10) (which was derived from the corresponding general eqn. (8)):

$$Q = \bar{v}\pi R^2 = \frac{n}{3n+1}\left(\frac{\Delta P}{2\eta_N L}\right)^{1/n} \pi R^{(3n+1)/n}$$

$$= \frac{n}{3n+1}\left(\frac{\Delta PR}{2\eta_N L}\right)^{1/n} \pi R^3$$

The term of the bracket is recognised as $\dot{\gamma}$, whence

$$Q = \frac{n}{3n+1}\dot{\gamma}\pi R^3$$

or

$$\dot{\gamma} = \frac{3n+1}{n}\frac{Q}{\pi R^3} \tag{13}$$

Introducing eqn. (12) into eqn. (13) it is seen that

$$\dot{\gamma} = \frac{\dot{\gamma}_N}{4}\frac{3n+1}{n} \tag{14}$$

the 'Rabinowitsch Correction'.

The Rabinowitsch Correction makes it possible to calculate $\dot{\gamma}$ in a power law liquid if n is known. In the Newtonian case $n = 1$, so that $\dot{\gamma}$ in eqn. (14) becomes identical with $\dot{\gamma}_N$.

From the definition of the power index n it is also obvious that eqn. (14) can be written:

$$\dot{\gamma} = \frac{\dot{\gamma}_N}{4}\left(3 + \frac{d\log\dot{\gamma}}{d\log\tau}\right) \tag{15}$$

In rectangular (wide slit) dies the appropriate Rabinowitsch Correction can be shown to be

$$\dot{\gamma} = \frac{2n+1}{3n}\dot{\gamma}_N = \frac{2n+1}{3n}\frac{6Q}{wh^2} = \frac{Q}{wh^2}\frac{4n+2}{n} \tag{16}$$

where w is the width and h is the height of the wide slit. The Hagen–Poiseuille equation for wide slit dies (Cartesian coordinate) is

$$Q = \frac{\Delta Pwh^3}{12\eta_N L} \tag{17}$$

and the general expression for power law liquids is

$$Q = \frac{\Delta P w h^3}{4 \eta_N L} \frac{n}{2n + 1} \tag{18}$$

In practice it is very useful in die design to be able to predict pressure drops by making ΔP the subject of Rabinowitsch-Corrected Hagen–Poiseuille equations which apply to wide slit or circular dies, as the case may be. When extruding polymer melts it has been found to be particularly beneficial to use tapered dies, since their use prevents flow defects at operating pressures which are substantially greater than those which would cause serious flow defects in dies with parallel-sided channels. The taper angles are generally less than $10°$. Attention has been drawn to this by Cogswell and Lamb[1] who also gave a mathematical treatment of the problem, and a similar treatment has been given by Plajer.[2] Lenk[3] has suggested a more rigorous treatment which is at the same time remarkably simple.

Four typical and common examples of linearly convergent dies are considered:

(1) vertically tapered wide slit dies;
(2) laterally tapered wide slit dies;
(3) wide slit dies with vertical as well as lateral taper;
(4) dies of a circular geometry (truncated right cones).

(1) *Pressure drop through a wide slit of constant width w and with a vertical taper angle θ*

A diagram of the geometry is given in Fig. 7.1. It is seen that

$$h = H_1 - L \tan \theta$$

$$\tan \theta = \frac{H_1 - h}{l} = -\frac{\mathrm{d}h}{\mathrm{d}l}$$

$$\mathrm{d}h = -\mathrm{d}l \tan \theta$$

$$\mathrm{d}l = \mathrm{d}h \cot \theta$$

In wide slit dies

$$\dot\gamma = \frac{2n + 1}{3n} \dot\gamma_N = \frac{2n + 1}{3n} \frac{6Q}{wh^2} \dagger = \frac{Q}{wh^2} \frac{4n + 2}{n}$$

$\dagger$ Note the use of the appropriate Rabinowitsch Correction for power law liquids in wide slit channels.

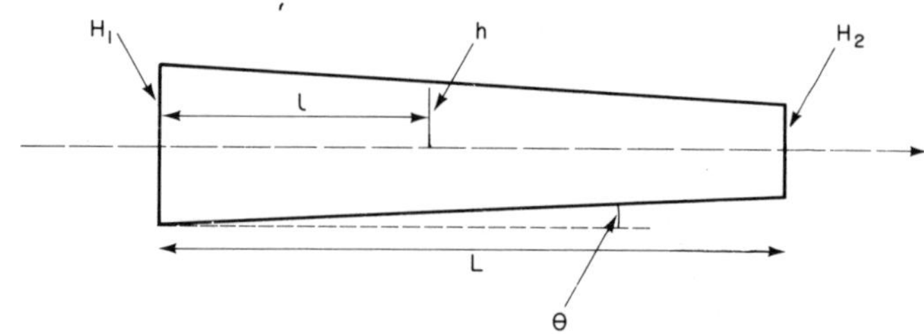

FIG. 7.1. Flow in a tapered wide slit die.

In flow through a slit die *without* taper

$$\tau = \frac{\Delta P h}{2L} = \eta \dot{\gamma}^n = \eta \left(\frac{Q}{wh^2} \frac{4n+2}{n} \right)^n$$

whence

$$\Delta P = \frac{2L_N}{w^n h^{2n+1}} \left(Q \frac{4n+2}{n} \right)^n \tag{19}$$

If the wide slit die is vertically tapered we consider the pressure drop dP for an infinitely small length dl in terms of dh and integrate with respect to h between the limits of H_1 and H_2, remembering that $dl = f(dh)$, namely

$$dl = -dh \cot \theta$$

$$dP = \frac{2dl\eta}{h^{2n+1}} \left(Q \frac{4n+2}{wn} \right)^n$$

and

$$\Delta P = -2\eta \cot \theta \left(Q \frac{4n+2}{wn} \right)^n \int_{H_1}^{H_2} h^{-(2n+1)} \, dh$$

$$= -2\eta \cot \theta \left(Q \frac{4n+2}{wn} \right)^n \left(-\frac{1}{2n} \right) h^{-2n} \Big|_{H_1}^{H_2}$$

Simplifying and taking the limits we get

$$\Delta P = \frac{\eta \cot \theta}{n} \left(Q \frac{4n+2}{wn} \right)^n H_2^{-2n} \left[1 - \left(\frac{H_1}{H_2} \right)^{-2n} \right] \tag{20}$$

(2) Pressure drop through a wide slit die with converging sides

This is an inverted fishtail die with bi-laterally equal sideways taper angle ϕ, and with constant height h. The width w reduces from the entrance width w_1 to the exit width w_2.

By trigonometry, as before

$$\tan \phi = -\frac{dw}{dl}$$

$$dw = -dl \tan \phi$$

$$dl = -dw \cot \phi$$

Starting from eqn. (19), we consider the pressure drop dP for an infinitely small length dl in terms of dw and integrate with respect to w between the limits of w_1 and w_2, remembering that $dl = f(dw)$, namely

$$dl = -dw \cot \phi$$

$$dP = 2\eta h^{-(2n+1)} \left(Q \frac{4n+2}{n} \right)^n dw \cot \phi\, w^{-n}$$

and

$$\Delta P = -2\eta h^{-(2n+1)} \left(Q \frac{4n+2}{n} \right)^n \cot \phi \int_{w_1}^{w_2} w^{-n}\, dw$$

$$= \frac{2\eta}{n-1} h^{-(2n+1)} \left(Q \frac{4n+2}{n} \right)^n \cot \phi\, w^{1-n} \Big|_{w_1}^{w_2}$$

Simplifying and taking the limits

$$\Delta P = \frac{2\eta}{n-1} h^{-(2n+1)} \left(Q \frac{4n+2}{n} \right)^n \cot \phi\, w_2^{1-n} \left[1 - \left(\frac{w_1}{w_2} \right)^{1-n} \right] \tag{21}$$

(3) Pressure drop through an inverted fishtail die with lateral taper angle ϕ and with a simultaneous vertical taper angle θ

By trigonometry, as before,

$$\tan \phi = -\frac{dw}{dl}$$

$$dw = -dl \tan \phi$$

$$dl = -dw \cot \phi$$

Starting from eqn. (19), we consider the pressure drop dP for an infinitely small length dl in terms of dh and dw and integrate with respect to:

(i) h between the limits of H_1 and H_2,
(ii) w between the limits of w_1 and w_2,

Polymer Rheology

remembering that dl may be expressed as $f(dh)$ *and* as $f(dw)$, namely

$$dl = -dh \cot \theta$$

and

$$dl = -dw \cot \phi$$

$$dP = 2\eta h^{-(2n+1)} \left(Q \frac{4n+2}{n} \right)^n (-dh \cot \theta) w^{-n} (-dw \cot \phi)$$

and

$$\Delta P = 2\eta \cot \theta \cot \phi \left(Q \frac{4n+2}{n} \right)^n \int_{H_1}^{H_2} h^{-(2n+1)} \, dh \int_{w_1}^{w_2} w^{-n} \, dw$$

Integration yields

$$\Delta P = 2\eta \cot \theta \cot \phi \left(Q \frac{4n+2}{n} \right)^n \left(-\frac{1}{2n} \right) \left(-\frac{1}{n-1} \right) h^{-2n} w^{1-n}$$

Taking the limits and simplifying, we finally obtain

$$\Delta P = \frac{\eta \cot \theta \cot \phi}{n(n-1)} \left(Q \frac{4n+2}{n} \right)^n H_2^{-2n} \left[1 - \left(\frac{H_1}{H_2} \right)^{-2n} \right] w_2^{1-n} \left[1 - \left(\frac{w_1}{w_2} \right)^{1-n} \right]$$

$$(22)$$

(4) *Pressure drop through dies having the shape of a circular truncated right cone*

The geometry of this die may be represented, *mutatis mutandis*, by a diagram similar to the one applicable to flow in a vertically tapered wide slit die (see part (1) above). In the present case, however, h, H_1 and H_2 are replaced by r, R_1 and R_2 respectively, and it is also understood that the taper angle θ is operative around the entire circumference of the die.

We can therefore write

$$r = R_1 - L \tan \theta$$

$$\tan \theta = \frac{R_1 - r}{l} = -\frac{dr}{dl}$$

$$dr = -dl \tan \theta$$

$$dl = -dr \cot \theta$$

In circular dies

$$\dot{\gamma} = \frac{3n + 1}{n} \dot{\gamma}_{\text{app}} = \frac{3n + 1}{n} \frac{4Q}{\pi R^3} \dagger$$

In flow through an *untapered* circular channel

$$\tau = \frac{\Delta P R}{2L} = \eta \dot{\gamma}^n = \eta \left(\frac{4Q}{R^3} \frac{3n + 1}{\pi n} \right)^n$$

and

$$\Delta P = \eta L 2^{2n+1} R^{-(3n+1)} \left(Q \frac{3n + 1}{\pi n} \right)^n \tag{23}$$

If the circular untapered (cylindrical) die is modified by tapering so that it becomes convergent and takes the shape of a truncated right cone, one has to consider the pressure drop dP for an infinitely small length dl in terms of dr and integrate with respect to r between the limits of R_1 and R_2, remembering that $dl = f(dr)$, namely

$$dl = -dr \cot \theta$$

$$dP = \frac{\eta \, dl 2^{2n+1}}{r^{3n+1}} \left(Q \frac{3n + 1}{\pi n} \right)^n$$

and

$$\Delta P = -2^{2n+1} \eta \cot \theta \left(Q \frac{3n + 1}{\pi n} \right)^n \int_{R_1}^{R_2} r^{-(3n+1)} dr$$

$$= -2^{2n-1} \eta \cot \theta \left(Q \frac{3n + 1}{\pi n} \right)^n \left(-\frac{1}{3n} \right) r^{-3n} \Big|_{R_1}^{R_2}$$

Taking the limits and simplifying:

$$\Delta P = \frac{2^{2n+1}}{3n} \eta \cot \theta \left(Q \frac{3n + 1}{\pi n} \right)^n R_2^{-3n} \left[1 - \left(\frac{R_1}{R_2} \right)^{-3n} \right] \tag{24}$$

At this stage we wish to draw attention to the symbol η in the various equations. The viscosities in wide slit dies and in dies with cylindrical and conical geometries are not, strictly speaking, identical. Rewriting them η''

$\dagger$ *Note:* the inclusion of the Rabinowitsch Correction distinguishes this treatment from that given by Cogswell and Lamb.

and η' respectively, one may take advantage of the empirical relationship between them which was used by Carley,[4] namely

$$\eta'' = 0{\cdot}91\eta'$$

Having thus identified η in eqns. (23) and (24) with η', we can therefore replace the η of eqns. (19) and (20) through (22) with $0{\cdot}91\eta'$. This makes it possible to apply viscosity data obtained from rheometers with circular channels to the calculation of pressure drops in wide slit dies.

It is clear that the method of calculating pressure drops may also be applied to dies in which the taper is curvilinear rather than rectilinear. Such dies are probably more expensive to produce, but they would have the considerable advantage of giving rise to a minimum of flow disturbance. The channel would be generated by rotating that portion of a quintic function of length and radius around an axis parallel to the tangent to the curve which is between the maximum and minimum of the curve.[5] Such a die would minimise the flow discontinuity at its entrance where the flow transition from a broad and sluggish stream to a narrow and fast stream occurs. The optimum gradualisation of this transition will have obvious benefits by enabling the extruder to increase the pressure (and hence the output) without exceeding the limit at which serious extrudate defects are encountered when ordinary dies are used.

In order that the pressure drop in such an advanced die may be calculated it would merely be necessary to determine the function of dl in terms of dr, followed by substitution in an equation for dP based upon eqn. (23) and integration between the limits of R_1 and R_2 in the same manner as that shown above.

In conclusion, two numerical solutions for calculating the pressure drop in dies with the shape of truncated cones are shown in Table 7.1.

TABLE 7.1

Case 1	*If:*	*Case 2*
10^4	η (poise)	3×10^4
2/3	power index n	$\frac{1}{2}$
1·0	Q (cm^3 s^{-1})	1·0
0·2	R_2 (cm)	0·25
5·0	L (cm)	6·0
5·71°, so that $\cot\theta = 10$	taper angle θ	9·5°, so that $\cot\theta = 6$
	Then:	
0·7	R_1 (cm)	1·25
$2{\cdot}92 \times 10^6$	ΔP (dyne cm^{-2})	$3{\cdot}9 \times 10^6$

REFERENCES

1. F. N. COGSWELL and P. LAMB, *Trans. & J. Plast. Inst.* (Dec. 1967).
2. O. PLAJER, *Plastverarbeiter*, **23**(6), 407 (1972).
3. R. S. LENK, *J. Appl. Poly. Sci.* (submitted 1977).
4. J. F. CARLEY, *SPE Journal*, **19**(9), 977 (1963).
5. R. S. LENK and G. SLATER, *J. Appl. Poly. Sci.* (to be submitted).

8

Pressure Drop in Wire Coating Dies. Two-Dimensional Flow in Extruder Screws

WIRE COATING DIES

In wire coating dies one has to consider two types of flow: the pressure flow from the extruder barrel through the cross-head; the drag flow caused by the velocity of the wire passing through the guide channel where it receives the coating. The volume rate of flow Q is the sum of the pressure flow Q_P and the drag flow Q_D. It is assumed that the height of the annular channel surrounding the wire H is small compared to the wire radius R_1, that the ultimate coating thickness after drawdown is h, that the wire velocity is V and that the pressure which supplies the melt to the wire guide is ΔP (Fig. 8.1). Treating the problem on the basis of Newtonian flow, we use the Hagen-Poiseuille equation for pressure flow in annular (wide slit) channels:

$$Q_P = \frac{wH^3 \Delta P}{12\eta L} \tag{1}$$

where ΔP is the difference between the pressure in the reservoir and atmospheric pressure.

The size of the wire, its velocity and the required coating thickness h determine the volume output which is the product of cross-sectional area and velocity, i.e.

$$Q = \pi V[(R_1 + h)^2 - R_1^2] = \pi V h(2R_1 + h) \tag{2}$$

We know that the drag flow $Q_D = V_z wH/2$, where w represents the mean value of the circumference of the annular space.

Combining these equations and solving for ΔP we get

$$\Delta P = \frac{6\eta V_z L(2h - H)}{H^3} \tag{3}$$

87

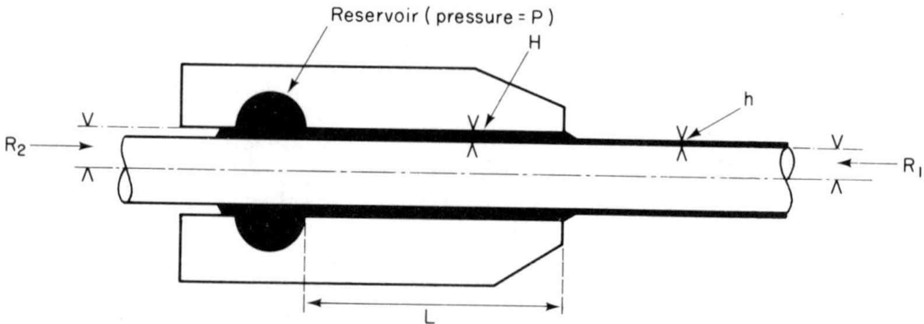

Fɪɢ. 8.1.   Diagram illustrating the basic principles of wire coating.

This equation shows:

(i)   that the pressure required in a wire coating die is directly proportional to the viscosity, the wire velocity and the die length;

(ii)   that zero pressure is needed in the reservoir if $h = H/2$. This is the optimum design condition because changes in the wire velocity or in the fluid viscosity will have no effect on the thickness of the coating.

This is a very simple solution to a rather complicated flow problem. In fact, it is no solution at all. There is no method available for calculating the pressures generated and the tension induced in the wire, and dimensions are obtained by trial and error using conical channels and by avoiding shear stresses which might produce melt fracture. The reason for this is that the polymer is not only non-Newtonian, but the flow is also axi-symmetric in a reducing section. Such a combination does not yield an analytical solution. A typical wire coating die is diagrammatically shown in Fig. 8.2.

Fenner and Williams[1] have pointed out that the use of digital computers enables one to obtain numerical solutions in terms of the characteristics for

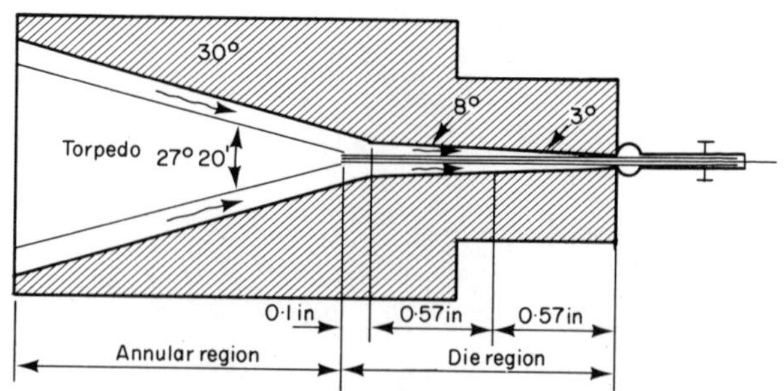

Fɪɢ. 8.2.   Typical wire coating die with torpedo (after Fenner and Williams[1]).

this type of flow for any given polymer. Much computing is required, but it needs to be done once only for a given polymer grade. The characteristic curves may then be used to predict the performance of a given die by graphical means. The numbers calculated do not constitute a complete answer. Although pressure profiles, shear rates and shear stresses may be calculated, these must then be related to the extruder characteristics and the rheological properties of the melt. There is no way of determining what is the optimum pressure profile. This makes it possible to compare good and bad dies and to determine the criteria for good design. The key is the determination of five dimensionless parameters and relating them to characteristic curves so as to obtain all parameters, integrating them and thus obtaining pressure profiles, wire tensions and maximum shear stresses.

The momentum equation is

$$\frac{1}{r}\frac{d(r\tau_{rz})}{dr} = \frac{dP}{dz} \tag{4}$$

Using the power law the velocity profile is obtained by numerical methods and computing. The results are expressed in dimensionless form.

The relevant die performance variables for a die with tapered torpedo include volumetric melt flow rate Q, extruder delivery pressure ΔP, wire tension F, and the maximum boundary shear stresses in the die and at the wire surface, in order to determine whether melt fracture is likely. These are defined in dimensionless form:

dimensionless flow rate

$$\Pi_Q = Q/WHV \tag{5}$$

dimensionless pressure gradient

$$\Pi_P = \frac{-(dP/dz)H}{\bar{\tau}} \tag{6}$$

dimensionless tension gradient

$$\Pi_F = \frac{1}{c\bar{\tau}}\frac{dF}{dz} \tag{7}$$

dimensionless die shear stress

$$\sigma_D = \tau_D/\bar{\tau} \tag{8}$$

dimensionless wire shear stress

$$\sigma_w = \tau_w/\bar{\tau} \tag{9}$$

where W is the mean circumference, V is the wire speed, τ_D and τ_w are the shear stresses at the die and wire, $\bar{\tau}$ is the mean shear stress at the mean shear rate V/H, i.e.

$$\bar{\tau} = \eta^0(V/H)^n \tag{10}$$

The die flow analysis is then followed by analysis of the annular flow and an overall performance analysis which includes the integration of the dimensionless parameters between the known limits. This involves numerical methods.

FLOW IN EXTRUDER SCREW CHANNELS

Let $N =$ velocity of the screw relative to the barrel in rpm, $w =$ constant channel width, $H =$ constant channel depth, $h =$ the depth of a layer of melt in laminar flow between the screw root ($h = 0$) and the top of the channel ($h = H$), $R =$ barrel radius (R is very much greater than H), and $\phi =$ flight angle of the screw channel.

The flight clearance and the consequent leakage flow are taken to be negligible. The coordinate system is that of a wide slit—therefore rectangular—in which x represents the direction of the flight width, y that of the flight height and z the main direction of forward flow. The velocities of flow are v_x, v_y and v_z respectively. Flow in the y-direction is taken to be negligible, but the sideways flow v_x is included in the analysis. Both v_x and v_z are functions of the positional coordinate y/h. The forward flow v_z has a velocity profile as in parallel plate flow. Zamodits and Pearson[2] have treated the problem as one of two-dimensional flow, viz:

the rheological equation used is the power law which is written

$$\tau = \eta\dot{\gamma}^n = c(\dot{\gamma}^2)^s\dot{\gamma} \tag{11}$$

where c is a constant and

$$s = (n - 1)/2 \tag{12}$$

Hence

$$\eta = c(\dot{\gamma}^2)^s \tag{13}$$

The shear rates for flow in the x- and z-directions are deformations acting perpendicularly to each other and are given, respectively, by

$$\dot{\gamma}_x = dv_x/dy$$

and

$$\dot{\gamma}_z = dv_z/dy$$

The total shear rate is the vectorial sum of its components:

$$\dot{\gamma}^2 = \dot{\gamma}_x^2 + \dot{\gamma}_z^2 = (dv_x/dy)^2 + (dv_z/dy)^2 \qquad (14)$$

When considering the stress tensor components one may neglect those in the y-direction, since flow in that direction is justifiably assumed to be negligibly small.

Of the x-direction components, however, only τ_{zx} is negligibly small, whilst τ_{xx} and τ_{yx} are not. The derivatives with respect to the axis along which the force is applied (x in τ_{xx}, y in τ_{yx}, z in τ_{zx}) must sum to zero, because nothing in the profile except the pressure depends on the distance along the z-direction of the channel, so that

$$(d\tau_{xx}/dx) + (d\tau_{yx}/dy) = 0 \qquad (15)$$

Now

$$\tau_{xx} = -P \qquad (16)$$

and using eqns. (11) and (14)

$$\tau_{yx} = \eta\dot{\gamma}_{yx}^n = c[(dv_x/dy)^2 + (dv_z/dy)^2]^s(dv_x/dy)$$

Hence

$$\left(\frac{dP}{dx}\right) = \frac{d}{dy}\{c[(dv_x/dy)^2 + (dv_z/dy)^2]^s(dv_x/dy)\} \qquad (17)$$

The analogous equation for the z-direction components of the stress tensor is

$$\left(\frac{dP}{dz}\right) = \frac{d}{dy}\{c[(dv_x/dy)^2 + (dv_z/dy)^2]^s(dv_z/dy)\} \qquad (18)$$

The boundary conditions for v_x and v_z are as follows:

$$\left.\begin{cases} \text{At } y = 0, v_x = 0 \quad \text{and} \quad v_z = 0; \\ \text{At } y = H, v_x = 2\pi RN \sin\phi \quad \text{and} \quad v_z = 2\pi RN \cos\phi \end{cases}\right\} \qquad (19)$$

Now, dP/dz is easily measured, but dP/dx is not. In order therefore to make use of eqns. (17) and (18) an additional boundary condition is needed. This is provided by remembering that there is assumed to be no leakage flow and therefore no output Q_x in the x-direction, so that

$$\int_0^H v_x \, dy = 0 \qquad (20)$$

Integration of eqns. (17) and (18) gives

$$\frac{1}{c}\int_{y_1}^{H}(dP/dx)\,dy = [(dv_x/dy)^2 + (dv_z/dy)^2]^s(dv_x/dy) = \frac{1}{c}\left(\frac{dP}{dx}\right)(y - y_1)$$

(21)

and

$$\frac{1}{c}\int_{y_2}^{H}(dP/dz)\,dy = [(dv_x/dy)^2 + (dv_z/dy)^2]^s(dv_x/dy)$$

$$= \frac{1}{c}(dP/dz)(y - y_z)$$

(22)

where y_1 is the level, in terms of h, where $(dv_x/dy) = 0$, and y_2 is the level, in terms of h, where $(dv_x/dz) = 0$.

Dividing eqn. (21) by eqn. (22), one obtains

$$\frac{(dv_x/dy)}{dv_z/dy} = \frac{(dP/dx)}{(dP/dz)}\frac{y - y_1}{y - y_2}$$

(23)

where

$$\frac{(dP/dx)}{(dP/dz)}$$

is a dimensionless quantity denoted in the following by P_0. Further dimensionless quantities are defined as follows:

$$y/H = Y$$
$$y_1/H = Y_1$$
$$y_2/H = Y_2$$

$$v_x^0 = \frac{v_x}{2\pi R N \sin\phi}$$

$$v_z^0 = \frac{v_z}{2\pi R N \cos\phi}$$

$$G = \frac{(dP/dz)H^{n+1}}{c(2\pi R N)^n}$$

If these dimensionless quantities are introduced into eqns. (21) and (22) one eventually obtains

$$v_x^0 = \frac{G^{1/n}}{\sin\phi}\int_0^y P_0(Y - Y_1)f(Y)\,dY$$

(24)

and

$$v_z^0 = \frac{G^{1/n}}{\cos\phi} \int_0^y P_0(Y - Y_2)f(Y)\,dY \tag{25}$$

where

$$f(Y) = [P_0^2(Y - Y_1)^2 + (Y - Y_2)^2]^{(1-n)/2n} \tag{26}$$

Substituting the boundary conditions into eqns. (23), (24) and (25), one obtains three equations in P_0, Y_1 and Y_2, and v_x and v_z follow from v_x^0 and v_z^0 respectively.

The total output of the screw is given by

$$Q = \int_0^H v_z\,dy/2\pi RHN \tag{27}$$

The constants c and s are obtained from rheological data, and G follows after determination of the pressure gradient dP/dz.

Numerical analysis by computer enables one to plot the 'screw characteristic' (output Q vs. pressure drop). In practice it is more convenient to plot Q vs. log G, where G is one of the dimensionless parameters designated above. All the other parameters (R, N, ϕ, w, H) are design constants. A family of curves is obtained in which each curve corresponds to an assumed value of n. This is illustrated in Fig. 8.3 which applies to an extrusion grade polyethylene and a screw with a flight angle of 30°.

It is obvious that the Newtonian assumption ($n = 1$) is likely to produce large errors, remembering that n commonly has values between 0·3 and 0·7. In fact the pressure required for a screw to give the required output is much less than that needed if the melt were Newtonian.

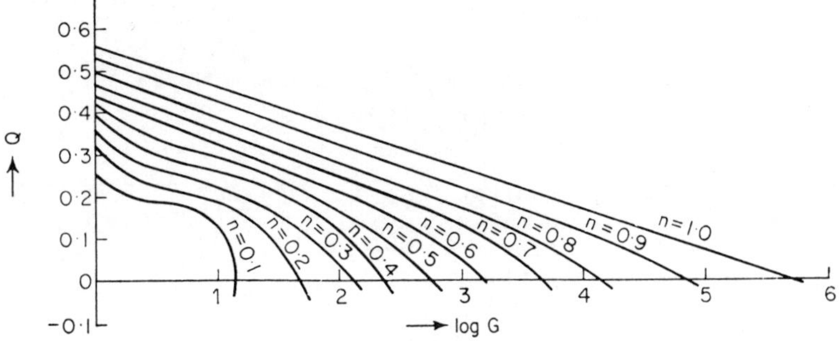

FIG. 8.3. Q vs. log G in an extrusion screw channel.

A *die characteristic* can also be included in a plot of Q vs. G, but the curve will pass through the origin, whatever value for n applies. If combined with the *screw characteristic* one obtains an overall operational diagram such as the one shown in Fig. 8.4.

The point of intersection of screw and die characteristics 'A' will indicate the operating output of the extruder. If a smaller die is substituted it is often

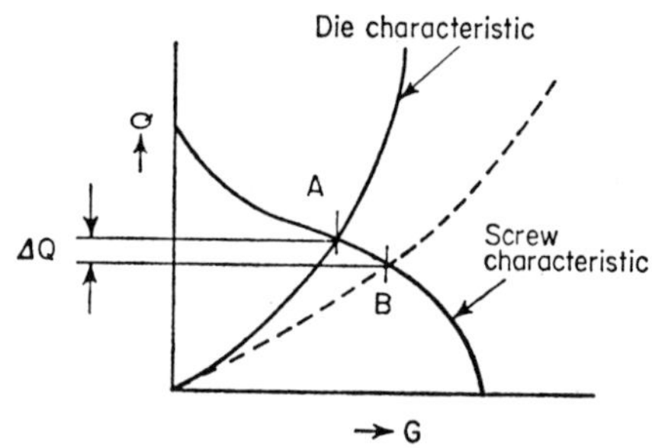

FIG. 8.4.   Combination of screw and die characteristics.

found that the output as indicated by point 'B' is only slightly affected. This is due to the fact that the screw characteristic in that particular region is rather flat, so that ΔQ is small. Only in a Newtonian liquid would the change in die cross-sectional area cause a strictly proportional change in net output.

The results from these calculations give answers which differ by only about 10% from observed values, a considerable improvement over methods which ignore the flow component v_x, since in the latter, errors of from 20 to 50% are common.

Strictly speaking, one should also allow for the temperature effects due to shear heating since they will affect the mean viscosity across the flow profile. This has been attempted by Gee and Lyon[3] and by Griffiths.[4]

REFERENCES

1. R. T. FENNER and J. G. WILLIAMS, *Trans. & J. Plast. Inst.*, 701 (Oct. 1967).
2. H. ZAMODITS and J. R. A. PEARSON, *Trans. & J. Plast. Inst.* (1965).
3. R. B. GEE and J. B. LYON, *Ind. Eng. Chem.*, **49**, 956 (1959).
4. R. M. GRIFFITHS, *Ind. Eng. Chem.* (*Funds*), **1**, 180 (1962).

9

Coextrusion

Coextrusion is a recent and highly sophisticated method for processing polymers. It has become rapidly established in many sectors of the extrusion field, but packaging is the dominating applications outlet, principally for film and sheet. Non-packaging applications include sheet for vacuum forming of refrigerator liners, luggage, boats, sanitary ware, pipe, wire and cable coating, profile and tubing, fibre and monofilament. Coextrusion can produce an entirely new range of products with special performance features, or with the prospect of considerable cost savings. An example of the latter would be a combination of virgin material on top, with reground material underneath.

Coextrusion equipment is generally limited to three systems:[1]

(i)   feedblocks, in which melt streams from different extruders are combined in a small cross-section before entering the die;

(ii) multi-manifold internal-combining dies, in which the various melt streams enter the die separately and join just inside the die orifice;

(iii) multi-manifold external-combining dies with completely separate manifolds for the different melts as well as distinct orifices through which they leave the die in order to join together just beyond it.

For blown film, multi-manifold internal-combining dies are preferred, mainly because there is a risk of spider lines and disruption of laminar flow of the layers when using feedblocks.

For pipe and tubing any system can be used, but feedblocks are said to be simpler and cheaper. In flat-die extrusion of film, sheet and coatings there is little to choose between the systems, and each has its advocates.

In wire and cable coating, multi-manifold dies are used exclusively.

Feedblocks are preferred for extruding heat sensitive materials, whilst multi-manifold dies are superior for combining streams of widely differing viscosities or temperatures. Feedblocks are said to be limited to combining materials with a viscosity ratio of 2:1 to 4:1, and temperature differences

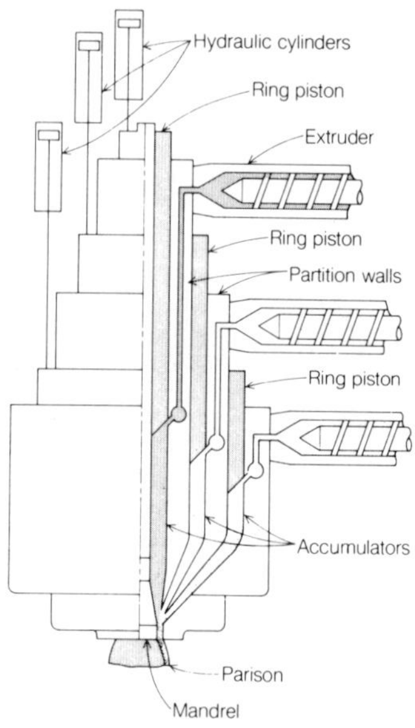

FIG. 9.1. Three-material, three-layer coextrusion accumulator head. (Reproduced by kind permission of *Modern Plastics International*, McGraw-Hill Inc.)

should not exceed about 30 °C; in multi-manifold internal-combining dies somewhat higher temperature differentials can be tolerated. Feedblocks are also limited by the power index of the different melts. Since the effective shear rate changes continuously in a feedblock/die system, the viscosities of the layers must change at comparable rates to maintain the original balance.

Intermittent extrusion-blow moulding, using independent concentric accumulators and annular piston rams, is a novel technique for producing multi-layer containers.[2] Figure 9.1 shows a three-material, three-layer coextrusion head. The basic design is similar to continuous coextrusion heads—except for the enlarged accumulating sections and corresponding ring pistons. These pistons are the key to intermittent coextrusion. The plasticated melts are accumulated within separate concentric reservoirs, each with its own independent temperature control system. The parison is formed against the mandrel by forcing the melt from the reservoirs by means of the ring pistons. Each of these is operated by an independent hydraulic cylinder; this allows the extrusion rate (and hence the pressure) of the various melts to be accurately controlled within wide limits, and feeding of one or more layers may be abruptly terminated whenever this is desired. Such on/off control is important for two reasons:

(i) expensive polymers can be positioned within the parison so that they are present in the container body, but not in the flash where they would be wasted;

(ii) the flash, consisting of a single material only, is easily recycled.

Coextrusion has been used for producing 'conjugate' fibres which consist of two semicircular components which resemble natural wool.[3-5] This resemblance is due to crimping which results from differential thermal contraction and consequent filament buckling during cooling.

Generally, the addition of a less viscous component reduces the pressure gradient. A 'pressure gradient reduction factor' (PGRF) is defined by

$$\left(\frac{\partial P}{\partial z}\right)_{A/B} \bigg/ \left(\frac{\partial P}{\partial z}\right)_{A}$$

where A is the more viscous and B the less viscous component. It was seen[6] that the PGRF remains virtually constant over wide ranges of volume-percentage compositions of the melts—presumably so long as there is sufficient of the lower viscosity component to ensure all-round lubrication. Using a power law model and a wide slit geometry, Han[7] derived equations for the volume output of each component as a function of the variables normally involved in single–phase extrusion and additionally as a function of the position of the interface in the cross-sectional die profile.

Owing to the lubrication effect which arises when the less viscous component is in contact with the die wall, the pressure gradient of a bicomponent system is greatly reduced when compared to that of a single phase system. This has long been exploited in increasing the pumping efficiency in oil pipes when a little water is added.

The shape of the interfacial boundary may vary as the two melts progress through the die, but there is a general rule that the less viscous component preferentially wets the die wall, whilst the more viscous component tends to become convex. This is to be expected since the less viscous component is *ipso facto* the one which flows more actively. The overall effect is that of an envelopment of the more viscous component. In order that this state be fully developed, it is necessary that the dwell time of the coextruding melts in the die be sufficiently long. This is more likely to be achieved when the channel is long and/or when the applied pressure is low or moderate. Han and Khan[8] confirmed this by studying the effect of length/diameter (L/D)

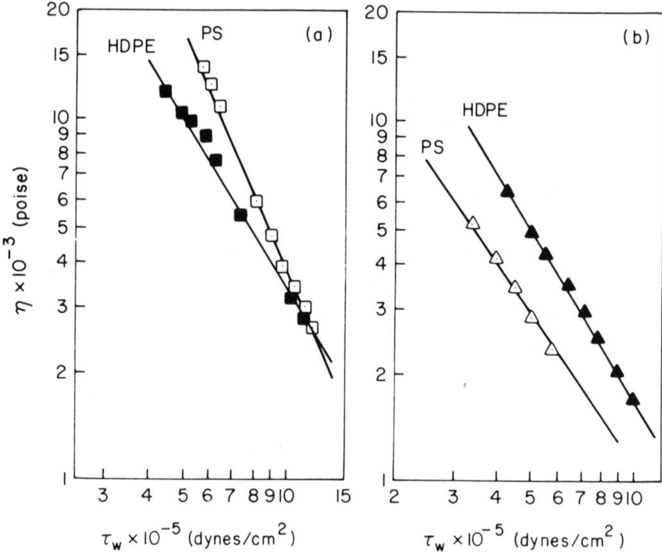

FIG. 9.2.   Viscosity vs. shear stress for PS (polystyrene) and HDPE (high-density polyethylene), used for coextrusion experiments. (a) Temperature: 200°C; (b) Temperature: 240°C. (Reproduced by kind permission of the author and Academic Press.[6])

ratios of circular dies on the interface shape of polystyrene (PS) and low-density polyethylene (LDPE) at a number of different output rates. At length/diameter ratios of 11 upwards the more viscous PS phase was completely enveloped by LDPE and assumed a circular shape after starting with a flat interface of two semicircular streams at the die inlet.[9] Han and Khan also carried out a theoretical study for determining the equilibrium interfacial configuration during stratified flow in rectangular ducts and provided theoretical backing for the view that viscosity will be far more important in deciding the interfacial geometry than elastic effects.

In a further experimental study Han coextruded PS and LDPE after feeding them into a circular die concentrically and eccentrically. Little change occurred when PS was the core material, but a dramatic (if gradual) reversal was observed when the lower-viscosity LDPE was at the centre initially, provided that the die was long enough for equilibrium to be established.

An interesting situation arises when the viscosity/shear stress curves of two polymer melts cross, so that the component which was the more viscous up to a given shear stress becomes less viscous than the second component beyond that shear stress. Such is the case in the system PS/HDPE (polystyrene/high-density polyethylene). At 200 °C (see Fig. 9.2(a)) the viscosity of PS exceeds that of HDPE below approximately 12×10^5 dynes cm^{-2}; but above that shear stress HDPE is the more viscous of the two melts. On the other hand, HDPE is the more viscous component throughout the entire available shear stress range at 240 °C (see Fig. 9.2(b)). When coextruding at 200 °C at a shear stress below 12×10^5 dynes cm^{-2} the core material is PS, but when the temperature was raised to 240 °C, reversal, i.e. encapsulation of the HDPE by the PS occurred.

Lee and White[10] and Southern and Ballman[11,12] used polymer systems which gave viscosity crossovers as the shear stress was increased. They duly observed that envelope reversal also occurred at the same points.

REFERENCES

1. M. H. NAITOVE, *Plastics Technology*, **23**(2) (1977).
2. A. IWAWAKI, K. KOJIMA and Y. SHITARA, *Mod. Plastics Int.*, **7**, 15 (1977).
3. W. E. SISSON and F. F. MOREHEAD, *Textile Res. J.*, **23**, 152 (1953); **30**, 153 (1960).
4. E. M. HICKS *et al.*, in *Man-made Fibres* (H. F. Mark *et al.*, editors), Vol. 1, p. 375, Wiley Interscience, New York (1967).
5. E. M. HICKS *et al.*, *Textile Res. J.*, **30**, 675 (1960).
6. C. D. HAN, *Rheology in Polymer Processing*, Academic Press, New York and London (1976).
7. C. D. HAN, *Rheology in Polymer Processing*, p. 264–70, Academic Press, New York and London (1976).
8. C. D. HAN and A. A. KHAN, *Trans. Soc. Rheol.* (in press).
9. C. D. HAN, *J. Appl. Poly. Sci.*, **17**, 1289 (1973).
10. B. L. LEE and J. L. WHITE, *Trans. Soc. Rheol.*, **18**, 467 (1974).
11. J. H. SOUTHERN and R. L. BALLMAN, *Appl. Poly. Symp.* No. 20, 1234 (1973).
12. J. H. SOUTHERN and R. L. BALLMAN, *J. Poly. Sci.*, *A-2*, **13**, 863 (1975).

10

The Effect of Melt Elasticity on Extrusion and Other Melt Processing Operations

It has long been known that grossly irregular extrudate is obtained at excessively high extrusion pressures, especially at low melt temperatures when the viscosity is very high. The gross irregularities suggest that an elastic effect is superimposed on the viscous effect. The stress history of the emerging extrudate must have involved a major flow disturbance—departure from laminar flow, turbulence, 'melt fracture'—and the die channel was too short to give the melt sufficient dwell time to enable it to relax and recover and to re-establish fully laminar flow. Tordella[1] has pointed out that the site of major disturbance must be the die inlet region, where a fat and sluggish flow is converted to a thin and fast flow. Kendall[2] studied the effect of the die inlet taper angle on the degree of distortion and found that the elastic stresses in the melt and the severity of the extrudate defects were much reduced when the taper was gentle. It was also possible to estimate the relaxation time of the melt, taking it as the minimum dwell time in the die which is necessary to eliminate the elastic memory and obtain good extrudate.

A highly polydisperse polymer will have a wide distribution of relaxation times, especially if it includes a substantial amount of low molecular weight fractions. In order to make good profile at a fast rate it is desirable to use low molecular weight polymer; on the other hand, this poses take-off problems because of the low melt viscosity and the poor mechanical properties of the extrudate. Optimisation, as always, demands a compromise between production efficiency and product quality. Optimisation differs from product to product.

In wire covering a long dwell time would cause thermal degradation of the melt with a consequent loss in electrical properties; moreover, production rates must be high. Since the mechanical strength of the extrudate is of secondary importance, polymers of much lower molecular weight can be used than would be the case in profile extrusion. A good cooling bath must be provided which may be a long trough with efficient

101

heat exchange. In wire coating dies (which have a cross-head geometry) flow disturbance is very liable to occur, which may be minimised by using a polymer of moderate molecular weight.

In fibre and monofilament extrusion the die diameters are small, the dies are short and the inlets tend to be square cut. Although extrudate distortion is very liable to occur, this is generally masked by the high draw rate which also orients the product and strengthens it in the draw direction. The strength may be further enhanced by subsequent cold-drawing. Nevertheless, as will be seen presently, high drawdown rates create their own particular problems which are again attributable to the elastic disturbance near the die entrance.

In tubular film extrusion, melt fracture is not normally encountered because the shear stresses are below that which marks the critical value for gross flow disturbance. However, Kendall managed to induce turbulence by using rather unusual operating conditions. Introducing a dyed marker masterbatch material between the screw end and the die entrance, he noted that the oscillatory movement of the marker line was out of phase with that of the undrawn extrudate itself. This behaviour is typical of elastic disturbance: when the disturbance was sufficiently severe the marker line fractured. In practice, melt distortion can be avoided by either reducing the pressure or by using a die of larger cross-section and increasing drawdown. The second remedy is only available for circular rod and when there is no objection to an increase in orientation.

When film is produced by chill-roll casting from a slit die there is little danger of flow disturbance because the polymer used is not usually of particularly high molecular weight and because temperatures are high and viscosities correspondingly low. The process is similar to monofilament extrusion, except that two-dimensional draw may be applied.

In extrusion blow moulding, flow disturbance is very prone to occur. To obtain the best possible mechanical strength, high molecular weight polymer is preferred, hence melt viscosities and stresses are high. The manifestation of defects is confined to the inside of the blown container because the mould walls smooth out the external irregularities, but they are not only undesirable in clear bottles, but are also liable to cause environmental stress cracking when a container with moulded-in stresses is exposed to an aggressive medium. For this reason extrusion blow moulding dies are usually fairly long dies.

In injection moulding the requirements of the material, mould design and operating conditions are highly conflicting and are therefore difficult to optimise. On the one hand a low viscosity is desirable for easy mould filling, because (a) the use of low temperatures would minimise degradation (or, conversely, because lower pressures could be used whilst keeping the temperature as it was), (b) weld lines and consequent weaknesses in the moulding are minimised, and (c) intricate mouldings—possibly including inserts—can be more readily achieved. On the other hand, high molecular weight polymer will give a product with the best mechanical properties, and faster cycle times are possible because the difference in temperature between the melt in the barrel and in the mould is great and the moulding can be ejected after a shorter mould-closed period, although this will also cause residual stresses in the moulding which may lead to ultimate distortion. Turbulent flow into the mould is not, of itself, undesirable since it promotes randomness and isotropy. Nevertheless, injection moulded articles are almost inevitably highly anisotropic.

Severs observed that at low shear stress a rod of extrudate could be snipped off cleanly at the die orifice—indicating good adhesion of the melt to the die wall. However, as the shear stress was raised above some critical value the entire die contents could be pulled out on giving the extrudate a smart tug. This demonstrates poor adhesion of the melt to the die wall. The problem of specific adhesion arises and will be considered again presently.

Some workers considered that the milder surface defects in extrusion arise at the die exit and are due to differential recovery of the highly strained extrudate skin and the less strained material in the body of the melt. But Tordella[3] preferred a notion involving slip-stick and found no evidence that initiation of roughness was due to differential elastic recovery between the various portions of the emerging polymer stream. Mieras[4] has shown that good quantitative agreement in assessing the elastic properties of polypropylene melts can be obtained between experimental techniques based on the measurement of die swell and on melt birefringence measurements. The elastic (recoverable) strain present in polymer melts shows itself clearly in the die swell of the emerging polymer stream. The swelling is independent of the die length at low output rates; at higher output rates it is less pronounced. This has been explained in various ways:

(i)   Re-randomisation of polymer chains which have become flow-oriented whilst passing through the die.

(ii) Transition from a near-parabolic velocity distribution at the die exit, to a constant velocity distribution some way down the line. The highly strained skin slows down the core, so that the cross-sectional area must increase.

Clegg[5] considers that both mechanisms play a part in *low shear rate* die swell. But neither accounts for the fact that at *high shear rates* (i.e. at high output rates) shorter dies produce larger swelling. This can only be explained as a memory effect of flow disturbance at the die entry. At low shear rates the upstream flow defect is mild and certainly nowhere near turbulence and melt fracture since the melt is below the Reynolds number. In order to examine this phenomenon further, Clegg studied the flow pattern by extruding rod containing a coloured core. This was accomplished by filling an inverted ram extruder barrel which was specially constructed for the laboratory investigation with natural polyethylene discs

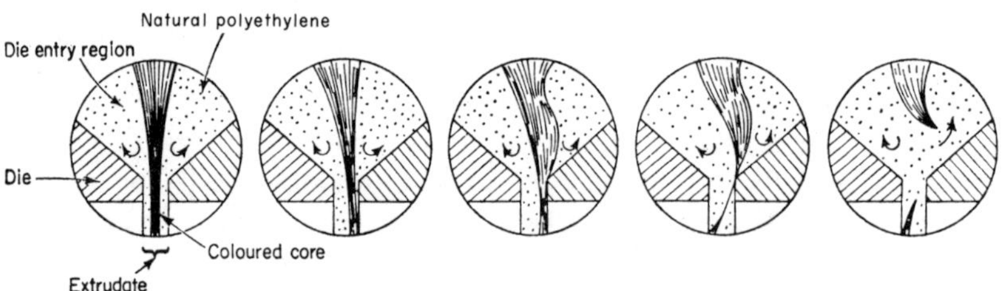

FIG. 10.1.   Flow phenomena at the entrance of an extruder die.

whose punched out centres had been replaced by a coloured masterbatch of the same grade. At low output rates a uniform extrudate was obtained in which the core remained central. At higher output rates the core moved in a vortex streamline with increasing sideways motion. At even higher output rates a critical point was reached when no core material was extruded at all for a brief interval. The extrudate showed a knobbliness or waviness whenever this occurred and the whole sequence of core–no core was repeated in regular cycles. This indicated that the disturbance had now reached the extremity of melt fracture and this was confirmed by a ciné film record of the events at the die entrance. Oscillatory flow was observed as the output rate increased (Fig. 10.1). Eventually the oscillation amplitude became so large that the hitherto continuous thread of colour core broke. The upper portion appeared to retract away and to follow the eddies whose course could be observed by means of poppy seed included in the uncoloured polymer. The effect was most pronounced with a square cut die inlet and was substantially diminished with a 12° half-angle inlet cone. With this tapered inlet the die length could be reduced by one-third and the shear rate increased by a decade before the critical stress was reached at which melt fracture occurred. This effect is absolutely analogous to the beneficial effect of using not only tapered inlets but even tapered dies, as has been seen earlier. Melt fracture, when it occurs, happens some small distance upstream from the die entrance rather than at the die entrance itself, in a region where the streamlines of the mainstream and those of the eddying currents adjoin and begin to interfere with each other at higher shear rates. Polymers which show flow irregularities at low output rates are those which have comparatively long relaxation times.

Elastic parameters depend on the die length whilst viscosity does not. Schreiber and Bagley found that some linear polyethylene fractions needed no end correction for effective die length (normally taken to be equivalent to six radii), whilst unfractionated polymers of the same average molecular weight as the fractions had appreciable end corrections. The two types also differed in extrudate swelling and in the rate with which swelling increased with increasing stress. Swelling was not only less with fractions, it also did not seem to matter whether the fractions were essentially monodisperse or not. Indeed reconstituted whole polymer after mixing of fractions did not restore the extrudate swelling observed before fractionation. Evidently factors other than molecular weight distribution were responsible for the differences in elastic properties, and these must have arisen as a direct consequence of the fractionation process itself. This was confirmed by Schreiber, Rudin and Bagley.[6]

Philippoff noted that solution treatment had a pronounced effect on polymer elasticity. He also observed a marked reduction in the birefringence of polystyrene solutions after filtration, whilst the viscosity remained unchanged. This led to the conclusion that fractionation and filtration in solution have the same effect, namely the elimination or reduction of the elastic response in melts obtained from the fractions subsequently, whilst the viscous response remained unaffected. Since the size of the flow units in polymer fractions and in whole polymers is much the same, it cannot be affected by solution treatment. What *can* be affected is entanglement. This may be regarded as transitory quasi-crosslinks which are loosened by solution treatment and 'combed out' by filtration. Fewer entanglements means less elasticity. Entanglements are largely restored in the melt over a period of time, but *some* elasticity is always present, if only because conditions at the die entrance promote entanglement. Evidently, melt elasticity depends greatly on the history of the polymer, including:

(i)   any solution treatment which will loosen and 'comb out' entanglements;

(ii) turbulent shear which will tend to promote entanglement but may also cause the opposite effect if it leads to mechanical chain scission;

(iii) thermal treatment which may cause chain shortening by degradation.

The viscous properties, on the other hand are unaffected for the reasons given above.

Schreiber and Storey[7] have taken up a suggestion by Busse[8] that molecular fractionation could occur along the wall of the die and that this may contribute to a reduction in extrudate swelling.

It is now recognised that both melt instability in the die entry region (ultimately leading to melt fracture) and slip-stick effects in the die itself play a part in causing extrudate defects. Cogswell and Lamb[9] investigated the criterion for the triggering of slippage. Starting from a consideration of the end correction for dies:

$$\tau_{true} = (P - P_0)r/2l$$

where P_0 is the pressure necessary to extrude through a die of zero length; it was found that P_0 is a function of the true shear stress τ_{true} and is largely independent of temperature. A linear plot showed a discontinuity in an otherwise linear plot at some critical shear rate at which distortion commences. If this distortion is due to slip-stick, as Benbow and Lamb[10] suggest, then the balance of axial forces in the die wall region may be considered with profit. The elastic stress is a function of P_0, say, $f_1(P_0)$ and it is always greater than the shear stress at the wall.[11] There will also be a static adhesion term, $f_2[(P - P_0)l'/l]$ which will be a function of the pressure at the particular distance from the die exit up to which dehesion extends (l'), a distance which, at the moment of triggering, will be zero since this commences at the die exit. $f_1(P_0)$ may also be a function of l' due to the relaxation of elastic strain during the passage of melt through the die. There may be additional forces Z acting from outside the die (haul-off). Cogswell and Lamb suggest that slip-stick is triggered when

$$f_1(P_0) - \tau \quad \text{exceeds} \quad f_2[(P - P_0)l'/l] + Z$$

and they continue to amplify this criterion as follows:

'The initial triggering occurs where the adhesion term is minimal, that is, at the die exit where $l' = 0$. However, P_0 increases more rapidly than τ, and at higher shear rates will be able to overcome greater adhesion; because of this the site of triggering will progress backwards toward the die entry as throughput is increased beyond the critical value. When this slipping occurs P_0 relaxes slightly and allows the polymer to re-adhere to the die wall. The result is a periodic effect. The periodicity and the mass per cycle M are related to the volume over which the elastic strain is relaxed as a result of one initiating slip event.'

Cogswell and Lamb go on to use the mass per cycle M as a measure of the severity of the flow defect and suggest that irregular distortion (gross melt fracture) occurs when the trigger site has moved upstream until it reaches the die entry ($l' \equiv l$), and when $f_1(P_0) - \tau$ must be greater than $f_2[(P - P_0)l'/l] + Z$, unless Z is unusually large as, for example in fibre spin haul-offs. For a constant shear rate and a given length/radius ratio, M is proportional to r^3. This implies that, for a fixed value of P_0, the volume of melt relaxed by a single slip-stick cycle occupies the same proportion of the die volume, whatever the radius. Thus M/r^3 constitutes a normalised measure of shear severity, so that the distorted extrudate from a small die is a true mini-reproduction of the distorted extrudate from a large die. On plotting M vs. the shear stress in polypropylene a smooth curvilinear relationship became evident which was, moreover, little affected by degradative treatment (i.e. molecular weight); this showed that volume and stress distribution effects are more important than molecular relaxation phenomena.

Cogswell and Lamb also drew attention to the fact that M decreases as the piston of a ram extruder approaches the die:

'At small piston heights the *radial* flow increases and the layer closest to the die wall receives a greater acceleration than earlier on and will therefore be more likely to trigger distortion. The central portion receives less acceleration, hence less elastic strain, and there is no change in the total pressure drop or output rate. The effect has been investigated as a function of initial piston height and thermal history which does not affect the result in any way. The evidence again points to the die entry region as the locus where the *bulk* effects responsible for distortion arise; triggering takes place in the boundary layer at the polymer/metal interface where the maximum elastic strain occurs.'

The conclusions reached can finally be summarised as follows:

(i)   the onset and severity of distortion depend on the recovery of elastic deformation;
(ii) the elastic deformation is imparted to the polymer melt during acceleration in the die entry region;
(iii) recovery is triggered by slippage at the die wall in a direction opposite to flow;

(iv)   triggering occurs where adhesion is least, i.e. at the die exit;

(v)   recovery may be prevented and melt distortion obviated by exerting large drawdown forces (fast haul-off).

One might add that since melt instability involves a mass/acceleration term a fundamentally dynamic situation arises which must lead to sinusoidal oscillation. The period of this oscillation is directly related to the shear stress and inversely related to the die length. The longer the die, the longer the dwell time of the melt in the die and the greater the probability that the dwell time inside the die is sufficient for relaxation to occur in imperceptible micro- rather than in macroscopic slip-stick events. The application of Z-forces has the effect of increasing the die length since it imposes restrictive boundary conditions *beyond* the die exit.

However, if drawdown is excessive, yet another situation arises. The melt-spinning of polymer fibres has been studied intensively by White and his colleagues and students at Tennessee[12] and by Han and his school in Brooklyn[13] and their flow visualisation and birefringence studies respectively represent most important contributions to the understanding of extrusion flow phenomena. Recently Lenk[14] showed that a phenomenon known as 'draw resonance' is just another aspect of the same overall elastic disturbance which arises at the die entrance and coalesces with effects caused further downstream and reacting backward. Draw resonance has

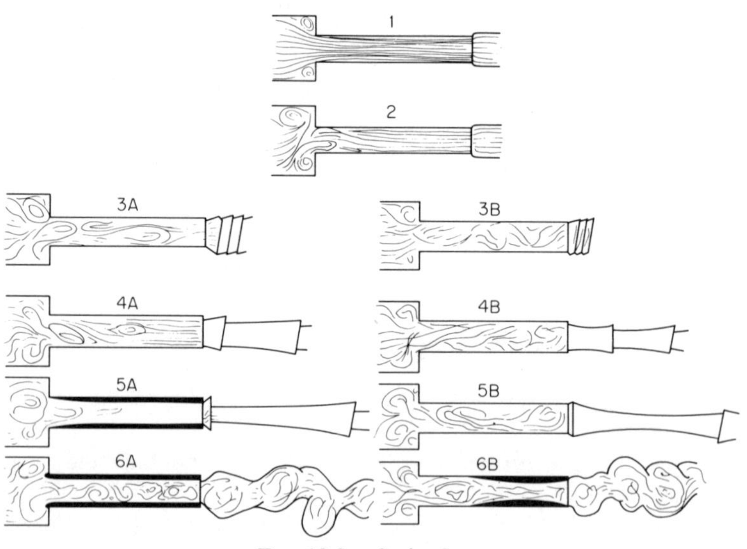

FIG. 10.2.   Series I.

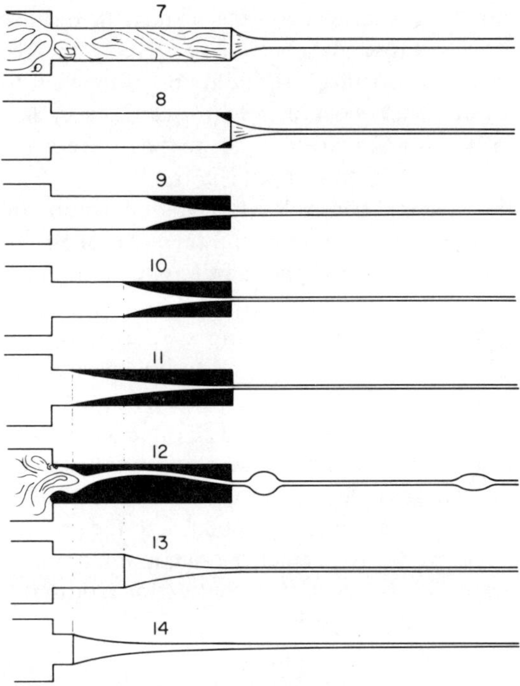

FIG. 10.3.   Series II.

been described as a periodic knobbliness in the fibre extrudate on applying excessive drawdown. The knobbles may be close together or a longer distance apart. Lenk illustrated the unified concept by means of two series of diagrams (Figs. 10.2 and 10.3).

In the first series, six diagrams represent a progressive increase in shear stress at zero drawdown or at a drawdown not exceeding that necessary for compensation of die swell.

Diagram 1 (Fig. 10.2) depicts stable laminar flow with eddying currents of stagnant flow at the shoulder of the die entrance. Drawdown is absent or slight, die swell being prominent. Diagram 2 shows the development of flow instability at the die entrance; the shear stress is not, as yet, excessive, so that the dwell time of the melt in the die is sufficiently long for melt relaxation to occur. This ensures that uniaxial laminar flow is re-established and that the elastic memory of the flow instability to which the melt has been subjected upstream is annealed out. The extrudate therefore shows no defect.

In diagram 3A the residence time of melt in the die is less than the relaxation time because of an excessive increase in the applied pressure (shear stress). This also affects the adhesion of the melt to the die wall and thus makes snapback possible. Snapback is a cyclic oscillatory melt retraction in an upstream direction due to the recovery of elastic strain. The greater that strain, the greater will be the retraction amplitude and the larger will be the mass per cycle M. This is shown by increasingly severe extrudate defects from surface mattness through 'orange peel' and 'sharkskin' to 'corkscrewing' and 'bambooing' (diagrams 4A and 5A). In diagram 6A the upstream snapback amplitude has become so large that it covers the entire die length right up to the site of primary melt flow instability. When this occurs, a blob of turbulent or stagnant material slips into the die and is carried some indeterminate distance downstream before the next slip-stick event is triggered. The cycles are now arhythmic and the corresponding variability in elastic recovery results in variable masses and shapes per cycle; gross distortion of the extrudate is observed and the extrudate ceases to bear much resemblance to the profile of the die orifice.

Whilst diagrams 3A–6A depict the maximum amplitudes of snapback cycles, diagrams 3B–6B show the other extreme of that cycle, namely the stage just prior to the triggering of snapback.

Figure 10.3 shows extrusion at constant pressure but at increasing drawdown rates. Drawdown not only reduces die swell for obvious reasons. It also orients the flow units in an axial direction, assists the return of the isovels to an axial distribution and opposes snapback. In so doing it brings about a steady state as shown in diagram 7. By the same token, the appearance of extrudate defects is either altogether prevented or—if present—disguised and rendered unidentifiable as the filament is stretched and smoothed out.

As drawdown is further increased the tension will rise to a magnitude sufficient to cause dehesion of melt from the die wall in an upstream direction starting at the die exit. The melt inside the die channel requires considerably less force to effect a reduction in cross-sectional area than the rapidly cooling (and therefore much more viscous) material further down the spin line. This progressively causes further dehesion of melt upstream as the rate of drawdown is increased (diagrams 8–11).

Dehesion in Fig. 10.3 differs from dehesion in Fig. 10.2 in that it is tension-induced and the strain cannot be relieved by an elastic snapback mechanism. Instead a non-cyclic state is established and the monofilament or fibre is still perfectly regular. However, the die length is *effectively* reduced (see diagram 10 which corresponds to the die length of diagram 13, and diagram 11 which corresponds to the die length of diagram 14). When drawdown becomes excessive, melt dehesion produces a die of effective zero length and the filament is drawn directly out of the unstable die entry region. At that moment an irregular blob will be drawn into the die, elastic relaxation will occur and the stresses leading to the next cycle will immediately begin to build up again. We thus witness a return to a cyclic sequence of melt dehesion and rehesion, but this time in the *absence* of snapback (diagram 12). This is the effect which Han *et al.*[15] describe as 'draw resonance'. It is observed when fibre is spun at excessively high drawdown, a condition which is easily reached in a process in which high drawdown rates are necessarily characteristic. It is possible to identify critical drawdown rates which will cause the appearance of regularly spaced nodules in the extrudate. It is a criterion for the appearance of 'draw resonance'. When it is exceeded, then the distance between nodules is steadily reduced, so that it may even cause the extrudate to assume a sawtooth appearance.

It is seen that melt fracture in the extrusion of rod and draw resonance during extrusion of fibre at excessive drawdown are *not* distinct flow instability phenomena, but that *both* are caused by elastic effects which arise near the die inlet. The difference is only in that in draw resonance (so-called) there is an abrupt appearance of cyclic knobbliness at excessive drawdown, whilst in the other case one observes a gradual escalation of severity of surface defects up to eventual gross irregularity at extreme pressures.

It is worth noting that the worst effects of extrusion defects can be avoided by tapering the die inlet and by slightly tapering the channel itself. It is also evident that the specific adhesion of polymer melts to the die metal at various stresses and temperatures would constitute a profitable field of study, as has already been recognised earlier by Cogswell and Lamb.[16]. Die swell and melt fracture have been studied by Vlachopoulos[17] with particular reference to the effect of molecular weight distribution. He concluded that the latter is very important both in melt fracture and die swell studies.

REFERENCES

1. J. P. TORDELLA, *J. Appl. Phys.*, **27**, 404 (1956); *SPE Journal*, **36** (Feb. 1956).
2. K. G. KENDALL, *Trans. & J. Plast. Inst.*, **31**(99), 49 (1963).
3. J. P. TORDELLA, *J. Appl. Poly. Sci.*, **7**, 215 (1963).
4. H. J. M. A. MIERAS, paper presented at the autumn conference of the British Society of Rheology, Shrivenham (1967).

5. P. CLEGG in *The Rheology of Elastomers*, p. 174, Welwyn Garden City Conference, Pergamon Press, Oxford (1958).
6. H. P. SCHREIBER, A. RUDIN and E. B. BAGLEY, *J. Appl. Poly. Sci.*, **7,** 887 (1965).
7. H. P. SCHREIBER and D. C. STOREY, *Polymer Letters*, **B3**(9), 723 (Sept. 1965).
8. W. F. BUSSE, *Physics Today*, **17,** 32 (Sept. 1965).
9. F. N. COGSWELL and P. LAMB, *Trans. & J. Plast. Inst.*, **35**(120), 809 (Dec. 1967).
10. J. J. BENBOW and P. LAMB, *SPE Trans.*, **3,** 7 (1963).
11. J. P. TORDELLA, *J. Appl. Phys.*, **27,** 454 (1956).
12. J. L. WHITE and Y. IDE, *Rheology and Dynamics of Fiber Formation from Polymer Melts*, Univ. Tenn. (Knoxville), May 1975, with 158 references, also *Appl. Poly. Symp.*, **27** (1975).
13. C. D. HAN, *Rheology in Polymer Melt Processing*, Chs. 5, 6 and 8, Academic Press, New York and London (1976).
14. R. S. LENK, A unified concept of melt flow instability during extrusion, *J. Appl. Poly. Sci.* (in press).
15. C. D. HAN, *Rheology in Polymer Processing*, Academic Press, New York and London (1976).
16. F. N. COGSWELL and P. LAMB, *Trans. & J. Plast. Inst.* (Dec. 1967).
17. J. VLACHOPOULOS, *Rheol. Acta*, **13,** 223 (1974).

11

The Rheology of Calendering

Assuming the melt to be in laminar flow, a complete analysis would answer the following questions:

Given the material constants, the geometry of the equipment and the selected operating variables, what will be:

(i)   the thickness of the foil or sheet produced;
(ii)   the temperature distribution in the material;
(iii) the rate of draw through the roll nip;
(iv) the force which tends to separate the rolls and acts on the calender bearings;
(v)   the power required to work the calender.

Assuming absence of elastic forces, incompressibility and Newtonian behaviour in the melt, Gaskell[1] has formulated an analysis which is relatively simple. One considers a model as depicted in Fig. 11.1 which represents the upper half of a symmetrical calender gap. A rectangular coordinate system is used, with x the forward direction and y the vertical axis. Because of incompressibility the continuity equation becomes

$$(\partial v_x/\partial x) + (\partial v_x/\partial y) = 0 \tag{1}$$

and the x-component of the momentum equation is

$$\rho\left[v_x\left(\frac{\partial v_x}{\partial x}\right) + v_y\left(\frac{\partial v_y}{\partial y}\right)\right] = -\left(\frac{\partial P}{\partial x}\right) + \eta\left[\left(\frac{\partial^2 v_x}{\partial x^2}\right) + \left(\frac{\partial^2 v_x}{\partial y^2}\right)\right] \tag{2}$$

The radius of both rolls is R and their surface speed V, whilst the gap at the nip is $2H_0$. Equation (2) is now simplified: the left-hand side represents the accelerational term. This may be ignored because flow is relatively slow and the viscous forces are large. As for the right-hand side, it is assumed that $(\partial v_x/\partial x)$ is small compared with $(\partial v_x/\partial y)$ so that it, too, may be neglected. This is reasonable because the geometry approximates that of parallel

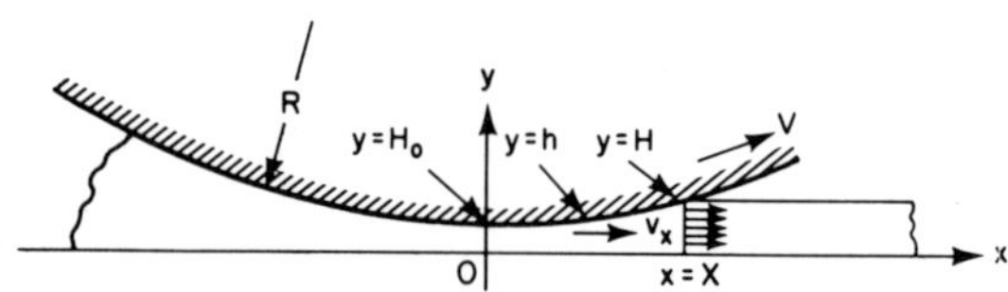

Fig. 11.1.   Diagrammatic representation of one half of the gap in a symmetrical calender.

plates if H_0 is small compared with R, and because the equation is only applied to the immediate region of the nip. Equation (2) thus becomes

$$(\partial P/\partial x) = \eta(\partial^2 v_x/\partial y^2) \tag{3}$$

It is also implicit in the assumptions that the pressure drop occurs only in the x-direction, since the other derivatives of pressure with respect to y and z are zero. Integration of eqn. (3) gives

$$\left(\frac{\partial v_x}{\partial y}\right) = \dot{\gamma} = \frac{1}{\eta}\left(\frac{\partial P}{\partial x}\right)y + c \tag{4}$$

and since $\dot{\gamma} = 0$ when $y = 0$, c_1 is also zero. A second integration and evaluation of the integration constant (using the boundary condition that $v_x = V$ when $y = h$) gives the velocity profile

$$v_x = V + \frac{y^2 - h^2}{2\eta}\left(\frac{\partial P}{\partial x}\right) \tag{5}$$

where h is the distance from the plane of symmetry to the roll surface at any given point along the x-axis. Because of the known roll curvature, h is a function of x.

The volumetric flow rate per unit width of rolls is

$$Q = 2\int_0^h v_x\,dy$$

Introducing eqn. (5) and integrating one obtains

$$Q = 2h\left[V - \frac{h^2}{3\eta}\left(\frac{\partial P}{\partial x}\right)\right] \tag{6}$$

It is now convenient to re-define x in terms of a dimensionless variable β:

$$\beta = x/(2RH_0)^{1/2} \tag{7}$$

Introducing this into eqn. (6) and making $(\partial P/\partial x)$ the subject of the equation we get

$$(\partial P/\partial \beta) = (2RH_0)^{1/2}\left(\frac{3\eta}{h^2}\right)\left(V - \frac{Q}{2h}\right) \tag{8}$$

Now, h is given as a function of x as follows:

$$h = H_0 + [R - (R^2 - x^2)^{1/2}]$$

A useful approximation of this is

$$h = H_0 + \frac{x^2}{2R} \tag{9}$$

and eliminating x between eqns. (7) and (9) it is seen that the dimensionless variable β is related to h by the equation

$$R/H_0 = 1 + \beta^2 \tag{10}$$

Selecting another parameter λ^2 as follows:

$$\lambda^2 = \frac{Q}{2VH_0} - 1 \tag{11}$$

which, according to Gaskell, is also equivalent to

$$\lambda^2 = X^2/2RH_0 \tag{12}$$

It is possible to introduce λ^2 into eqn. (8) and obtain:

$$\left(\frac{\partial P}{\partial \beta}\right) = \frac{\eta V}{H_0}\sqrt{\frac{18R}{H_0}}\left[\frac{\beta^2 - \lambda^2}{(1 + \beta^2)^2}\right] \tag{13}$$

The significance of λ is that it is identical with β when $x = X$, i.e. at the point when the sheet loses contact with the rolls (see Fig. 11.1). X can be immediately obtained by measuring the sheet thickness as it leaves the calender. The pressure drop through the calender is obtained by integrating eqn. (13)

$$P = \frac{\eta V}{H_0}\sqrt{\frac{9R}{32H_0}}[g(\beta, \lambda) + c] \tag{14}$$

The integration constant is evaluated from the boundary condition that $P = 0$ when $\lambda = 0$ and is reasonably approximated by $c = 5\lambda^3$.

Equation (13) shows that $(\partial P/\partial \beta)$ is zero when $\beta = \pm\lambda$ and that the pressure is at a maximum when $\beta = -\lambda$ and at a minimum when $\beta = +\lambda$.

The function $g(\beta, \lambda)$ is complicated, but it has two roots of interest, namely those where the pressure is just zero. These points represent positive and negative values of $g(\beta, \lambda)$ respectively. It is obvious that the positive point is the point where the laminate leaves the rolls ($\beta = \lambda$) and the negative point can be designated as $\beta = -\beta_0$. At these points eqn. (14) shows that $g(\beta, \lambda) = -5\lambda^3$. This means that there exists a unique relationship between λ and β_0 which may be graphically represented (Fig. 11.2).

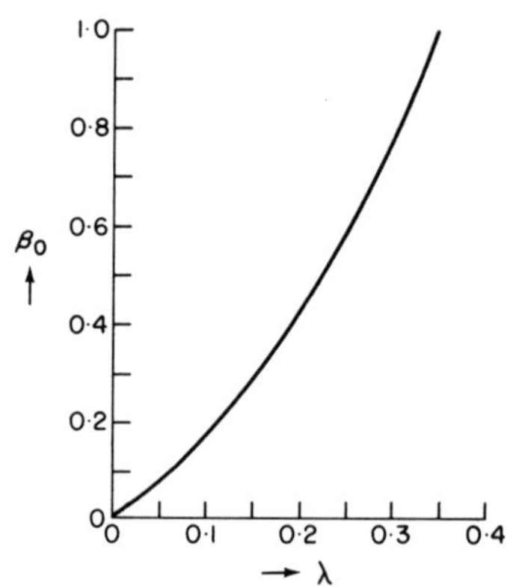

FIG. 11.2.   Relationship between λ and β_0.

The maximum pressure is obtained from eqn. (14) by setting $\beta = -\lambda$:

$$P_{\max} = P(-\lambda) = \frac{5\lambda^3\eta V}{H_0}\sqrt{\frac{9R}{8H_0}} \tag{15}$$

The dimensionless pressure profile is the ratio of eqns. (14) and (15) viz:

$$P/P_{\max} = P(\beta)/P(-\lambda) = \frac{1}{2}\left[1 + \frac{g(\beta, \lambda)}{5\lambda^3}\right] \tag{16}$$

For the special case where $\beta = 0$, i.e. at the nip, $g(\beta, \lambda) = g(0, \lambda) = 0$ and eqn. (16) takes the value of $\frac{1}{2}$. *The pressure at the nip is exactly half the maximum pressure.*

The effect of λ on the pressure profile is clearly seen when all the constants in eqn. (14) are collected together and combined with eqn. (16) to give

$$P = K\lambda^3 \frac{P}{P_{\max}} = K\lambda^3 P' \tag{17}$$

Figure 11.3 shows a plot of P vs. β for three values of λ. As λ increases the

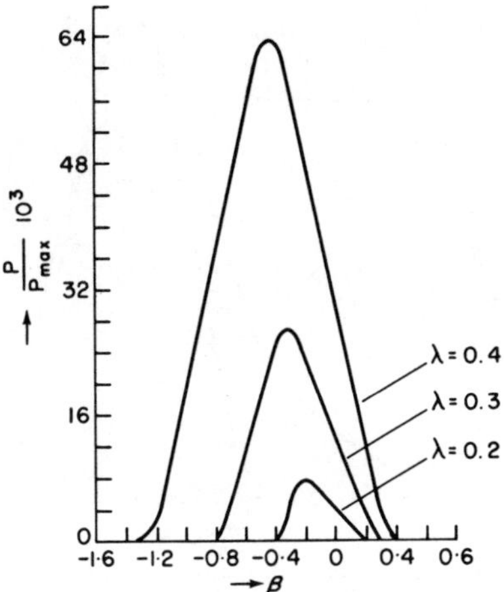

FIG. 11.3.   Dimensionless pressure profile for arbitrary selected values of λ.

profile broadens and the maximum pressure increases. Because of the third power dependence of P_{max} on λ, small changes in λ cause large variations in P_{max}. Thus, if λ is doubled, P_{max} increases eightfold, and if λ goes from 0·28 to 0·32, P_{max} increases by nearly 50%!

Although λ has a profound effect on the slope of the pressure profile, it is only *experimentally* determinable. If a pressure profile is available, λ can be determined either from the zero pressure points, from the maximum pressure point, or from the nip pressure, depending on which gives the best overall fit with the experimental data. Bergen and Scott[2] measured pressure profiles, using strain gauge pressure transducers on the surface of a 10 in diameter roll. Comparison of the data with the theoretical curve showed a good fit for $\beta > -\lambda$, but not such a good fit for $\beta < -\lambda$ (Figure 11.4, a grade of plasticised PVC). The curves coincided at P_{max} because λ was assigned the value of β which is appropriate for P_{max}. The main reason for non-coincidence—especially on the left of the plot—is due to the non-Newtonian nature of the melt. The theory assumes η to be constant at all points, but since the melt is pseudoplastic, η is smaller in the nip region, so that the pressure must build up earlier than required by Newtonian flow theory.

By combining eqns. (5) and (13), transforming x to β and rearranging, one obtains a dimensionless velocity profile

$$v_x/V = \frac{2 + 3\lambda^2[1 - (y/h)^2] - \beta^2[1 - 3(y/h)^2]}{2(1 + \beta^2)} \qquad (18)$$

     Polymer Rheology

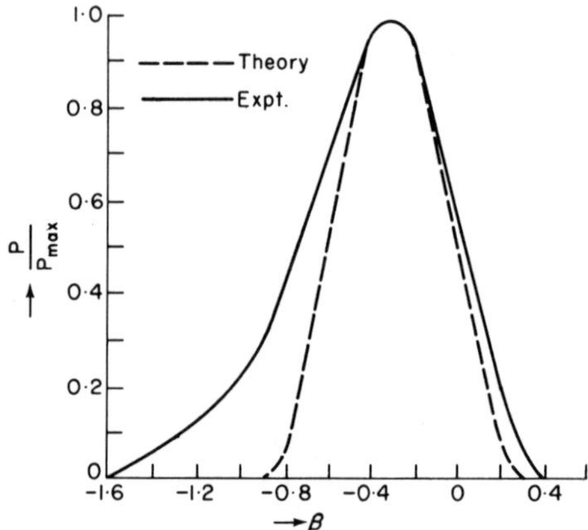

FIG. 11.4.   Comparison of theoretical and experimental pressure profiles (after Bergen and Scott[2]).

which shows that v_x/V is a function of β, y/h, and λ, where the first two are the positional coordinates.

Assuming $\lambda^2 = 0\cdot10$, one can depict the velocity profiles for various values of β in terms of λ, as is exemplified in Fig. 11.5. The following points should be noted:

(i)   When β lies between $-\lambda$ and $+\lambda$ the pressure gradient is negative and the velocity profile is concave—pressure flow occurs in the forward direction.

(ii)   When β is *less* than $-\lambda$ the pressure gradient is positive and forward flow is retarded—the velocity profile is concave.

(iii)   As β decreases further, a point is eventually reached when the velocity at the mid-plane will be zero. This stagnation point, designated β^*, is found by setting v_x/V in eqn. (18) equal to zero, giving

$$(\beta^*)^2 - 3\lambda^2 - 2 = 0$$

This shows that β^* is a function of λ. When $\lambda^2 = 0\cdot1$, the stagnation point is at $\beta^* = -4\cdot8\lambda$.

(iv)   Even further back, when $\beta < \beta^*$, the velocity components have both positive *and* negative values: towards the mid-plane melt moves away from the nip because the velocity component is positive. Near the roll surfaces it is positive and the melt moves *towards* the nip. As a result there is partial circulation of the melt in this region.

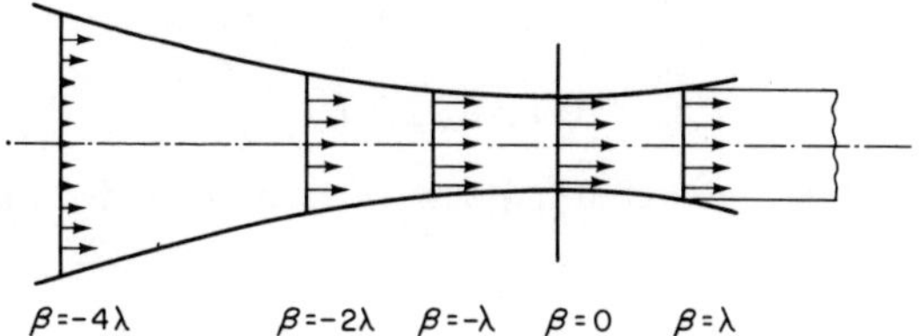

FIG. 11.5.   Velocity profile for $\lambda^2 = 0\cdot10$.

The shear rate in the melt is obtained by differentiating v_x as given in eqn. (18) with respect to y, giving

$$\dot{\gamma} = \left(\frac{\partial v_x}{\partial y}\right) = \frac{1}{h}[\partial v_x/\partial(y/h)] = \frac{3V(y/h)}{h}\left(\frac{\beta^2 - \lambda^2}{1 + \beta^2}\right) \tag{19}$$

and since $h/H_0 = 1 + \beta^2$ (eqn. (10)),

$$\dot{\gamma} = \frac{3V}{H_0}\frac{y}{h}\left[\frac{\beta^2 - \lambda^2}{(1 + \beta^2)^2}\right] \tag{20}$$

At the roll surface, where $(y/h) = 1$ and the shear rate is $\dot{\gamma}_h$ we have

$$\dot{\gamma}_h = \frac{3V}{H_0}\left[\frac{\beta^2 - \lambda^2}{(1 + \beta^2)^2}\right] \tag{21}$$

To determine the power p required to drive each roll, one takes the product of:

(i)   the roll velocity at the surface, V;
(ii)   the roll length along the axis, W;
(iii)   the melt viscosity η;
(iv)   the sum of all the shear rates $\dot{\gamma}_h$ along the x-axis where the melt is in contact with the rolls.

There are two rolls, therefore

$$p = 2VW\eta \int \dot{\gamma}_h \, dx \tag{22}$$

and on transformation to the dimensionless variable β we get

$$p = 2VW\eta \sqrt{2RH_0} \int_{-\beta_0}^{\lambda} \dot{\gamma}_h \, d\beta \tag{23}$$

                     Polymer Rheology

Substituting for $\dot{\gamma}_h$ according to eqn. (21) and integrating one obtains

$$p = 3WV^2\eta\sqrt{2R/H_0}\,f(\lambda) \tag{24}$$

where $f(\lambda)$ is a lengthy expression which will be presented graphically a little further on.

To determine the hydrostatic pressure of the melt which exerts a thrust tending to separate the rolls, one takes the product of:

(i)   the roll velocity at the surface, V;
(ii)   the roll length along the axis, W;
(iii)   the sum of all the pressures P at the roll circumference along the x-axis where the melt is in contact with the rolls.

$F = VW\int P\,\mathrm{d}x$ which, on transforming x to the dimensionless variable β according to eqn. (14) becomes

$$F = W\sqrt{2RH_0}\int_{-\beta_0}^{\lambda} P\,\mathrm{d}\beta \tag{25}$$

Integration yields

$$F = \frac{3\eta VRW}{4H_0}\,q(\lambda) \tag{26}$$

where the function $q(\lambda)$ is defined by another complicated equation which need not be given explicitly, since it is depicted, together with $f(\lambda)$ in Fig. 11.6.

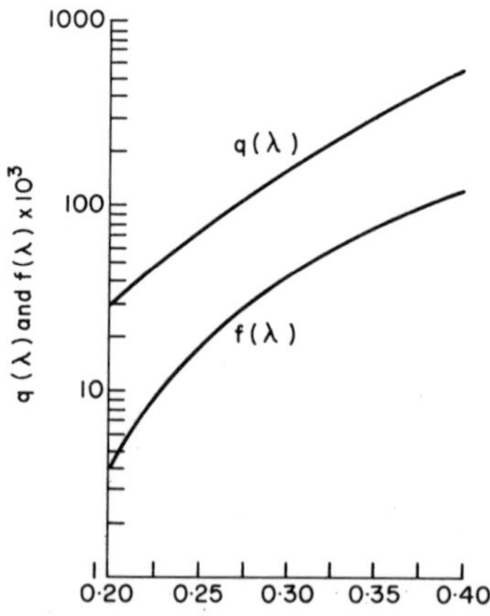

FIG. 11.6.   Plot of $q(\lambda)$ and $f(\lambda)$ as a function of λ.

Example

A calender has a radius R of 10 cm, an axial width W of 100 cm, a roll separation H_0 of 0·01 cm and it runs at a peripheral speed of 40 cm s^{-1} producing a sheet of 0·0218 cm thickness. The melt viscosity is 10^4 p. Calculate the maximum pressure developed in the material; the power required to operate the calender; the roll separating force; and the volumetric calendering rate.

Solution: λ can be calculated because the thickness of the calendered sheet (and hence the point of separation from the rolls in terms of x) is given: at the point of separation $h = H$ and $H/H_0 = 1 + \lambda^2$, where H represents the half-thickness of the calendered sheet and H_0 is half the roll separation at the nip. Putting in the given data, $\lambda = 0·30$.

$P_{\max}$ is now obtained from eqn. (15) giving 181×10^6 dynes cm^{-2}.

The roll separating force is obtained from eqn. (24), but $f(\lambda)$ is first read off from Fig. 11.6 at $\lambda = 0·3$, where it is found to be 0·043. Hence p is found to be 922 ergs s^{-1} or 0·92 kW.

The roll separating force is obtained from eqn. (26), but $q(\lambda)$ is first read off for $\lambda = 0·30$ in Fig. 11.6. It is found to be 0·16, hence F is found to be $4·8 \times 10^9$ dynes.

The volumetric calendering rate $Q = 2VHW = 87·2$ cm^3 s^{-1}.

Extension of the hydrodynamic analysis of calendering to non-Newtonian melts. Essentially the same procedure is used, making the same simplifying assumptions as before, except that the viscosity is now also a function of shear rate. For the pressure gradient the following equations are obtained:

For $\beta^2 > \lambda^2$

$$\frac{\partial P}{\partial \beta} = K \frac{(\beta^2 - \lambda^2)^n}{(1 + \beta^2)^{2n+1}} \tag{27}$$

for $\lambda^2 > \beta^2$

$$\frac{\partial P}{\partial \beta} = K \frac{(\lambda^2 - \beta^2)^n}{(1 + \beta^2)^{2n+1}} \tag{28}$$

where the coefficient K is defined by

$$K = \left(\frac{2n+1}{n}\right)^n \left(\frac{V\eta^0}{H_0\tau^0}\right)^n \left(\frac{\tau^0}{H_0}\right) \tag{29}$$

Example

Reconsider the calendering process evaluated in the previous example. The polymer melt is now taken to be a power law liquid with power index $n = 0.5$ and with a standard reference viscosity $\eta^0 = 10^4$ p at the standard shear rate of $\dot{\gamma}^0 = 1\,\text{s}^{-1}$. K is found to be 12.6×10^7 dynes cm^{-2}. P is obtained by integration of eqn. (28). Using graphical integration techniques and taking $\lambda = 0.30$, as before, it is found that

$$\int_0^1 \lambda\,\mathrm{d}(\beta/\lambda) \cong 0.71$$

and $P_{\text{nip}} = 8.08 \times 10^6$ dynes cm^{-2}, and P_{max} is twice that value, namely 16.2×10^6 dynes cm^{-2}. Note that this is about one-tenth of the Newtonian with a *constant* viscosity of 10^4 p! Even under shear conditions which are mild compared with those encountered in extrusion it is thus seen that the effect is the reduction of the Newtonian viscosity by a decade when the power index is 0.5, a very common sort of value for pseudoplastic polymer melts.

REFERENCES

1. R. E. GASKELL, *J. Appl. Mech.*, **17**, 334 (1950).
2. J. T. BERGEN and G. W. SCOTT, *J. Appl. Mech.*, **18**, 101 (1951).

12

Stretching Flows

F. N. Cogswell

*Senior Research Physicist, Imperial Chemical Industries Limited,
Plastics Division, Welwyn Garden City, Herts AL7 1HD, UK*

INTRODUCTION

Polymer melts resist flow. In viscous flow the energy producing the flow is dissipated as heat. During flow some energy may be stored, giving rise to elastic effects, and, should the stress exceed some critical value, fracture may occur. When we are considering simple fluids then the viscosity may be clearly defined. The earlier chapters have shown that the viscosity of a polymer melt depends strongly on temperature, pressure and rate of deformation. This chapter introduces one further variable: geometry.

Simple liquids are isotropic; their properties are the same in all directions. Polymers are, by definition, long molecules and when a polymeric system is deformed those molecules may become oriented. Depending on the degree of anisotropy introduced, so the properties of the polymer become different in different directions. Consider a handful of matches in a random array: they will have the same resistance to deformation in any direction. Now order the matches as they would pack into a matchbox; there is now a qualitative difference in the resistance to deformation, dependent on the direction in which you try to push the matches.

If such a simple system produces such differences how much more must we expect the properties of complex polymer melts to depend on the way in which we deform them and the extent to which they are ordered. The concept of stretching as well as shearing flows allows us to account for this difference. However, far from adding complexity this chapter will show that this single extra concept allows us to rationalise a whole range of phenomena which occur during processing.

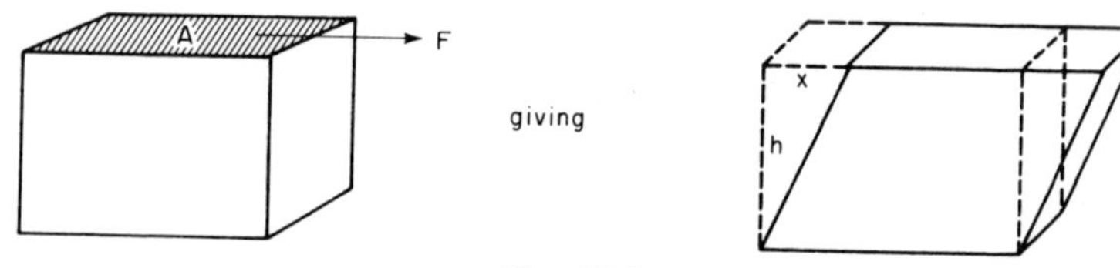

Fig. 12.1.

The Differences between Shearing and Stretching Flows

In a shearing flow the stress acts tangentially as shown in Fig. 12.1, where shear stress, σ_s, is defined as F/A, shear strain, γ, as x/h, and rate of shear, $\dot{\gamma}$, as $(1/h)(dx/dt)$. The viscosity in simple shear

$$\eta = \frac{\sigma_s}{\dot{\gamma}}$$

In a stretching flow the stress acts normal to the cross-section as shown in Fig. 12.2, where tensile stress, σ_E, is defined as F/A, tensile strain ε, as $\delta l/l$,† and rate of strain $\dot{\varepsilon}$ as $(1/l)(dl/dt)$ *which* $= v/l$ (if one end is fixed).

The stretching flow viscosity

$$\lambda = \frac{\sigma_E}{\dot{\varepsilon}}$$

For simple materials Trouton[1] used a simple geometrical construction to derive $\lambda = 3\eta$. However we have shown that polymer melts are not simple

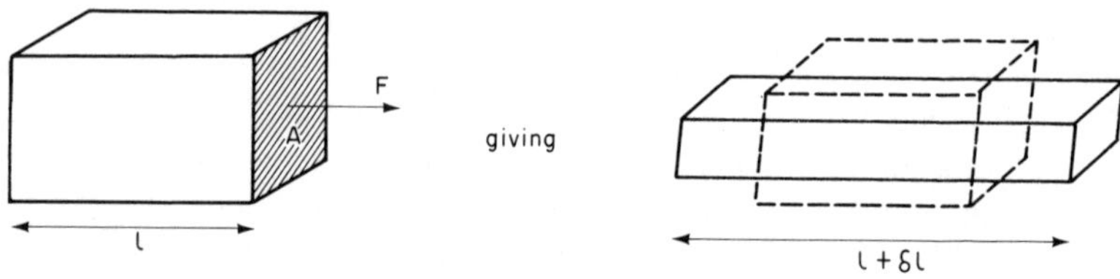

Fig. 12.2.

† *Note:* if the deformation is large

$$\varepsilon = \int_{L_0}^{L_1} \frac{dl}{l} = \ln \frac{L_1}{L_0}$$

where L_0 is the initial length and L_1 the final length. This is sometimes called the Hencky measure of strain or true strain as opposed to engineering strain $(L_1 - L_0)/L_0$.

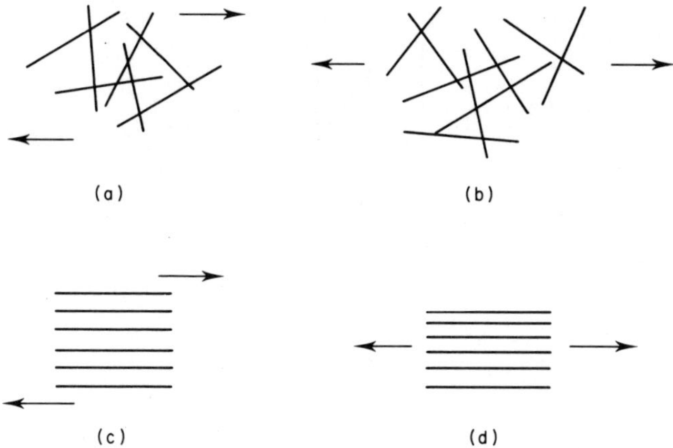

FIG. 12.3.   (a) Shear; (b) tension; (c) shear; (d) stretch.

materials: their molecules may be ordered. Consider again the match stick analogy. First the random array as shown in Fig. 12.3(a) and (b). We experience little qualitative difference in the response to the different deformations. But now consider the ordered matches in Fig. 12.3(c)—it becomes very easy to shear the matches in the plane in which they are ordered; but as seen in Fig. 12.3(d), if we try to stretch them it is impossible. This is an analogy only, but it serves to illustrate the qualitative differences which we may reasonably expect. For those readers who read Chapter 5, this qualitative difference has been formulated by Lodge.[2] Our experience with the matches combined with the guidance from theoreticians suggests that while we are familiar with the concept of shear thinning we need not be appalled by the possibilities of tension stiffening.

The rheological response of polymer melts combines three concepts: viscosity, elasticity and fracture. To represent the viscoelastic response it is convenient to consider a simple model system as shown in Fig. 12.4. The Maxwell model comprises a dashpot having a viscous resistance λ in series with a spring of modulus E. Such is the model that if a stress is applied for a time the material will respond as shown in Fig. 12.5, where the recoverable, elastic strain

$$\varepsilon_R = \frac{\sigma_E}{E}$$

and

$$\dot{\varepsilon} = \frac{\sigma_E}{\lambda}$$

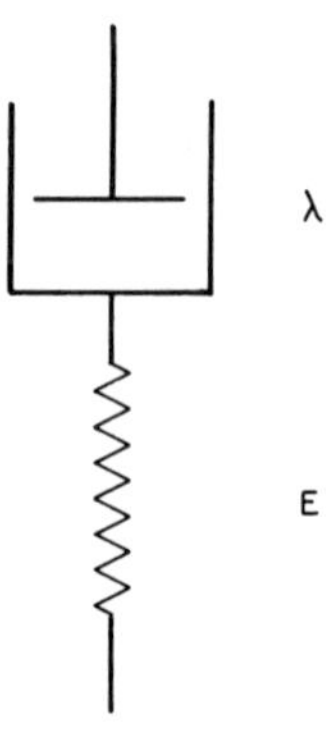

Fig. 12.4.

This model is a simplification of the real response but it serves as an adequate one for many practical purposes. Chapter 14 gives a more detailed treatment of models.

We have deduced that the parameters of this model may depend on the orientation of the molecules and that in turn will depend on the stress to which they are subjected. Any model will break if the stress to which it is subjected is too high and our model is no exception. In fact polymer melts break readily; the maximum stress which they can sustain is about $10^6\,\mathrm{N\,m^{-2}}$. Compare that value against the strength of a solid polymer at about $10^9\,\mathrm{N\,m^{-2}}$. This lack of strength is less remarkable when it is compared with that of virtually all other common liquids—consider how easy it is to fracture water by taking the top off a bottle of lemonade. The

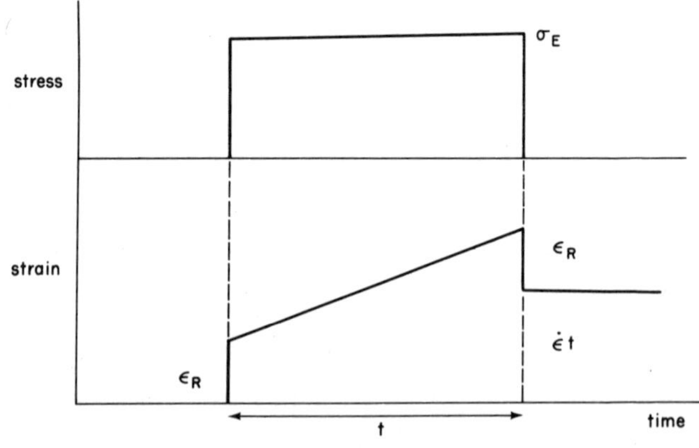

Fig. 12.5.

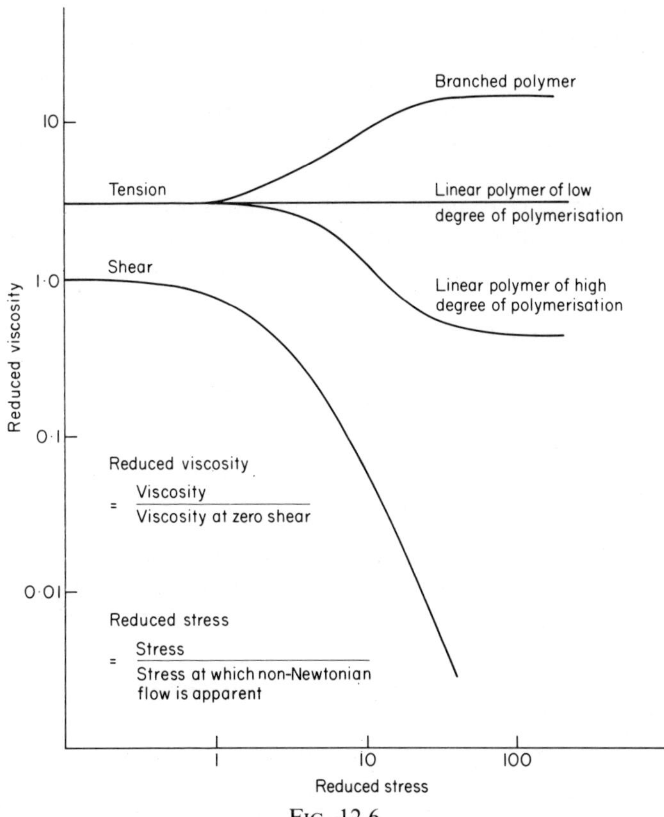

FIG. 12.6.

elastic response of all polymer melts is similar to that of rubber: at low stress
deformation is proportional to stress but at high stress level there is a
tendency towards a limiting elastic strain. For most polymer melts that
elastic limit is a draw ratio of about 7:1. The viscous response is more
complex. We have seen how the shear viscosity of all polymer melts
decreases with increasing stress. As a first approximation the stretching flow
viscosity is independent of stress and equal to three times the zero shear
viscosity. If there is a high degree of polymerisation the viscosity will tend to
decrease as the stress level is increased—though less rapidly than the shear
viscosity. There is one important exception to this rule. If the polymer is
branched then the stretching flow viscosity will show a tendency to increase
with increasing stress level. The stretching flow response of typical polymer
melts is indicated in Fig. 12.6.

In measuring the viscosity of liquids it is easy to subject them to a prescribed shear field and this is achieved in many commercial rheometers, while it is extremely difficult to achieve a well defined stretching flow. If, for example, we hang a weight on a filament the filament stretches. But, as it does so, the cross-section reduces so that the stress level (force/cross-sectional area) increases and the stretching accelerates. Despite this difficulty the gradual recognition of the importance of stretching flows has meant that much effort has been deployed in this field, techniques are being developed, and data will become more readily available in the future.

Having appreciated the difference between shearing and stretching flows we must now seek a simple test by which we can recognise when a flow contains a stretching component. The simple test is to ask, 'Are the streamlines always parallel?' If they are, we are dealing with a simple shearing flow and almost certainly an artificially contrived one at that. If they are not always parallel—and in almost all practical processing systems this will be the case—then there is a stretching flow component. That component may be geometrically small but the stress level which it produces is the product of the deformation and the resistance to deformation. In Fig. 12.6 we note that the resistance to stretching flow may be more than a hundred times greater than the resistance to shearing flow. The implication is that the stretching flow component need be only 1 % of the total deformation to be significant, even dominant, in its effect. The stretching flow component should never be ignored.

SOME EXPERIMENTS WITH STRETCHING FLOWS

Some meaningful experiments can be carried out with a Melt Flow Index tester and two samples of linear and branched polyethylene. If possible choose two samples of the same Melt Flow Index and for preference extrusion grade materials:

(1) Extrude a 100 mm length of each sample and let them hang under their own weight for a predetermined time, then lay them flat until they solidify. Check the change in cross-section—it will be about the same for each material.

(2) Extrude a shorter length and then stretch it slowly as far as you can. The branched sample, low-density polyethylene, will draw uniformly but the linear polymer may tend to draw unevenly.

(3) With the same short sample lengths draw them rapidly to about

twice their original length and release. Note how the samples retract, this retraction is an indication of the extent to which the molecules were oriented.

(4) Snatch the extrudate from the die lips. The branched polymer will fracture in a superficially brittle fashion but the linear polymer will fail in a ductile manner by necking down to a fine thread.

(5) Extrude the melt as rapidly as possible. The extrudate will be distorted if you have pushed hard enough. Note, however, the chaotic structure of the branched polyethylene while the linear sample gives a more regular distortion. Compare these different distortions with the difference in failure during the snatch test.

These are qualitative examples of the phenomena of stretching flows.

STRETCHING FLOWS IN PRACTICE

To illustrate the significance of stretching flows we select from the range of practical processes some of those systems where the behaviour in stretching flows is critical to the success of the process. The list is not intended to be exhaustive but only to give an idea of the sort of effects which may be observed so that they may be more readily recognised in other situations. Regrettably this section cannot be considered to contain statements of unchallenged fact. In an area such as stretching flows which has only recently attracted significant academic interest there are bound to be differences of opinion, sometimes quite irreconcilable ones. What follows are personal opinions based on empirical studies complemented by balancing the theories and observations of other investigators.

Free Surface Flows
Free surface flows include fibre spinning and film blowing where the surface is able to adopt any conformation between set starting and finishing points. Vacuum forming and blow moulding are other examples. In nearly all such processes the thickness uniformity of the product is of considerable importance. Large surface areas may be involved in such flows but in general surface tension effects in polymer melts are small by comparison with viscous resistance. In most practical free surface flows there are other important factors affecting processability such as heat transfer and crystallisation. Very thin sections are often involved so that heterogeneities may be especially important. Such factors interact strongly with the following rheological effects.

Of all practical flows *fibre spinning* is the most obvious stretching flow. We stretch melts to make fibres firstly to make very thin sections and secondly to introduce a preferred orientation. What is required of a fibre spinning process, as of all processes, is that it should yield a uniform product at as high a rate as possible. The first limitation is that the fibre may break. In practice such rupture of very thin sections is usually associated with heterogeneities. However, all polymer melts have an upper limit of

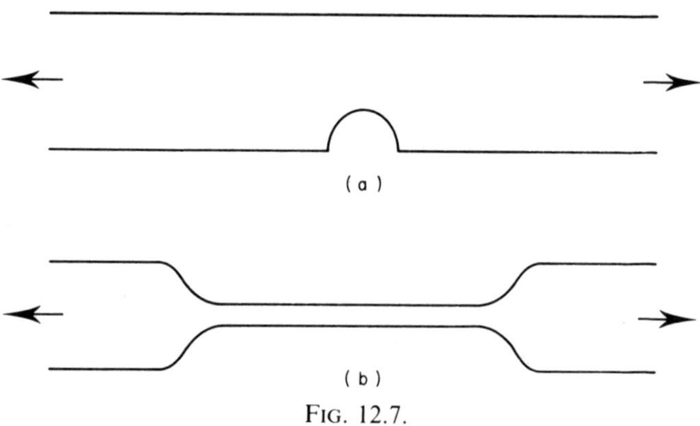

Fig. 12.7.

tensile stress which may not be exceeded without rupture—there is always a limit to speed. In a material with a low resistance to deformation the build-up of stress towards that limit will be less rapid and such materials will be preferred for spinning. The criterion of uniformity is less easy to meet. All real systems contain non-uniformities and the question which we must ask is do such non-uniformities grow (bad) or are they suppressed (good) during stretching flows. If we consider a non-uniformity being stretched (see Fig. 12.7(a)) it is clear that the thinner section will be under a higher stress than the material on either side of it. If the resistance to deformation decreases as the stress increases then that defect will grow forming a neck (see Fig. 12.7(b)). However if the resistance to deformation is greater in the more highly stressed region then the non-uniformity will be diminished. The resistance to deformation is a compound of elastic and viscous response. The viscosity may be tension stiffening or tension thinning but the resistance to elastic deformation always increases with stress so that the more elastic the deformation the more likely the process is to give a uniform stretch. More rapid deformations tend to be more elastic. The ratio of viscosity/modulus for a melt gives the units of time. Such a measure is sometimes called the 'natural time' of the material. By dividing the natural time scale for the material by the time scale of the process we obtain the 'Deborah Number' (Judges v. 5). A Deborah Number greater than unity will be favoured for stability. In fibre spinning there is a conflict between stability which is achieved by increasing the speed, and rupture where probability is increased by increasing the speed. No wonder that such processes have narrow tolerances for option running.

In *extrusion blow moulding* the extruded tube or 'parison', may tend to sag under its own weight causing an uneven reduction in thickness and an increase in length. A limited amount of such sagging may be accommodated, but if there is too much the process will be inoperable. The stress causing the sagging depends on the mass and the cross-sectional area of the parison. The mass of the parison is the product of length, l, cross-section, A, density, ρ, and acceleration due to gravity, g, so that the stress becomes

$$\sigma_E = \frac{lA\rho g}{A} = l\rho g$$

The rate of strain, $\dot{\varepsilon}$, is proportional to the sag velocity, v, divided by the length, l. We may now write down the rheological equations

$$\sigma_E \propto l$$

$$\dot{\varepsilon} \propto \frac{v}{l}$$

$$\text{viscosity } \lambda = \frac{\sigma_E}{\dot{\varepsilon}} \propto \frac{l^2}{v}$$

As the parison sags the cross-section decreases, the stress increases, and the sagging accelerates. To keep the sagging below a critical value the viscosity of the parison must be increased in proportion to the square of the parison length. But even this is not enough, for longer parisons will also take longer to extrude allowing more time for sagging. If we assume that parison handling time is proportional to parison length then to achieve comparable handling between long and short parisons we must increase the tensile viscosity in proportion to the cube of the parison length. As a first approximation tensile viscosity is inversely proportional to Melt Flow Index. If we can just handle 1 m long parisons from a polymer of MFI 1·0, the Melt Flow Index required to make a 2 m parison will be 0·125. The choice may be critical, for the need to employ a very low fluidity polymer to resist sagging will bring with it high power consumption in the extruder and possibly excessive heat generation reducing the efficiency of the process.

In many blow mouldings the tubular parison is inflated into a non-tubular mould. Note that pressure inside any shape of parison will first inflate it to a tube. In a square section moulding the melt will first strike the face and then flow into the corner, thinning down as it does so (see Fig. 12.8). If the resistance to deformation is tension stiffening there will be a tendency to draw materials out of the sides and into the highly-stressed corner making a more uniform product, but, if it is tension thinning the

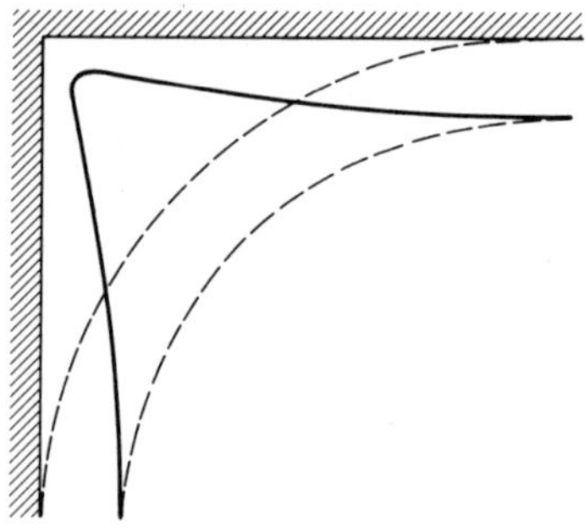

Fig. 12.8.

reverse flow may occur and the corners will be very thin. The criteria are the same as for stability in fibre spinning: the more rapid inflation should give the more uniform product.

In *film or sheet casting* from a strip die, the edges tend to 'neck-in' so that the final width is less than that of the die. Furthermore the edges of the sheet may be thicker than the centre. When we draw a polymer melt so that it extends by a draw ratio D then, because the volume remains constant, the thickness contracts. In a fibre both dimensions will change by the ratio $(1/D)^{1/2}$ but in a wide sheet there is no opportunity for lateral contraction except at the edge. The edge of the sheet does not know that it is part of a sheet rather than a fibre so that we may take a qualitative three element picture as shown in Fig. 12.9.

In a *film or sheet extrusion die* the width is usually at least 100 times the thickness and the edge effect would extend over a significant width. Geometric factors such as the distance over which the draw occurs and the amount of draw greatly affects 'neck-in': the greater the draw distance the more the 'neck-in'. For viscous systems 'neck-in' increases with the haul-off speed if the draw height and output rate are held constant. As the haul-off rate is increased the time scale of the process is reduced and the polymer melt will appear to become more elastic in its response. As the melt becomes elastic the 'neck-in' tends to reduce so that in the rapid processing of highly elastic melts—like low-density polyethylene—'neck-in' actually decreases as the haul-off rate is increased. We again see the importance of elasticity in stabilising a stretching flow.

In melt *film blowing* the stretch takes place in two directions. As with film drawing we may readily establish the importance of the elastic response in stabilising the process to give a uniform thickness. The elastic response in

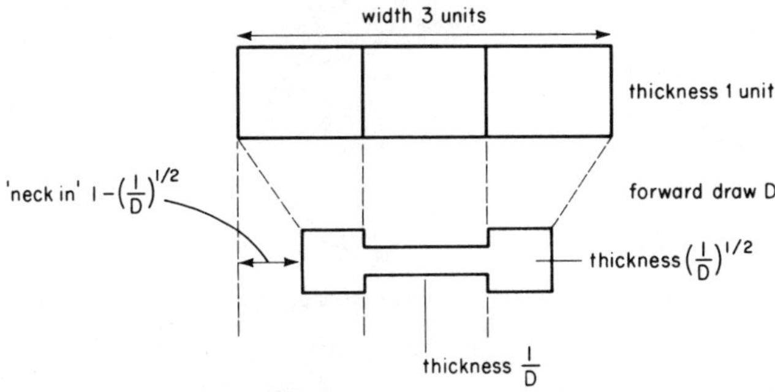

FIG. 12.9.

the film blowing process orders the molecules and that ordering determines the anisotropy of the final product. In the extrusion of the film the need to keep the extrusion pressure within moderate bounds and the demands of engineering tolerance effectively limit the die gap to about 500 μm. If it is intended to make film of 20 μm thickness it is clear that we require a 25:1 areal draw. In principle this could be achieved by a 5 × 5 draw ratio, but if this course were taken we would have a bubble delicately balanced on a very narrow base with very little tension along the line. Such a bubble would be subject to all the vagaries of draughts and might tend to be unstable. To reduce the susceptibility to such extraneous factors an extra tension is added along the machine direction so that bubble extensions of 8 (machine direction) × 3 (transverse direction) tend to be much more practical—with a corresponding anisotropy of the product. The processor has many tools at hand such as venturi curtains and mechanically constraining the shape of the bubble to aid him in achieving the desired balance of properties.

We have considered examples of uniaxial and biaxial stretching. The next stage is triaxial. When a thick section of polymer cools from the melt to a solid, there is a change of density. The volume decreases by as much as 30% in crystalline polymers. The outside solidifies first forming a skin. At first this skin is able to shrink or distort to accommodate the volume shrinkage of the inside but as it becomes thicker its rigidity increases until it produces a tension in the melt as seen in Fig. 12.10. This tension takes the form of a negative pressure. Our melt can only withstand a limited tensile stress and when that is exceeded the stress must be relieved. The only way in which relief can be obtained is through *cavitation*. The hydrostatic tension which causes cavitation is only of the order of ten atmospheres. However, if we are in a situation where cavitation is occurring, then by the same argument the

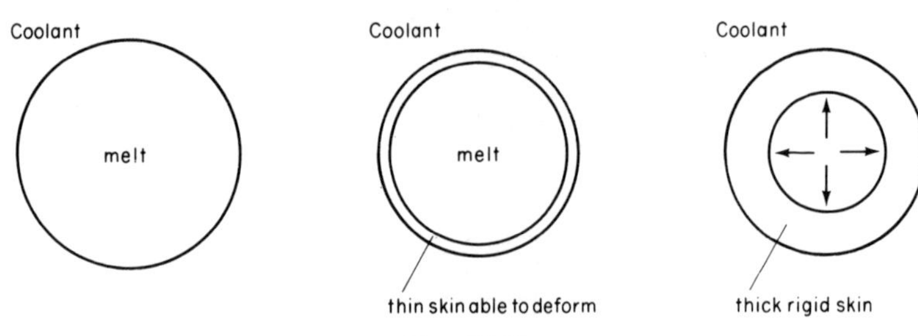

FIG. 12.10

application of modest external pressures may significantly reduce the problem. Cavitation is a problem but we make great use of the brittleness of melts by introducing chemical or physical blowing agents to produce *foaming*. In many systems foaming is initially suppressed by a modest external pressure and a sudden release of that pressure can assist in providing a homogeneous foam. As the foam bubbles grow the material in the cell wall is stretching so that it is again the stretching flow rheology which dominates cell growth and the possible fracture of the cell walls to give an open cell foam. In a foam, larger bubbles tend to grow at the expense of smaller ones leading to uneven cell size distribution. As with other processes a tension stiffening and more elastic response tends to add stability.

These are some of the areas where stretching flows are of primary importance in determining the acceptability of a product. In virtually all cases the practical process is complicated by other factors such as heat transfer, crystallisation or nucleation by inhomogeneities. A practical study of such flows must take these factors, as well as the rheology, into consideration.

Constrained Flows

The field of constrained flows includes all those cases when the boundary is defined by processing machinery. The areas of most interest in polymer processing are the flows in mixing systems, through dies, and in moulds. Most such flows are driven by a pressure gradient and the melt tends to adhere to the boundary wall and flow rapidly along the central channel— similar to the flow in the centre of a river, which is faster than at the banks. When such a flow occurs the layers near to the bank are sheared and in constrained flows shear is usually the dominant factor. However, in practical flows the streamlines are rarely parallel, and, when they are not, a stretching flow is also occurring.

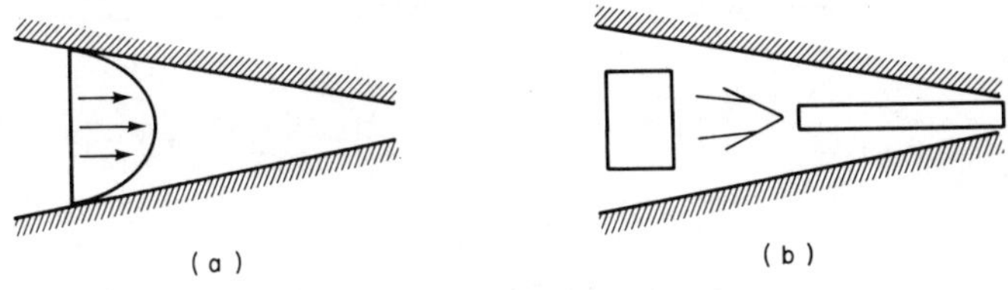

FIG. 12.11.

In convergence from the barrel of an extruder to a die the flow may be considered as having two components (see Fig. 12.11(a) and (b)). If the taper is very narrow then there will be very little stretch but the shearing flow component is high: the reverse occurs in a wide angle. Each flow requires a pressure to maintain it (see Fig. 12.12). By plotting out the two components and adding them, we may arrive at a total pressure drop and note that at some angle there will be a minimum pressure drop. Now suppose that the convergence has a wider angle than that for minimum pressure. In this case we may suggest that the melt will construct its own boundary leaving a slowly re-circulating 'dead zone' where the taper should have been to give a

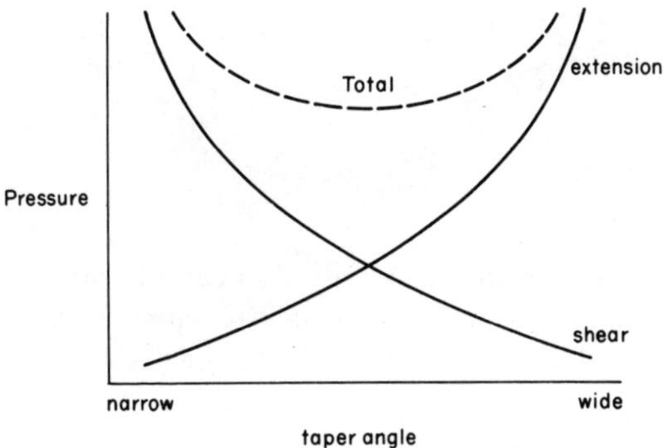

FIG. 12.12.

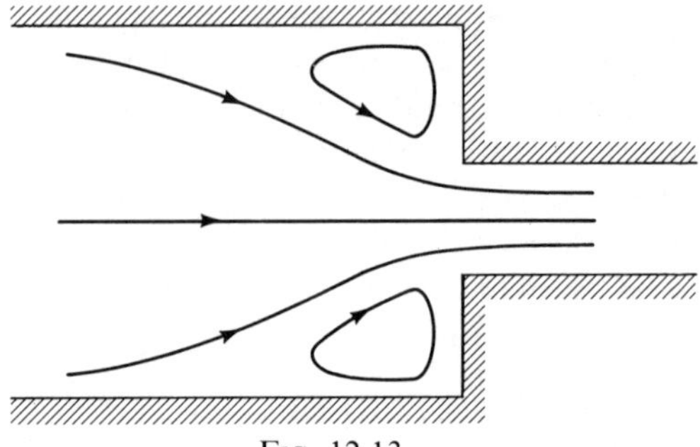

FIG. 12.13.

minimum pressure (Fig. 12.13). The size of the 'dead zone' will be greater the greater the difference between the resistance to shear and elongation. Referring back to Fig. 12.6 we note that the 'dead zone' will increase as rate is increased and that it will tend to be large in branched polymers. The 'dead zone' is an undesirable feature in practical processing. Material which accumulates in such zones will stay there until it degrades and such degraded material then contaminates the flow. The stretching flow, because it tends to be highly elastic, may produce excessive swelling or if the breaking stress is exceeded flow defects can result. The optimum design of taper dies requires minimising such deleterious effects while maintaining the extrusion pressure as low as possible.

In *calendering* the flow is more complex since the boundary is not only constrained but also moving.

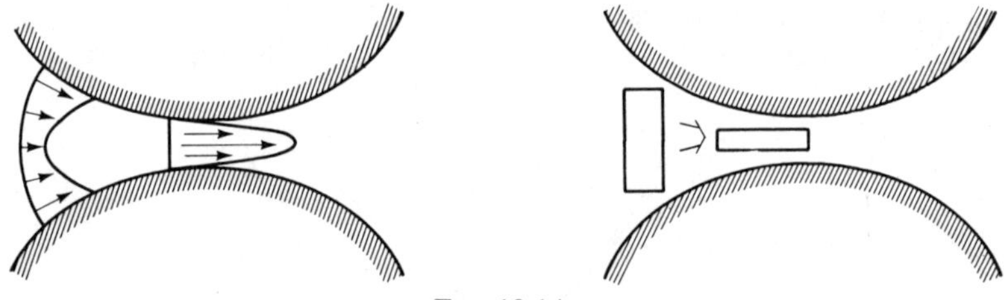

FIG. 12.14.

In Fig. 12.14 note how increasing the diameter of the rolls for a given nip gap and throughput reduces the rate of stretching in the nip. The product quality from small calenders is not necessarily a good guide to operation on the larger scale.

During *injection moulding* through a central gate the flow spreads out in a disc shape. A cheese cut from such a disc shows that the deformation history

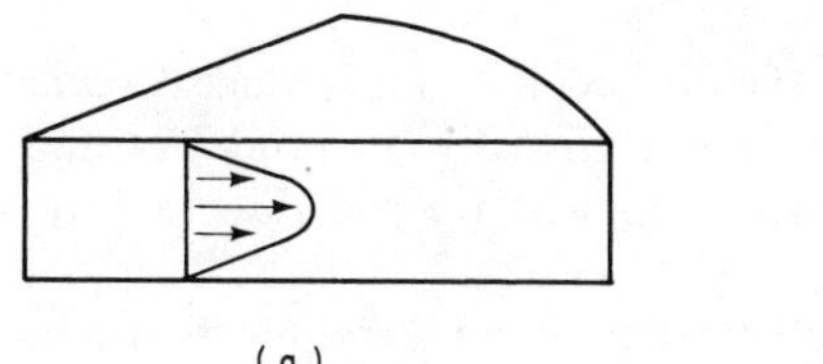

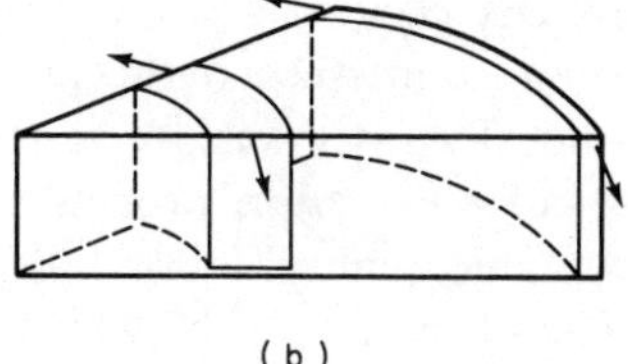

FIG. 12.15.   (a) Shear; (b) extension.

includes both shear and extension (see Fig. 12.15). The shearing flow is localised in the surface and will produce a radial orientation there while the stretching flow produces a transverse orientation along the central plane. In a thin disc moulding it is desirable that the two orientations should be balanced so that in the inevitable shrinkage that follows moulding the contraction will be the same in both directions. Should they not be equal the plate could become saddle shaped (excessive radial shrinkage) or dished (excessive transverse shrinkage) (see Fig. 12.16). There are other factors, in particular mould design and cooling, which contribute to shrinkage but in critical mouldings only a narrow range of moulding conditions will give satisfactory results.

Some of the most important of the 'working' flows in processing may contain significant stretching components. An important function of a screw extruder is to provide mixing and homogenisation. Practical mixing flows, by definition, involve deliberate interruption or distortion of the streamline pattern. In flow through porous beds, such as are used as filter packs, the path is tortuous. Such flows almost certainly contain an element of stretching behaviour which has yet to receive an adequate theoretical study.

In complex constrained flows the shearing component will usually be geometrically larger. Those shearing flows will tend to determine the major viscous effects, such as pressure drop and heat generation, associated with quantity of output. The stretching flow, however, is more likely to produce intense elastic effects or even fracture and will generally determine those factors which control the quality of the product.

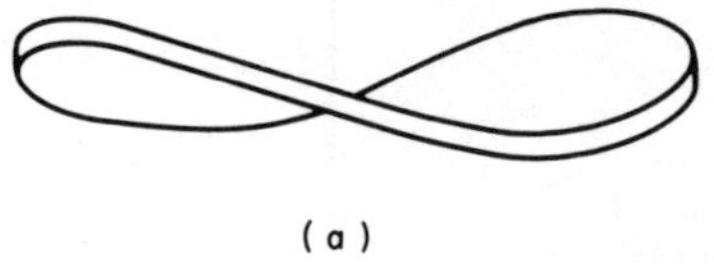

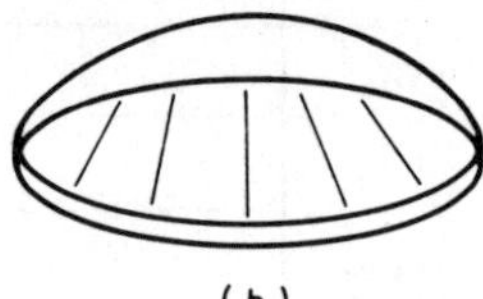

FIG. 12.16.   (a) Saddle shaped; (b) dished.

Flow Defects

An important aspect of the quality of a product is its appearance. In general a smooth glossy surface is required but virtually all processes and in particular extrusion processes, have a limited output rate to give an acceptable quality.

The phenomenon by which the extrudate issuing from a die becomes grossly distorted is sometimes termed '*melt fracture*'. The term is aptly chosen since today it is generally considered that such defects occur when the tensile stress, in the convergent flow at the die inlet, builds up to a value where stable stretching flow is not possible. In some cases rupture of the streamlines may be the only way of relieving the stress. When this occurs the flow pattern is disturbed.

Tapering the die to reduce the stretching flow component can delay the onset of distortion considerably. Indeed in good industrial practice the flow rates employed may be a hundred times those which would give 'melt fracture' in a poorly designed die. (Melt fracture is discussed in more detail in Chapter 10.)

Many polymers *crystallise* and those which do invariably crystallise more readily if the molecules are ordered by being subjected to a stretching flow and to high pressure. So effective are these agents that melts which in their absence may supercool by 50 °C can be induced by the convergent flow at a die entry to crystallise 50 °C above their normal melting point. Such effects can produce a variety of symptoms of which the most startling is complete cessation of flow.

The die exit region also produces a stretching flow. Inside the die the melt adheres to the wall, having zero velocity at that point. But downstream of the exit the surface velocity is finite (see Fig. 12.17). The result is a sudden acceleration or stretching of the surface at the die exit. If the acceleration is too rapid the surface may rupture giving a rough texture sometimes called '*sharkskin*'. This defect is usually treated by locally heating the die lips to

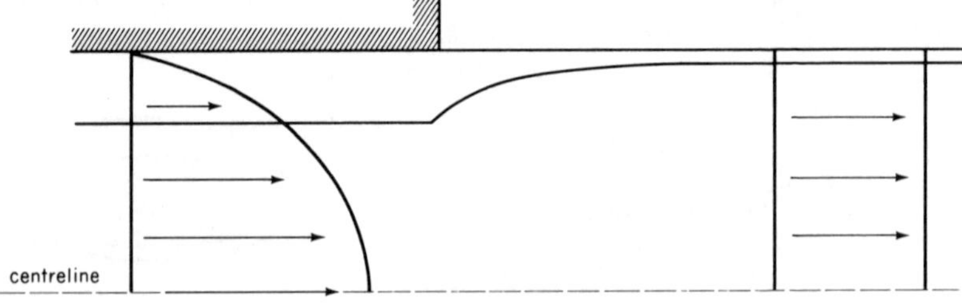

FIG. 12.17.

reduce the viscosity, and so reducing stress level in that critical surface region.

REFERENCES

1. F. TROUTON, Proc. Roy. Soc. A77, 426–40 (1906).
2. A. S. LODGE, *Elastic Liquids*, Academic Press, 114–18 (1964).

BIBLIOGRAPHY

1. J. DEALY, *Poly. Eng. Sci.*, **11,** 433–45 (1971). (General review of both theory and practice in stretching flows.)
2. F. N. COGSWELL, *Trans. Soc. Rheol.*, **16,** 3, 383. (A review of rheometers for studying stretching flows.)
3. F. N. COGSWELL, *Appl. Poly. Symp.*, **27,** 1–18 (1975). (A survey of rheological experiences with extensional flows.)
4. H. M. LAUN and H. MUNSTEDT, *Rheol. Acta*, **15,** 10, 517–24 (1976). (Probably the most detailed study of the stretching flow response of a single melt.)

13

The Rheology of Injection Moulding

I. T. Barrie

Research Physicist, Imperial Chemical Industries Limited,
Plastics Division, Welwyn Garden City, Herts AL7 1HD, UK

INTRODUCTION

The business of injection moulding and the study of melt rheology are both very demanding and totally absorbing activities. Historically speaking, the practitioner of either has had little time or energy to spare to be concerned with the activity of the other. Yet it is now widely recognised that there is an important community of interest between the moulder and the rheologist. The object of this chapter is to identify this community of interest, define the common problems, and establish a common language.

VISCOSITY DATA

The experimental data usually cover a wide range of flow rates, making log scales essential. The effect of increasing the melt temperature is mainly that of lowering the apparent viscosity over the entire shear rate or shear stress range, with little change in its shape.

As a first approximation we may use the following shear rate relationships:

$$\dot{\gamma} = \frac{4Q}{\pi r^3} \tag{1}$$

for a circular channel, radius r and volume flow rate Q, and

$$\dot{\gamma} = \frac{6Q}{wh^2} \tag{2}$$

for a rectangular channel, height h and width w. Analogous expressions can be derived for other geometries. Typically the shear rates lie in the range of 30 to $3000\,s^{-1}$ for practical purposes.

Because polymer melts are non-Newtonian a power law approximation can be used relating shear rate and shear stress, which involves the power index n and a viscosity coefficient which is appreciably affected by temperature and pressure. As has been seen in earlier chapters, the power law can be expressed in terms of apparent viscosity in logarithmic form and a best straight line can be fitted which approximates the experimental data, representing them very well over a range of two or three decades.

For a given material it is generally possible to use the power law model over the range of temperatures which is of practical significance to the injection moulder. This is represented by a family of virtually straight lines with identical slope n, so that the lines are parallel for various temperatures.

The numerical value of the viscosity coefficient represents the apparent viscosity only at unity shear rate in the power-law model. It should be noted that:

(i)   the approximation is valid only over the chosen range of shear rates to which it was fitted;

(ii)   that the *viscosity coefficient* is not a *viscosity* and does not have the units of viscosity, except when $n = 1$.

The temperature dependence of the viscosity coefficient k is given by an Arrhenius type equation

$$k_T = k_o \exp(-\alpha(T - T_o)) \tag{3}$$

where α is the temperature coefficient best fitting the experimental data with reference to a value k_o at a corresponding temperature T_o. Using the definition of the shear stress, namely $\Delta P r/2l$, or $\Delta P h/2l$ for cylindrical and rectangular channels respectively, it is easily seen that the pressure drop ΔP is given by the following equations:

$$\Delta P = \frac{2\left(\dfrac{4}{\pi}\right)^n klQ^n}{r^{3n+1}} \qquad \text{(cylindrical channels)} \tag{4}$$

and

$$\Delta P = \frac{2(6)^n klQ^n}{w^n h^{2n+1}} \qquad \text{(rectangular channels)} \tag{5}$$

In practical moulding situations two further channel geometries are

TABLE 13.1
TYPICAL SHEAR VISCOSITY INDEX VALUES

Material	Shear viscosity index (at shear rate of $10^3\ s^{-1}$)								
	Temperature °C								
	150	170	190	210	230	250	270	290	310
LD Polythene (MFI 20)	1·15	0·85	0·65	0·50	0·40	0·30	0·25	0·20	
PVC (plasticised)	1·65	1·00	0·60						
PVC (rigid)[a]		3·60	3·10	2·70					
EVA (MFI 2)		2·2	1·75	1·45	1·15	0·95			
Polypropylene (copolymer)			1·15	1·05	0·85	0·75	0·65	0·60	
Polypropylene + 25% Glass			1·45	1·25	1·10	0·95			
Polyacetal			2·9	2·0	1·4				
Polystyrene (high impact)				1·40	1·15	0·95			
SAN				2·10	1·75	1·30	0·90		
ABS				2·60	1·95	1·40			
Acrylic (PMMA)				6·10	3·00	1·50	0·60		
'Noryl' (modified PPO)						2·40	2·00	1·60	
Polycarbonate						7·90	5·70	3·60	1·90
Nylon 6:6[a]							1·15	0·80	0·55

[a] The approximation $n = 1/3$ cannot be used.

Note: To obtain shear viscosity values at $10^3\ s^{-1}$ shear rate in SI units, multiply by 100; in poises, multiply by 1000.

encountered for which the isothermal equations are given below. These geometries are:

(a)   a tapering channel of circular cross-section with initial radius r_1 and final radius r_2, over a length, l;

(b)   a 'spreading disc' flow geometry between parallel plates a distance x apart and fed from a central hole of radius r_0. The flow length is R, i.e. instantaneous radius of the disc.

The corresponding pressure drop equations for isothermal shear flow are:

(a)
$$\Delta P = \left(\frac{4}{\pi}\right)^n \frac{2klQ^n}{3n(r_1 - r_2)}\left[\left(\frac{1}{r_1}\right)^{3n} - \left(\frac{1}{r_2}\right)^{3n}\right] \tag{6}$$

and

(b)
$$\Delta P = \left(\frac{6Q}{2\pi}\right)^n \frac{1}{1 - n}\frac{2k}{x^{1+2n}}[R^{1-n} - r_0^{1-n}] \tag{7}$$

Constructing a Simple Comparative Table of Polymer Melt Viscosities
The melt flow behaviour of most polymers used in the injection moulding industry can be represented by using a power-law index value of $n = 1/3$. Notable exceptions to this generalisation are Nylons and rigid PVC for which n is nearer to 1/2–3/4. Using $n = 1/3$, the best-fit values of k as a function of temperature are determined for materials and grades of commercial interest. Table 13.1[1] shows typical comparative data.

The viscosity coefficients given in Table 13.1 have the approximated values of the melt viscosity at $10^3\,\mathrm{s}^{-1}$: (a) in SI units when multiplied by 100, and (b) in poise when multiplied by 1000. The same coefficients give k in SI units when multiplied by 10^4 and in cgs units when multiplied by 10^5.

PRACTICAL DATA FOR ISOTHERMAL MELT FLOW IN CIRCULAR-SECTION CHANNELS

In an injection moulding machine the polymer melt is prepared and stored in the injection barrel before being delivered to the mould at high speed through heated delivery channels. These channels are usually of circular cross-section with at least one highly convergent tapered section, forming the injection nozzle of the machine. For some mould designs it is necessary to extend the heated delivery channels between the nozzle exit to the mould entry point(s) by means of 'hot runners'.

Practical Data on Pressure Loss in Melt Delivery Channels

Figure 13.1 shows the dimensions of one such nozzle-and-runner arrangement used for experimental measurement of pressure drops; Section A represents the barrel end-cap and nozzle and Section B the hot-runner. Sections A and B were both maintained at melt temperature.

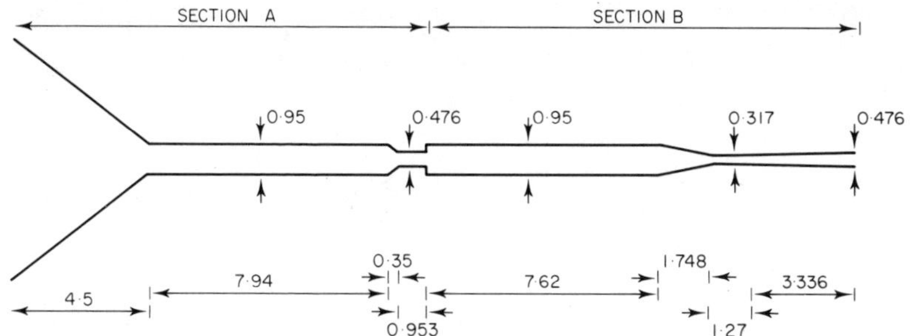

FIG. 13.1   Machine nozzle and hot-runner dimensions (in cm).

Using a commercial grade of polypropylene at 200 and 260 °C, the oil pressure acting on the injection ram was measured for various rates of melt discharge in the range 40 to 300 cm^3 s^{-1}. This discharge into air was made (a) through the nozzle only (i.e. Section A in Fig. 13.1), and (b) through the nozzle, runner and sprue (i.e. Sections A and B together in Fig. 13.1). The data are plotted in Fig. 13.2. By subtraction of curve A from curve (A + B) the component of pressure dropped in Section B alone was deduced; this is plotted in Fig. 13.3 for each melt temperature.

These measured pressure losses can be compared with the pressure losses calculated from simple shear flow eqns. (4) and (6). The comparison is given in Figs. 13.4 and 13.5 using the approximation $n = 1/3$, $k_{200\,°C} = 1·1 \times 10^5$ and $k_{260\,°C} = 0·7 \times 10^5$ SI units.

The estimation of pressure drops from simple shear equations alone is clearly inadequate, particularly for highly convergent channels as in Section A. There is an obvious case for quantitative knowledge of melt behaviour under conditions of deformation other than by shear.

Elongation Flow Behaviour of Polymer Melts

The 'elongational flow' concept can be seen from Fig. 13.6, by the shape distortion of an elemental volume through a convergent channel. Several workers have studied the subject in recent years[2,3] involving some refined measuring techniques; the subject is also treated in Chapter 12. The result is

Polymer Rheology

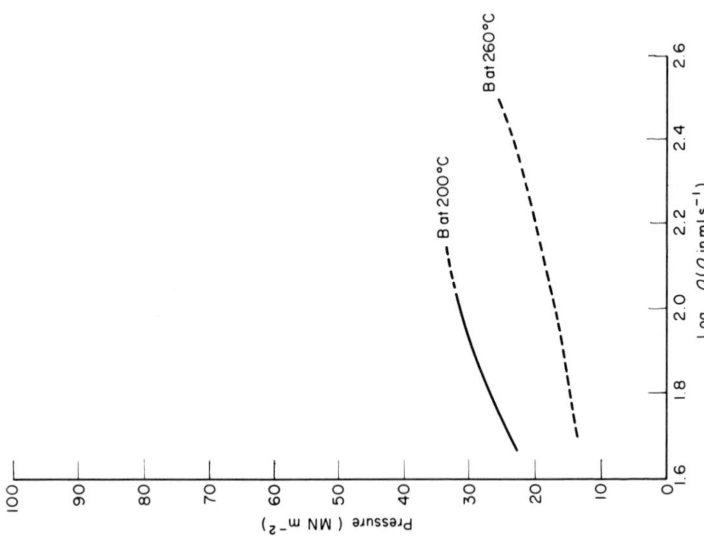

Fig. 13.3. Hot-runner pressure losses (Section B only).

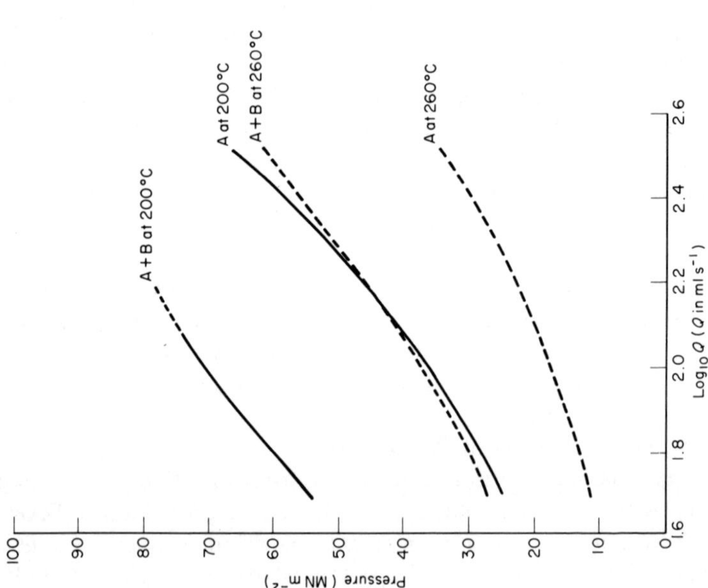

Fig. 13.2. Delivery channel pressure losses.

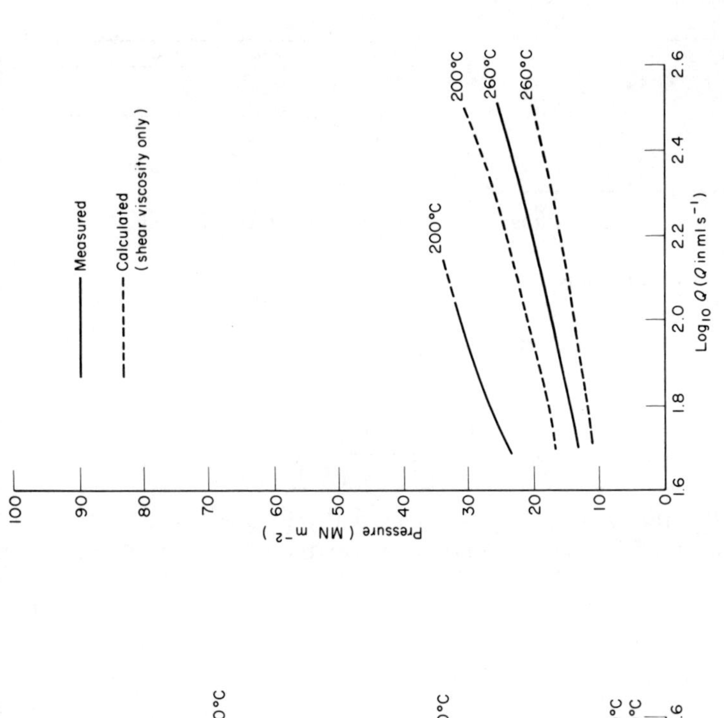

FIG. 13.5. Comparison of measured pressure loss with calculated simple shear loss in hot-runner (Section B).

FIG. 13.4. Comparison of measured pressure loss with calculated simple shear loss in injection barrel end-cap and nozzle (Section A).

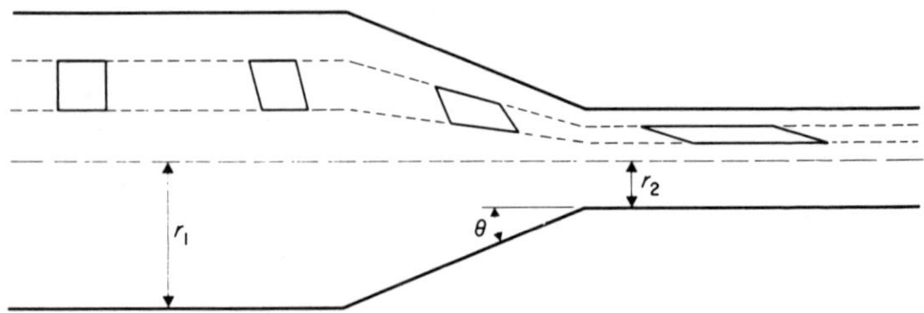

FIG. 13.6.   Elongational flow in a convergent channel.

the generation of equations for a 'tensile viscosity' component of the form:

$$\sigma_E = \lambda_T \frac{\dot\gamma}{2} \tan\theta$$

and

$$\Delta P_E = \frac{2}{3}\sigma_E\left[1 - \left(\frac{r_2}{r_1}\right)^3\right] \tag{8}$$

where ΔP_E is the pressure drop associated with the elongational stress σ_E, λ_T is a temperature-dependent material constant (the 'tensile viscosity') and $\dot\gamma$ is the shear rate.

The contribution of the tensile viscosity to the pressure drop in the delivery channels can be calculated if λ_T is known. Typical values of λ_T for some polymers[1] are given in Table 13.2.

Using the polypropylene values $\lambda_{200\,°C} = 3600\ \mathrm{N s\,m^{-2}}$ and $\lambda_{260\,°C}\ 1200\ \mathrm{N s\,m^{-2}}$ the tensile viscosity contribution to pressure losses in the delivery channels in Fig. 13.1 are shown in Figs. 13.7 and 13.8. The effects of tensile viscosity (shaded areas) have been added to those of shear viscosity previously calculated. The agreement between the measured and calculated pressure losses is now excellent in Section B of the channel but there is still much of the pressure drop in Section A unaccounted for. This shortfall is explicable in terms of friction between the injection screw and the barrel.

Frictional Losses between Injection Ram and Barrel
If the melt pressure in the injection ram is calculated from the oil pressure applied to the ram, the pressure loss due to friction between the injection ram/screw and the barrel should be allowed for.

In practical situations this loss is usually significant but not dominant. Experimental evidence suggests that the frictional component accounts for between 5 and 20 $\mathrm{MN\,m^{-2}}$.

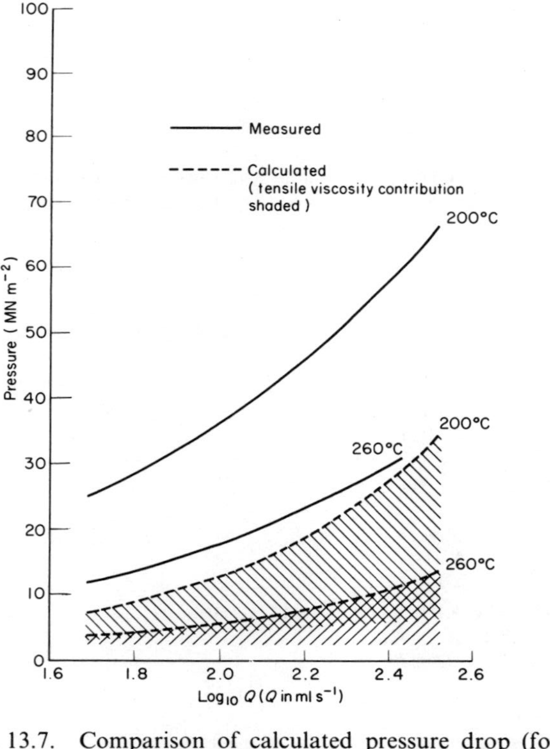

FIG. 13.7. Comparison of calculated pressure drop (for tensile and shear effects) with practical data. Section A (barrel end-cap and nozzle).

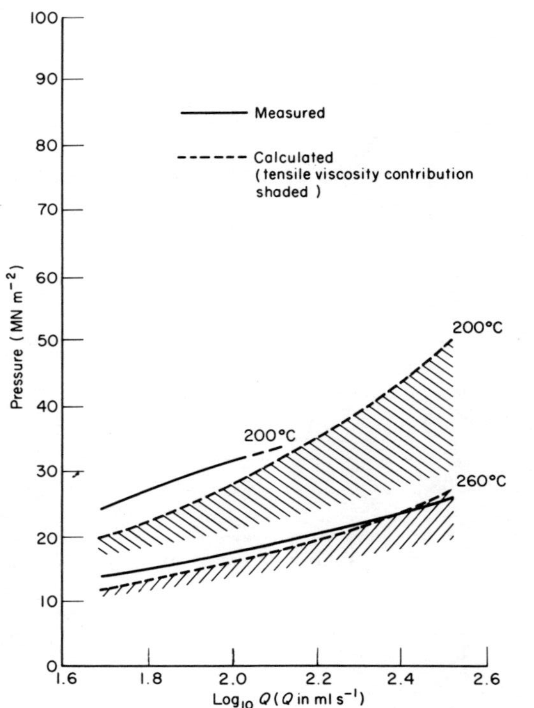
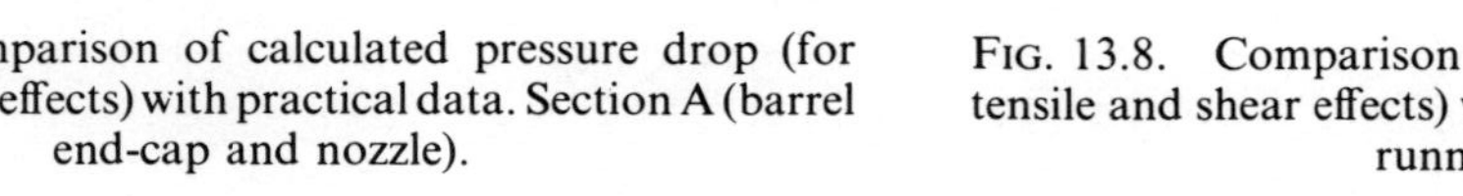

FIG. 13.8. Comparison of calculated pressure drop (for tensile and shear effects) with practical data. Section B (hot-runner and sprue).

TABLE 13.2
TYPICAL EXTENSIONAL VISCOSITY INDEX VALUES
(IF THE CALCULATED TENSILE STRESS EXCEEDS $2 \times 10^6\,\mathrm{N\,m^{-2}}$ THE MELT WILL RUPTURE)

Material	*Extensional viscosity index (at shear rate of $10^3\,s^{-1}$)*								
	Temperature °C								
	150	170	190	210	230	250	270	290	310
LD Polythene (MFI 20)	150	90	50	30	18	11	7	4	
PVC (plasticised)[a]	70	30	10						
PVC (rigid)[a]		600	200						
EVA (MFI 2)		800	450	250	150	100			
Polypropylene (copolymer)			43	30	21	15	10·5		
Polyacetal			35	20	14				
Acrylic (PMMA)				500	140	40	11·5		
Nylon 6:6							2·7	0·5	

[a] The values given are very approximate.

Note: To obtain extensional viscosity values at $10^3\,s^{-1}$ shear rate in SI units, multiply by 100; in poises, multiply by 1000.

Effects of Shear Heat Generation and Pressure on Viscosity

Heat is generated in shear flow. Unless this is dissipated through the channel walls, an increase in the average melt temperature results. For thermodynamic reasons this increase is proportional to the total pressure drop and amounts to around 6 °C per 1000 psi for most polymers. A temperature rise of around 1·5 °C per 1000 psi occurs on compression but this is reversible and as the material decompresses cooling by the same amount occurs. Thus, the temperature rise during melt flow is dominated by shear heating.

The generation of pressure also densifies the melt and causes an increase in viscosity. This viscosity increase can be considered as equivalent to that brought about by cooling the melt; i.e. although no actual temperature decrease occurs (and in fact the reverse situation has already been discussed), the material behaves as though it were cooler. At the present time data for the pressure dependence of viscosity is only available for relatively few polymer melts; in terms of temperature change equivalents, it is -6 °C per 1000 psi for polypropylene, -4 °C per 1000 psi for Nylon 66 and -2 °C per 1000 psi for PMMA. (Note: $1000\,\text{psi} = 6·9 \times 10^6\,\text{N m}^{-2}$).

Thus, for pressure losses in shear flow conditions, the effects of pressure and shear heating are opposite and tend to cancel out. For polypropylene, these opposing factors do roughly cancel each other, whereas for nylon and PMMA the indications are that shear heating effects are the greater.

As a general rule a greater error is introduced by allowing for only shear heating effects *or* the variation of viscosity with pressure, rather than by ignoring both. (Note: viscosity data are usually quoted for low pressure conditions of, say, below $10\,\text{MN m}^{-2}$.)

PRACTICAL CALCULATIONS FOR MOULD FILLING (NON-ISOTHERMAL FLOW)

The previous section has dealt with quantitative estimates of pressure losses incurred in delivering a polymer melt from the injection barrel to the mould entry point. Attention is now turned to pressure and other considerations for melt flow in cooled and relatively thin-sectioned walls. In this case there is one simplification in the calculations, that is the fact that elongational flow effects are quantitatively negligible in all practical cases. There is also one major complication, namely the fact that the melt is continually cooling and freezing on the mould cavity walls during the filling operation.

Pressure Requirements for Mould Filling

If isothermal conditions are applied to mould filling calculations, the basic flow equations for 'spreading disc' and uniform channel situations could be used to obtain fair pressure-drop estimates for most practical cases.

Figure 13.9 shows the effect of a cooled mould on the pressure vs. flow rate curve. The isothermal pressure curve falls steadily with falling flow rate. For a cooled mould, however, there exists a minimum flow rate, $Q_{\min}$, corresponding to a complete freezing of the melt in the cavity just at the instant of filling the mould. At rates below $Q_{\min}$ the mould cannot be filled even with a very high melt pressure head. As Q increases above $Q_{\min}$ the freeze-off effects diminish and the cavity pressure at the instant of filling falls. At the other extreme, at very fast filling rates, the melt has little time to lose heat to the cavity walls and the situation approaches the isothermal case. The actual pressure/rate curve is continuous between these two extremes leading to the shape shown in Fig. 13.9. Of particularly practical interest is the shallow minimum in the curve implying a relatively constant pressure requirement for a range of flow rates.

It is little surprise, therefore, to find that:

(i)   moulders generally do not consider the injection rate as an important variable for mould-filling problems except for thin-section mouldings;

(ii) injection moulding machines are generally designed with injection rates fast enough to operate in the pressure-minimum region, but no faster.

A moulder is particularly interested in the minimum pressure drop in the mould cavity for two reasons. Firstly, by minimising pressure gradients, the stress level frozen into the mouldings is minimised giving maximum dimensional stability in subsequent storage and use. Secondly, lower pressures mean lower clamp forces to prevent mould-opening, which for large projected-area mouldings can be a problem.

A detailed theoretical calculation of the P_{o} vs. Q curve for a cooled mould of the simplest geometry would involve highly sophisticated computer programming and expensive computation facilities. Even then present day knowledge is still insufficient to give complete and accurate answers. Given that approximations and assumptions have to be made the question is one of balance between simplification and adequacy.

One simplified approach to estimating non-isothermal flow behaviour in moulds[4] is based on the empirical observation in a centre-gated disc moulding that the frozen polymer layer is approximately uniform on the

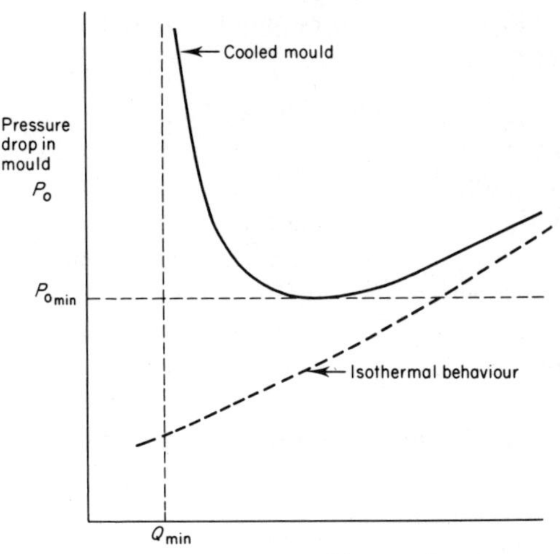

FIG. 13.9.   Isothermal and non-isothermal pressure drop in a mould cavity.

mould surfaces at the instant of mould filling. This means that the melt flow is still occurring in an approximately uniform thickness cavity, but one which is thinner than the original cavity. Thus, if the original (empty) cavity had a thickness x, and a thickness Δx of frozen materials had formed on each cold surface, then the effective remaining flow path is given by:

$$x_{\text{eff}} = x - 2\Delta x \tag{9}$$

The thickness of each frozen layer, Δx, is a function of the time taken to fill the mould, t, the mould temperature, θ, the melt temperature, T, the melt freeze-off temperature, T_o, and the thermal diffusivity of the polymer melt, x.

Experimental observation of the variation of Δx with t suggests a relationship

$$\Delta x = Ct^{1/3} \tag{10}$$

where C is the constant of proportionality which includes θ, T, T_o and x.

By analogy with heat flow in simpler situations, the value of C is given approximately by the empirical equation:

$$C = 2a^{1/2}\left(\frac{T_o - \theta}{T - \theta}\right) \tag{11}$$

where $a = $ the heat diffusion coefficient of the melt.

In summary the assumptions for a simplified treatment of pressure drops in a non-isothermal case are:

(i)   material freezes on the mould surfaces in a uniform layer;

(ii)   this layer increases with mould-filling time according to a cube-root law (eqn. (10));

(iii) the build-up of frozen material depends on mould and melt temperatures and the material freeze-off temperature and heat transfer properties according to eqn. (11);

(iv) isothermal flow equations (e.g. eqns. (5) and (7)) can be applied using the reduced, effective cavity thickness given by eqn. (9).

Estimates of Cavity-Filling Pressures for a Centre-Gated Disc Mould

For the sake of simplicity a centre-gated, circular disc cavity will be used to illustrate the derivation of the pressure equations and for comparisons with practical observations.

Using eqns. (11) and (10) in (9)

$$x_{\text{eff}} = x - 2Ct^{1/3} \tag{12}$$

$$= x - 4a^{1/2}\left(\frac{T_{\text{o}} - \theta}{T - \theta}\right)t^{1/3} \tag{13}$$

The cavity filling time t is obtained from the filling rate, Q, and the cavity volume, V:

$$t = \frac{V}{Q}$$

For a circular disc cavity, radius R and thickness x,

$$V = \pi R^2 x$$

$$t = \frac{\pi R^2 x}{Q} \tag{14}$$

At the instant of filling the cavity the flow conditions are assumed to be as shown in Fig. 13.10. Experimental evidence suggests that the model is reasonable for cavity thicknesses between 2 and 5 mm.

The pressure drop from the centre of a 'spreading disc' flow situation in a cavity of thickness x_{eff} to a radius R is given by (eqn. (7)):

$$P_{\text{o}} = \frac{2K}{x_{\text{eff}}^{4+2n}}\left(\frac{6Q}{2\pi}\right)^n \frac{1}{1-n}(R^{1-n} - R_{\text{o}}^{1-n})$$

Two assumptions greatly simplify the algebraic equation without affecting

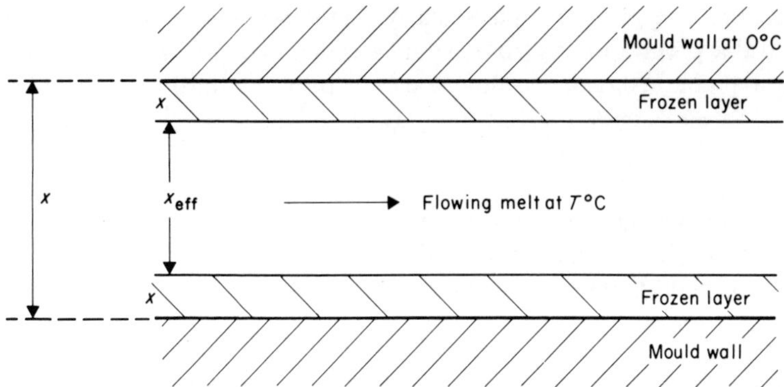

FIG. 13.10. Model for flow conditions in a cooled cavity.

the logical argument. These are that $R \gg R_0$, i.e. the inlet hole is very small compared with the disc radius, and $(6/2\pi)^n$ is approximately unity. Therefore

$$P_0 \simeq \frac{2KQ^n R^{1-n}}{x_{\text{eff}}^{1+2n}(1-n)} \tag{15}$$

Substituting for x_{eff} using eqns. (14), (12) and (9), a relationship is obtained between the pressure drop, P_0 and the flow rate, Q, in terms of the material rheological and thermal properties (K, n and C) and the cavity geometry, R and x. This equation is:

$$P_0 = \frac{2KQ^n R^{1-n}}{(1-n)x^{1+2n}\left[1 - 2C\left(\dfrac{\pi R^2}{Qx^2}\right)^{1/3}\right]^{1+2n}} \tag{16}$$

When practical values are used for K, n, C, R and x the resulting curve of P_0 vs. Q is as shown previously in Fig. 13.6.

As already discussed the practical moulder is particularly concerned with minimising the cavity pressure gradients. The value of $P_{0\text{min}}$ can be obtained from eqn. (16) by differentiating the right-hand side with respect to Q and equating to zero. This gives

$$P_{0\text{min}} = f(n)KC^{3n}\left(\frac{\pi^n R^{1+n}}{x^{1+4n}}\right) \tag{17}$$

where

$$f(n) = \frac{2^{3n+1}}{1-n}\left(\frac{1+5n}{3n}\right)^{3n}\left(\frac{1+5n}{1+2n}\right)^{1+2n} \tag{18}$$

The form of eqn. (17) is particularly useful containing, as it does, three separable groups. Firstly f(n) is a pure member; secondly KC^{3n} contains the material characteristics of flow and heat transfer; thirdly

$$\left(\frac{\pi^n R^{1+n}}{x^{1+4n}} \right)$$

depends upon the cavity geometry.

All three factors contain the power-law index n.

The general form of this equation is

$$P_o = f(n)\mu_T G(n) \tag{19}$$

and only the 'material' term μ_T is temperature-dependent. Equation (17) can be further simplified for the case where $n = 1/3$ is a good approximation to the material flow behaviour in this case.

$$P_o \simeq 35 K_T C_T \left(\frac{1 \cdot 5 R^{4/3}}{x^{7/3}} \right) \tag{20}$$

An Index of Mouldability for Polymer Melts

One result of the analysis in the previous section was the isolation of one

TABLE 13.3

THERMAL PROPERTY VALUES FOR SOME MOULDING-GRADE MATERIALS

Material	*Thermal property*	
	Apparent freeze-off temperature (fast-cooling) °C	*Thermal diffusivity* $m^2 s^{-1}$ $(\times 10^{-8})$
LD Polythene (MFI 20)	90	11
PVC (plasticised)	90	7
PVC (rigid)	140	12
EVA (MFI 2)	115	11
Polypropylene (copolymer)	135	11
Polypropylene + 25% Glass	135	11
Polyacetal	135	9
Polystyrene (high impact)	130	8
SAN	150	9
ABS	140	9
Acrylic (PMMA)	160	9
'Noryl' (modified PPO)	180	10
Polycarbonate	200	10
Nylon 6:6	240	13

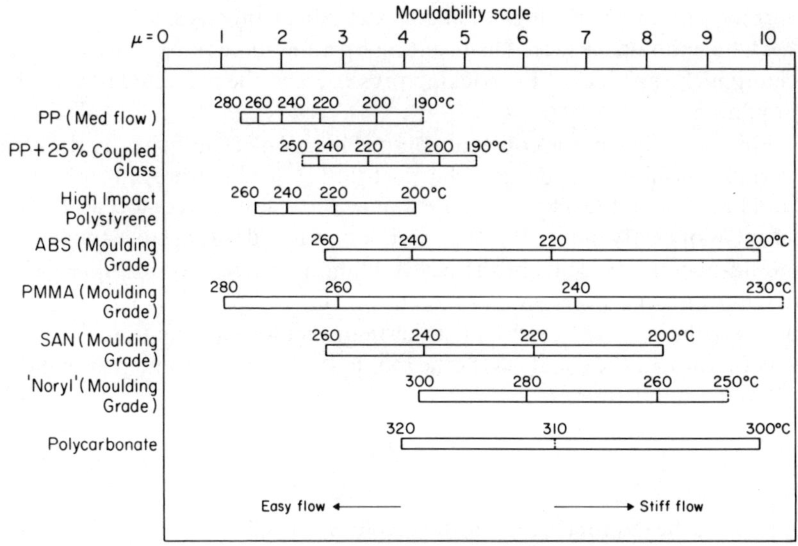

FIG. 13.11.   Comparative mouldability of materials.

group of factors in the cavity-pressure equation related entirely to material parameters; these were given the symbol μ in eqn. (19). The value of μ at any temperature is a quantitative indication of the relative mouldability of a range of materials sharing the same value of n. Strictly, the mould temperature, θ, also affects the value of μ at any one melt temperature; however θ appears linearly in both the numerator and denominator in the estimate of C (eqn. (11)) and does not strongly influence the value of C for mould temperatures between 30 and 70 °C, which are generally encountered in practice.

Two further material parameters are needed for the calculation of μ values; these are the apparent freeze-off temperature and the thermal diffusivity. Typical values of these parameters for some common moulding materials are given in Table 13.3.

Figure 13.11 shows the approximate values of μ vs. temperature for some common moulding materials for which $n \simeq 1/3$.

Estimation of Clamping Requirements

The period during a moulding cycle which usually makes most demands on the mould clamping requirement is the mould-packing phase. At this time the packing pressure, applied after the mould is filled in order to eliminate surface sink-marks during cooling, is virtually acting hydrostatically over the whole mould surface. The resulting opening force to be resisted by the clamp is the product of the packing pressure and the projected area of the moulding.

Occasionally when a 'constant volume injection' technique is used with a 'vertical flash' mould design, the post-filling, packing-pressure phase is eliminated and the maximum demands on clamp forces occur at the moment of cavity filling. It is therefore sometimes desirable to estimate the mould-opening force from the cavity filling pressure. For a centre-gated circular disc and other simple geometries this can be calculated from the pressure distribution given by (isothermal) rheological equations.[4-6] The general form of the equation for the mould-opening force, and hence of the clamping requirements, is

$$F = \frac{P_o A}{\beta_n} \tag{21}$$

where F is the required clamping force and A is an effective projected area of the moulding, β_n is a numerical term greater than unity and dependent on the power-law index n. In most practical cases the value of β_n lies between 1·5 and 2·5; the more centrally gated, flat, and symmetrical the moulding the higher is the β value.

For the circular disc mould considered earlier the clamping force equation is

$$F_{min} = \frac{f(n)}{\beta_n} KC^{3n} \left(\frac{\pi^{n+1} R^{5-n}}{x^{1+4}} \right) \tag{22}$$

and for $n = 1/3$, $\beta_n = 4$

$$F_{min} = 13\pi K_T C_T \left(\frac{R^{10/3}}{x^{7/3}} \right) \tag{23}$$

The Feasibility Diagram for Mould Filling

Injection moulders are faced every day with judgements and decisions concerning the choice of material and machine for mouldings—hopefully in the process of design, but all too often for mouldings already frozen in design, or worse still for moulds already made. The material is chosen primarily to optimise the performance and functional usefulness of the final article, and only secondarily to ensure a minimum of production problems when manufactured by the thousand or, in some cases, by the million.

One fundamental problem, especially for large-area or relatively thin-walled mouldings, consists of ensuring that the mould can be filled with melt from the chosen injection point(s) using the chosen machine. A moulding costed for one machine but which can only be moulded on a larger one causes problems all round.

A moulder will instinctively ask questions such as:

(i)   On which machine(s) can the mould be hung (platen size limitations)?

(ii)   Which machine has sufficient shot capacity?

(iii)   Has the machine adequate ratings of pressure, clamp and injection rate to ensure consistent filling of the mould with the chosen material?

Rheological Limitations as Defined by Maximum Flow Ratio
The traditional answer to (iii) above, which involves the melt rheology, is to consider the 'flow ratio' for the material, i.e. the expected ratio of maximum flow length, L, to cavity thickness, x. A material with a flow ratio of 150, for example, would require a cavity thickness of at least $L/150$ to be filled, where L is the greatest melt flow distance occurring from any injection point. This crude rule-of-thumb has embodied many injection moulders' total concern with melt rheology for several decades.

Figure 13.12 shows what these feasibility criteria amount to, using a plot

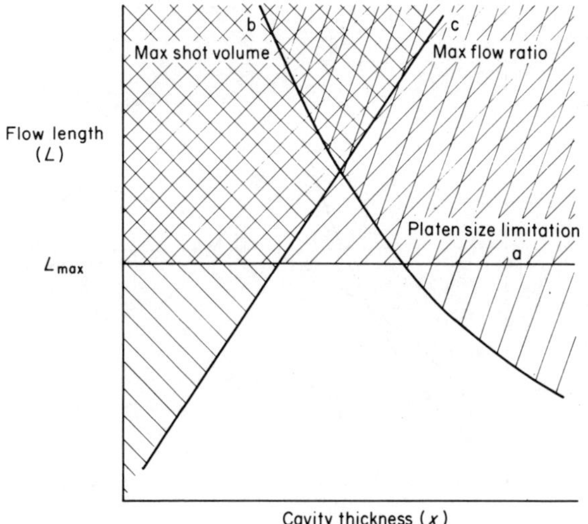

FIG. 13.12.   Elementary feasibility diagram.

of flow length vs. thickness. For example, if we consider a 2×1 rectangular cavity, gated at the centre, the platen dimensions will define the largest mould with a cavity of such a shape which can be hung. The maximum flow length is the semi-diagonal of the rectangle, plotted as curve 'a'. Flow lengths greater than L_{max} for the given shape are not feasible (shaded area).

Shot size limitations define a further curve, 'b', which excludes larger and thicker rectangles in the appropriate shaded area.

Finally a 'maximum flow ratio' value for the material is given by the line 'c' for which L/x is constant. This defines a further boundary between what is feasible and what is not (again shaded).

Applying all three limiting criteria together gives an area in Fig. 13.9 which defines the dimensions of feasible mouldings of the stated shape.

Feasibility in Terms of the Flow Equations Previously Derived
The application to the feasibility diagram of the flow equations derived for estimating maximum cavity pressure and clamp requirements eliminates the need to rely on the single, rule-of-thumb, value for flow ratio.

Any injection moulding machine has a specified 'maximum melt pressure' value, P. This usually refers to the maximum pressure which may be generated under hydrostatic conditions. During mould filling, the injection drive pumps are power-limited to providing only about two-thirds

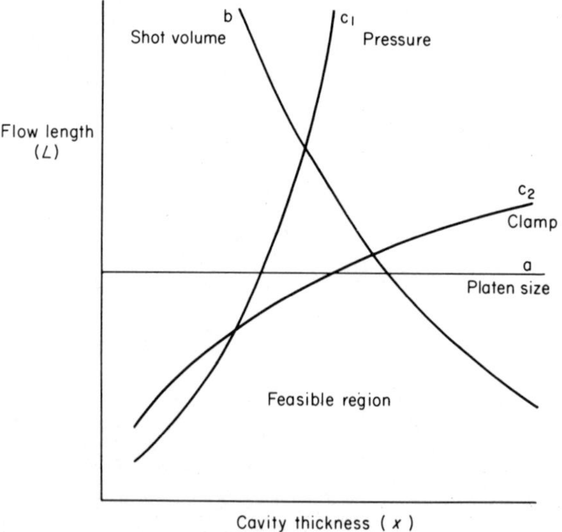

FIG. 13.13.   Refined feasibility diagram.

of the maximum hydrostatic pressure when injecting at the maximum rate. This 'pressure-bank' under dynamic conditions is drawn on to overcome the delivery pressure drop from injection barrel to mould entry, ΔP, as well as the maximum cavity pressure during filling, P_o.

Taking typical practical values, P is around 140 MN m^{-2} (20 000 psi) and ΔP can be limited by choice of delivery channel dimensions to around 35 MN m^{-2} (5000 psi). This means that the maximum pressure available for filling a cavity can be as low as 2/3 of 140 MN m^{-2} less 35 MN m^{-2}, i.e. P_o is around 55 MN m^{-2} (8000 psi) for design purposes.

Assuming a fixed design value for P_o, and injection at the optimum rate, eqn. (20) can be used to calculate the L, x values for the feasibility boundary for any given material at a given temperature. (An assumption being that the pressure required to fill a given shape is the same as that needed to fill an escribing circular cavity with the injection point at the centre and a radius equal to maximum flow length, i.e. $L = R$ in eqn. (20).) This curve is plotted in Fig. 13.13 as 'c_1', and represents the pressure limitations for a rectangular shape as discussed above. Curves 'a' and 'b' are the same as in Fig. 13.12.

The curve 'c_2' arises from a consideration of a limiting clamping pressure, F_{max}, in eqn. (23). The precise location of boundary curves c_1 and c_2 is temperature-dependent in each case.

When the material choice is highly dependent on moulding feasibility, as

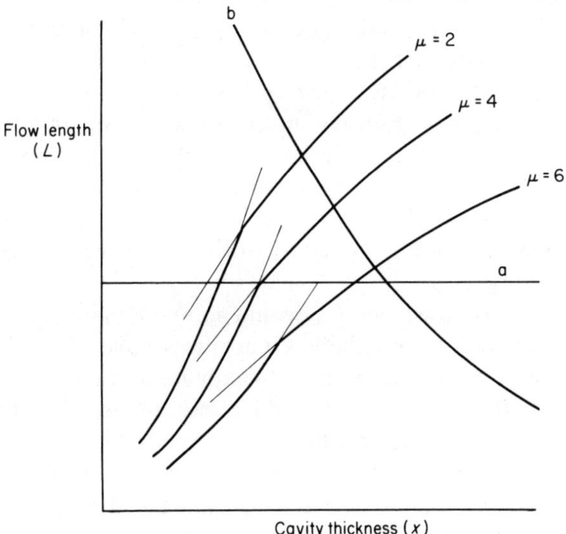

FIG. 13.14.   Feasibility envelopes for various mouldability indices.

in the case of large area panels for example, a feasibility diagram can be constructed initially in terms of mouldability indices (see Fig. 13.11) as shown in Fig. 13.14.

The resultant series of 'feasibility envelopes' clearly shows the limitations of a given machine, moulding a particular shape, as the material properties vary. The machine limitations represented by curves 'a' and 'b' remain fixed.

DISCUSSION

To the injection moulder and mould designer the melt flow behaviour is only one of many problems that are encountered every day. They can rightly claim that they have got by sufficiently well for some decades drawing on their experience and the use of a few rules of thumb when estimating flow distances and clamp pressures. Some trimming of delivery channel dimensions, gate sizes and even cavity dimensions is considered acceptable during the commissioning trials for a mould, even if that mould may have to be taken back to the machine several times. From the moulder's point of view polymer melt rheologists have failed to provide simple answers to what appear to be simple questions; with few exceptions the traditional reliance on a mixture of trial-and-error and expertise has not been bettered by complicated equations and computer programs.

However, there are two factors of growing importance which put the traditional approach under strain. Firstly, there are now attractive markets for large injection mouldings such as business machine panels, sundry panels for the automotive industry and even complete boat hulls and decks. Secondly, there is the considerable growth in injection mouldable engineering plastics such as the nylons, polycarbonate, poly(butylene terephthalate) and modified poly(phenylene oxide), and these materials have their own flow problems such as rapid freeze-off (nylons) or high apparent viscosities.

On the other hand, the rheologist would point out that polymer melts are non-Newtonian viscoelastic fluids; that the simple (Newtonian) flow equations cannot be applied because the apparent viscosity is shear-rate-dependent and that, owing to the viscoelastic nature of polymer melts significant tensile properties are often encountered which add to the energy required to change the configuration of the fluid. Furthermore, significant reversible and irreversible changes in temperature are brought about by the compression and decompression of the melt, as well as by the shear heat generated in the high-shear-rate region near the channel walls. These complications are fairly well understood and can in many cases be adequately quantified. Taking them all fully into account involves non-trivial computer programs and computations. Finally, all quantitative attempts to analyse the melt flow in injection moulds must also account for the build-up of frozen material on the cold cavity surfaces. Factors such as polymer-to-metal heat transfer, mould cooling efficiency and even crystallisation kinetics would loom large in any rigorous analysis.

In this chapter the significant parameters have been identified and used to estimate those quantities which are of prime importance to the injection moulder and mould designer. In particular, approximate relationships between material properties, machine design parameters and mould geometry (for simple shapes) have been given. The operating equations have been simplified as much as possible using a number of reasonable assumptions, empirically established relationships and mathematical models. Greater refinement and better accuracy could no doubt be achieved at every stage—but always at the cost of simplicity. It has been the aim throughout to communicate a quantitative understanding of what is happening to the melt in the injection moulding process with a minimum of mathematical complexity. Any over-simplification of rheological matters may be remedied by reference to the literature.

REFERENCES

1. F. N. COGSWELL, private communication.
2. F. N. COGSWELL, *App. Poly. Symposia*, **27**, 1 (1975).
3. J. DEALY, *Poly. Eng. Sci.*, **11**, 433 (1971).
4. I. T. BARRIE, *Plastics & Polymers*, **38**, 47 (February 1970).
5. I. T. BARRIE, *Plastics & Polymers*, **37**, 463 (October 1969).
6. I. T. BARRIE, *SPE Journal*, **27**, 64 (August 1971).

14

Deformation in the Solid State—Small Strains

A polymer will respond in one of three ways to an applied stress:

(i)   In a rapid elastic manner characterised by a high modulus which corresponds to bond stretching and the deformation of bond angles. This applies to both amorphous polymers below T_g and to crystalline polymers.

(ii)   By viscous flow characterised by a low modulus which corresponds to the irreversible slippage of flow units past one another. This occurs in amorphous polymers above T_g and generally in all molten polymers.

(iii)   In a rubber–elastic manner characterised by a low modulus and by largely *reversible* slippage of flow units. This may result in several hundred percent elongation. As the flow units straighten out from their randomly coiled configuration, so they orient themselves along the stress axis. This process is retarded by the internal frictional (viscous) forces of the material.

The responses are illustrated by a model of springs and dashpots (Fig. 14.1). At low temperatures the viscosities of the dashpots η_2 and η_3 are high and the polymer can only respond in a rapid elastic manner which is defined by the spring modulus G_1.

At higher temperatures the viscosity of dashpot η_2 will become apparent, and with it the second spring modulus G_2. If this is the dominant behaviour, then the polymer will be soft and rubbery rather than rigid.

At still higher temperatures when the viscosity η_2 is low, the stress is quickly transmitted to the second dashpot of viscosity η_3 and when this becomes the dominant response, then irreversible flow occurs with little elasticity, since a restoring force (a spring in parallel with that dashpot) is effectively absent.

In reality all three processes operate, with one or other dominating. The mixed response is referred to as 'viscoelasticity'. At any one temperature the

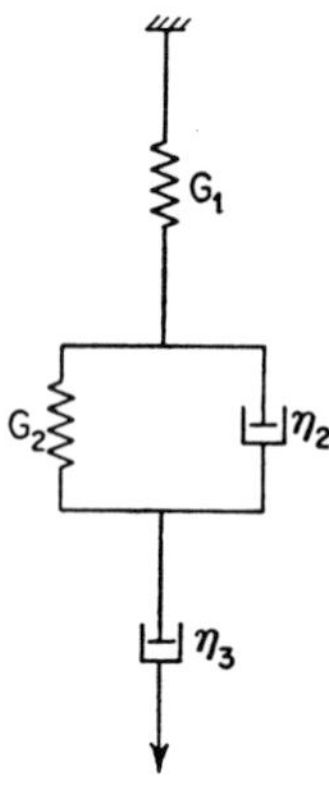

Fig. 14.1.

extent to which each mechanism operates depends on the *rate* of stress application. Viscosity cannot show up under impact conditions; under prolonged loading viscosity will dominate and manifest itself as creep.

The spring and dashpot model is useful as a conceptualising device, but it is crude and oversimplified. It suffers from two major limitations:

(i)   the dashpots suggest Newtonian flow although Newtonian flow in polymers occurs only at very low shear stresses;

(ii)   it applies only to amorphous polymers.

If we confine ourselves mainly to amorphous polymers, then we must also limit the applied deformations to those which are small enough to justify the assumption that we are still operating within the linear viscoelastic region. It is difficult to be certain that any given stress, even a small one, is small enough to fall within that region, so that conclusions on the mechanical properties of materials which are based upon a linear viscoelastic theory are always suspect. However, an analysis based on classical linear viscoelasticity will serve as a starting point.

The models of viscoelasticity were invented in the 19th century when few man-made polymers were known and when their structure was still unrecognised. These models add nothing new and everything that can be deduced from them is also available from phenomenological data; but they do give a fair picture of the significance of internal parameters of state. These parameters are represented by springs and dashpots. A spring stores the energy which has been expended in effecting deformation; a dashpot dissipates that energy. The internal parameters of *real* substances indicate

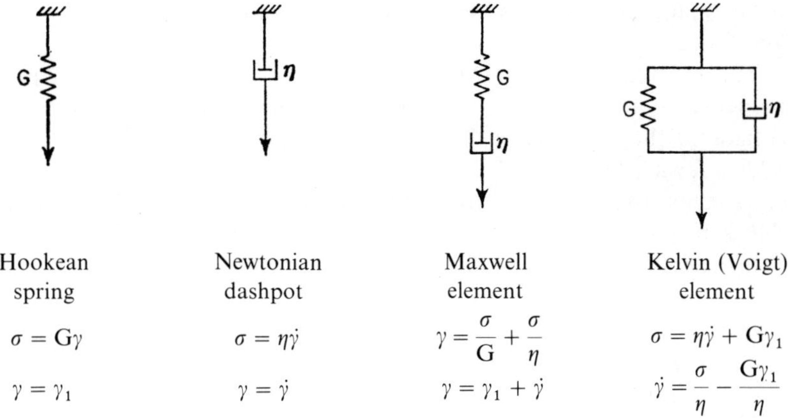

Hookean spring	Newtonian dashpot	Maxwell element	Kelvin (Voigt) element
$\sigma = G\gamma$	$\sigma = \eta\dot{\gamma}$	$\gamma = \dfrac{\sigma}{G} + \dfrac{\sigma}{\eta}$	$\sigma = \eta\dot{\gamma} + G\gamma_1$
$\gamma = \gamma_1$	$\gamma = \dot{\gamma}$	$\gamma = \gamma_1 + \dot{\gamma}$	$\dot{\gamma} = \dfrac{\sigma}{\eta} - \dfrac{G\gamma_1}{\eta}$

Fig. 14.2.   Key: σ = stress; γ = total strain; γ_1 = strain on the spring; $\dot{\gamma}$ = strain rate; η = (Newtonian) viscosity; G = spring modulus.

to what extent the substances store and dissipate the energy applied by an external stress. The simplest mechanical models are shown in Fig. 14.2.

It is noted that σ is used for stress. In the liquid state we have used the symbol τ. We now wish to use the symbol τ for a different function, as will become evident presently.

Strictly speaking, the models of Fig. 14.2 should also include an inertial element, as, for example, in Fig. 14.3(a); this may be omitted in large deformations of plastics since damping is substantial; but in small deformations, damping is less and inertial effects may result in sinusoidal oscillations.[1] Figure 14.3(b) shows this more clearly because the dashpot is omitted. When a force is applied to the inertial Hookean spring, its subsequent release will lead to a yoyo-like oscillation which, in the absence

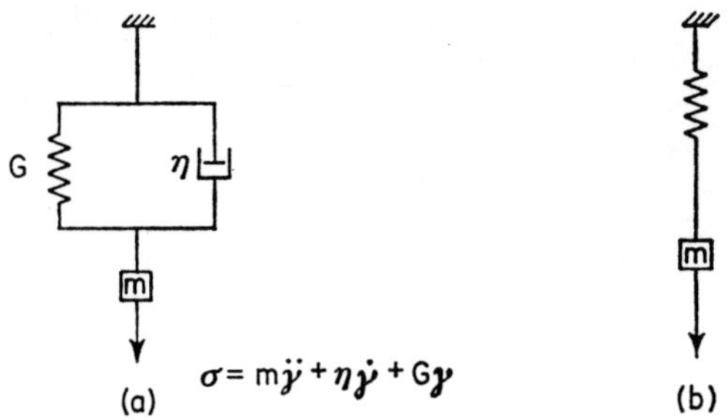

Fig. 14.3.   Kelvin (Voigt) model with inertia ($\ddot{\gamma}$ = strain acceleration).

of damping, produces a sinusoidal time/strain curve with constant amplitude.

The difference between Maxwell relaxation and Kelvin retardation models is as follows: when springs and dashpots are arranged in series (Maxwell) they each bear the entire stress, but their deformations are additive; when springs and dashpots are arranged in parallel (Kelvin) they each suffer the same deformation, but they divide the stress among themselves.

In order that the physical meaning of η/G for Maxwell and Kelvin models should not be confused, it is denoted as the *relaxation time* and *retardation time* respectively. The symbol τ is used whenever the two cannot be confused, but where this risk exists τ_J is used for retardation time and τ_R for relaxation time.

The relaxation time may be obtained from stress relaxation experiments at constant strain. The retardation time is obtainable from a creep experiment at constant stress or from a recovery experiment following a creep experiment on removing the stress.

The Kelvin element is a good example of a viscoelastic body. If the model is stressed, part of the stress energy is stored in the spring while the remainder is dissipated in the dashpot and causes the deformation to be retarded. On releasing the stress, recovery will occur due to the elasticity of the spring, but this is, again, retarded by the dashpot. The time which the system requires to reach equilibrium is the retardation time τ_J.

Under a constant stress the Kelvin element shows creep. The deformation obeys the equation

$$J = J_0(1 - \exp(-t/\tau_J))$$

where J is the creep compliance at time t, and J_0 the creep compliance at the time of stress application. The compliance J is the reciprocal of the modulus G. If the applied stress is cyclic ('dynamic'), then the storage and loss compliances are given by the Debye equations:

Storage $$J'(\omega) = J_0 \frac{1}{1 + \omega^2 \tau_J^2}$$

Loss $$J''(\omega) = J_0 \frac{\omega \tau_J}{1 + \omega^2 \tau_J^2}$$

where the frequency ω has the dimensions of reciprocal time.

The Kelvin element is suitable for representing the retardation of a constant stress, but not for the relaxation of a constant strain. For the latter the Maxwell element needs to be considered.

If a Maxwell element is deformed to a constant strain at zero time only the spring is initially strained and stores the entire strain energy. But as time passes, so the dashpot flows under the influence of the restoring force of the spring which doles out the stored energy to be dissipated by the dashpot. The stress will relax and approach zero monotonically. The monotonic decay obeys the exponential equation

$$G(t) = G_0 \exp(-t/\tau_R)$$

where τ_R again has the units of time and represents the 'relaxation time'. When $\tau_R = t$ the modulus will have decayed to one-eth of its initial value. Since $G = \sigma/\gamma$ and remains the same for all values of t in an experiment in which the strain is to remain constant, we can also write

$$\sigma(t) = \sigma_0 \exp(-t/\tau_R)$$

For dynamic strains the analogous Debye equations for storage and loss moduli are:

Storage
$$G'(\omega) = G_0 \frac{\omega^2 \tau_R^2}{1 + \omega^2 \tau_R^2}$$

Loss
$$G''(\omega) = G_0 \frac{\omega \tau_R}{1 + \omega^2 \tau_R^2}$$

One of the advantages of dynamic tests lies in the fact that storage and loss components can be readily separated. This becomes apparent when the Debye equations are used in plotting $\ln(t/\tau_J)$ and $\ln(t/\tau_R)$ vs. J and G respectively (Fig. 14.4). Separate curves are obtained for the storage and loss modulus. The sum of these two curves represents the complex dynamic compliance (or modulus) and the hysteresis loop between the complex and storage compliance (or modulus) therefore represents the 'hysteresis loss'. Note that parts (a) and (b) of Fig. 14.4 are mirror images of each other.

Before leaving one- and two-parameter models let us consider the stress as a function of time when a constant arbitrary strain is applied at zero time and released at time t (Fig. 14.5).

Looking back at Fig. 14.2 which includes the mathematical stress and strain relations for each type, it is clear that these cannot be used for Maxwell and Kelvin elements as they stand because flow is part of the total deformation in both and flow depends not only on the stress, but also on the duration of that stress. One must therefore first write down an expression

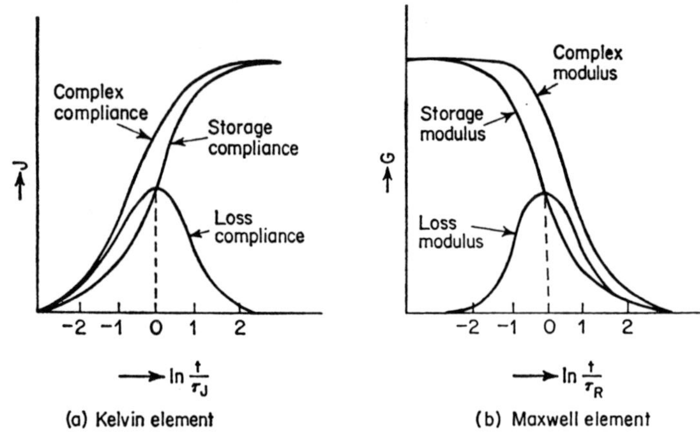

$$\text{Fig. 14.4.}$$

which gives the deformation *rate* in the Maxwell element. The flow displacement γ_2 is already a time derivative ($\dot{\gamma}$, by definition, is $d\gamma_2/dt$, which, according to Newton's Law, is σ/η). The elastic deformation γ_1 obeys Hooke's Law and is given by σ/G. On differentiation we obtain

$$\frac{d\gamma_1}{dt} = \frac{1}{G}\frac{d\sigma}{dt}$$

and summing the two derivatives, the total deformation rate is seen to be

$$d\gamma/dt = d\gamma_1/dt + d\gamma_2/dt = \frac{1}{G}\frac{d\sigma}{dt} + \frac{\sigma}{\eta}$$

This fundamental differential equation determines the mechanical response of a material to a constant shear stress or strain. At constant strain $d\gamma/dt$ is zero and the equation reduces to

$$\frac{1}{G}\frac{d\sigma}{dt} + \frac{\sigma}{\eta} = 0$$

which can then be integrated[2] to give

$$\sigma = \sigma_0 \exp(-Gt/\eta) = \sigma_0 \exp(-t/\tau_R)$$

This is the law of Maxwellian decay in which the ratio η/G is the relaxation time τ_R. Since γ is constant, $\sigma/\sigma_0 = G/G_0$ and the law of Maxwellian decay can also be written

$$G = G_0 \exp(-t/\tau_R)$$

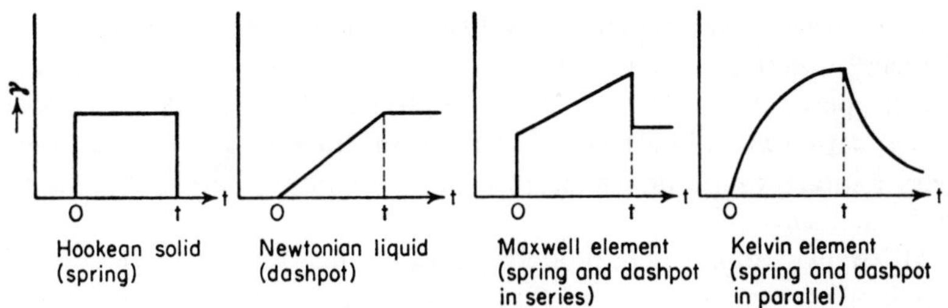

FIG. 14.5.

The behaviour of a Maxwell body in tension is similar to that in shear and the relaxation times will be the same, provided that the material is incompressible, because the tensile modulus and viscosities can then be assumed to be $3G$ and 3η respectively and their ratio is again τ_R.

Again looking back at Fig. 14.2, the basic differential equation for the Kelvin element can be made apparent from

$$\sigma = \eta\dot{\gamma} + G\gamma_1$$

by rewriting it

$$\frac{d\sigma}{dt} = \eta\frac{d\gamma_2}{dt} + G\frac{d\gamma_1}{dt}$$

which at constant stress can be set to zero and integrated,[2] giving

$$\gamma = \frac{\sigma}{G}\frac{1}{1 - \exp(-Gt/\eta)}$$

where η/G is τ_J, the retardation time so that

$$\gamma = \frac{\sigma}{G}\frac{1}{1 - \exp(-t/\tau_J)}$$

Upon removal of the stress the sample slowly returns to its original shape where $\gamma = 0$, following the experimental function

$$\gamma = \gamma_0\exp(-t/\tau_J)$$

Thus, after constant stress, a retarded elastic specimen recovers monotonically at a rate which is determined by its retardation time, just as at constant deformation a Maxwell body monotonically relaxes its stress at a rate which is determined by its relaxation time.

The relative importance of the elasticity/flow mechanism of response thus depends not only on G and η, but also on the experimental time scale t.

Thus, in a Maxwell model $(\sigma = \sigma_0 \exp(-t\tau_J))$, if t is large compared with the relaxation time τ_R, σ will approach zero. If it is small compared with τ_R, σ will approach σ_0. The same will apply exactly to the Kelvin model $(\gamma = \gamma_0 \exp(-t\tau_J))$, in a recovery experiment, taking γ_0 as the maximum deformation (at zero time) and substituting the appropriate symbols in the first statement.

Maxwellian decay can be plotted against t/τ_R in several ways. If $\log(\sigma/\sigma_0)$ is plotted vs. t/τ_R we get a straight line, as expected. But if instead we plot

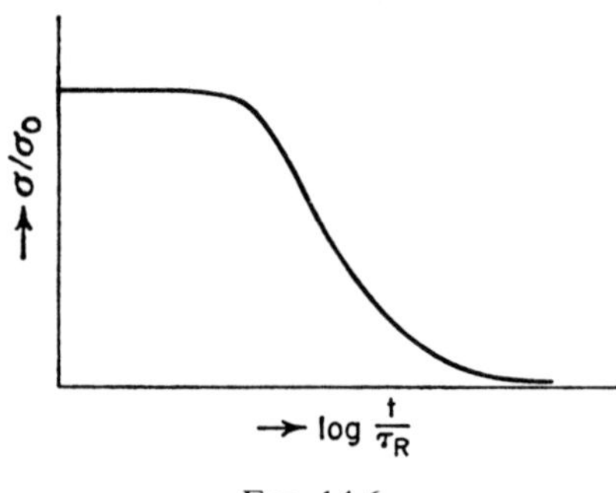

Fig. 14.6.

σ/σ_0 vs. $\log(t/\tau_R)$ we obtain a curve of characteristic sigmoid shape, where a change in relaxation time τ_R does not alter the shape but merely displaces it along its time axis (Fig. 14.6). The curve is therefore universal for all materials which obey the Maxwell law of relaxation. Relaxation time is a function of free volume and therefore varies with temperature. By carrying out relaxation experiments at different temperatures it is possible to determine the corresponding relaxation times and to obtain portions of a curve which need only be shifted along the abscissa (Fig. 14.6) to reveal a complete master curve. The shift constant depends upon the relaxation time. This is known as the time/temperature *superposition principle* which makes it possible to predict long-term relaxation behaviour at constant temperature from a study of short-term behaviour of a material at various temperatures.

It has been seen that a model with at least four parameters is required to obtain even a rough approximation of polymer behaviour. The model depicted in Fig. 14.1 resembles the Kelvin type because it has a spring and a dashpot in parallel. An alternative representation is shown in Fig. 14.7(b) which resembles the Maxwell type because it has two Maxwell elements in parallel. The relaxation and retardation times are ratios of a spring

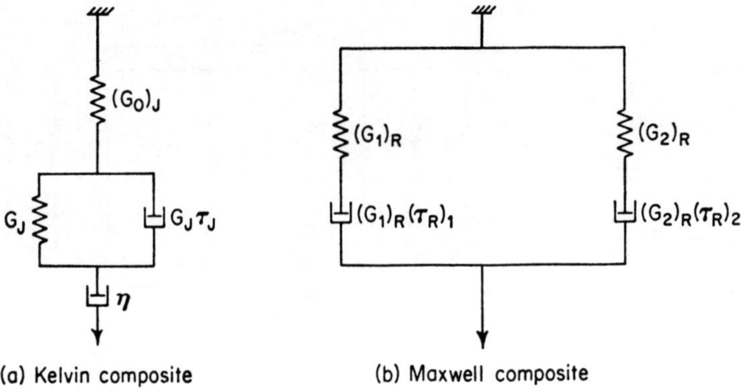

$$\text{FIG. } 14.7.$$

modulus and a dashpot viscosity ($\tau = \eta/G$). Each viscosity—apart from the lone dashpot in the Kelvin composite (Fig. 14.7(a))—can therefore be represented as a product $G\tau$. Both the Kelvin and Maxwell composites are equally representative; they are mechanically equivalent. The Kelvin composite is preferred for calculating the time-dependent strain at constant stress (creep) whilst the Maxwell composite is preferred for the calculation of the time-dependent stress at constant strain (relaxation). If the parameters $(G_1)_R$, $(G_2)_R$, $(\tau_R)_1$, $(\tau_R)_2$ of the Maxwell composite and $(G_0)_J$ of the Kelvin composite are known, then the retardation time τ_J of the Kelvin composite can be calculated from the relationships

$$(G_1)_R = (G_0)_J \frac{\dfrac{1}{(\tau_R)_1} - \dfrac{1}{\tau_J}}{\dfrac{1}{(\tau_R)_1} - \dfrac{1}{(\tau_R)_2}}$$

$$(G_2)_R = (G_0)_J \frac{\dfrac{1}{(\tau_R)_2} - \dfrac{1}{\tau_J}}{\dfrac{1}{(\tau_R)_1} - \dfrac{1}{(\tau_R)_2}}$$

where $(\tau_R)_1$ and $(\tau_R)_2$ are relaxation times and τ_J is a retardation time.

The simplification implicit in four-parameter models must now be adjusted in order to represent the actual behaviour of plastics more accurately. It just is not true that plastics are fully characterised by one or two relaxation or retardation times. The large numbers of possible

                                Polymer Rheology

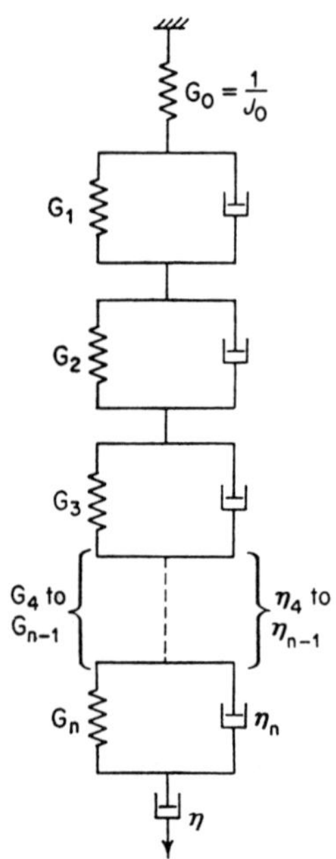

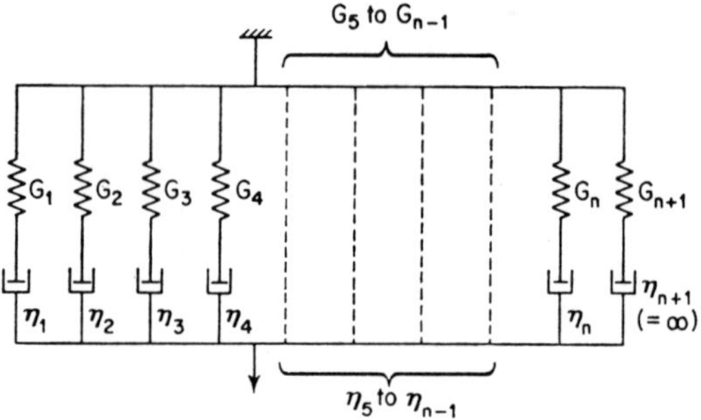

FIG. 14.8.   *Note:* the dashpot η_{n+1} must have infinite viscosity, so that the spring G_{n+1} ensures that a finite stress remains after infinity time.

configurations which are available to flow units must be reflected in a corresponding number of relaxation or retardation times. Some of the configurations will be more probable than others and in general the relaxation or retardation times will cluster around a central value in a Gaussian manner. The first step in arriving at a distribution function lies in constructing generalised Maxwell and Kelvin composites as shown in Fig. 14.8.

The generalised Kelvin model has an instantaneous elasticity parameter G_0, a Newtonian flow parameter η and n different Kelvin elements, each with its particular retardation time (τ_J), and spring and dashpot constant; there are thus $(2n + 2)$ parameters present. The creep behaviour is given by the equation

$$J = J_0 + \sum_{i=1}^{n} J_i(1 - \exp(-t/(\tau_J)_i)) + t/\eta \tag{1}$$

The generalised Maxwell model consists of $(n + 1)$ Maxwell elements arranged in parallel, each with its particular relaxation time $(\tau_R)_i$ and spring and dashpot constant. Of the $(n + 1)$ relaxation times n correspond to the retardation times in the generalised Kelvin model and the additional one represents the flow in the generalised Kelvin model. Again, $(2n + 2)$ spring and dashpot parameters are present. The stress relaxation behaviour is given by the equation

$$G = G_\infty + \sum_{i=1}^{n+1} G_i \exp(-t/(\tau_R)_i) \tag{2}$$

We now extend the generalised Kelvin and Maxwell models from a finite number of $(2n + 2)$ springs and dashpots to an infinite number of springs and dashpots, so that their characteristic parameters vary continuously with their retardation or relaxation time. This results, in each case, in an infinite network characterised by a continuous function of a single independent variable, namely:

(1) the distribution function of retardation times (retardation spectrum) for the continuous Kelvin model;
(2) the distribution function of relaxation times (relaxation spectrum) for the continuous Maxwell model.

The distribution functions can be fitted from experimental data after plotting the compliance or modulus against

$$\ln \frac{t}{\tau_J} \qquad \text{or} \qquad \ln \frac{t}{\tau_R}$$

for retardation and relaxation respectively. The resultant plot is shown in Fig. 14.9. The plot also includes the theoretical curve obtained from the

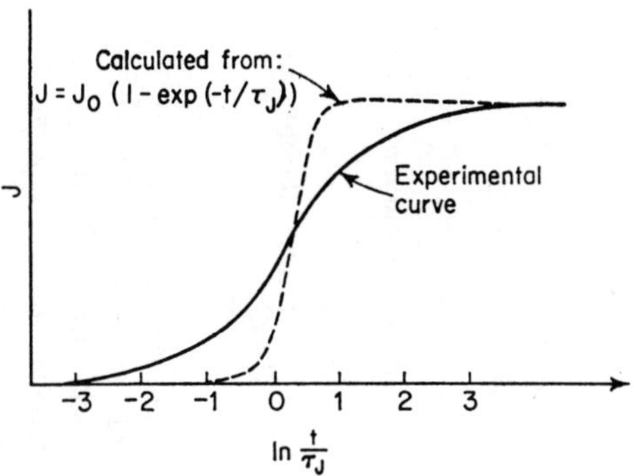

Fig. 14.9.

TABLE 14.1

Kelvin		*Maxwell*
$$J = J_0 + \sum_{i=1}^{n} J_i(1 - \exp(-t/\tau_J)) + \frac{t}{\eta}$$	Equations for generalised finite models	$$G = G_0 + \sum_{i=1}^{n+1} G_i \exp(-t/\tau_R)$$
the retardation times of the individual units $f(\tau_J)\,\mathrm{d}\tau_J$ will be between zero and infinity		the relaxation times of the individual unit $g(\tau_R)\,\mathrm{d}\tau_R$ will be between zero and infinity

Replacement of the summation sign $\qquad\qquad$ Replacement of the summation sign

$$\sum_{i=1}^{n} J_i \text{ by the integral } \int_0^\infty f(\tau_J) \qquad\qquad \sum_{i=1}^{n+1} G_i \text{ by the integral } \int_0^\infty g(\tau_R)$$

The retardation spectrum $f(\tau_J)$ gives the contribution (to the equilibrium compliance of viscoelastic deformation) per unit interval of the time scale of the retardation processes with retardation times around τ_J

$\qquad$ Alfrey's definition[5] of the functions $f(\tau)$ and $g(\tau)$

The relaxation spectrum $g(\tau_R)$ gives the contribution (to the equilibrium elastic modulus) per unit interval of the time scale of the relaxation processes with relaxation times around τ_R

$$J = J_0 + \int_0^\infty f(\tau_J)(1 - \exp(-t/\tau_J)\,\mathrm{d}\tau_J + \frac{t}{\eta}$$

Equations for *infinite models*

$$G = G_\infty + \int_0^\infty g(\tau_R)(\exp(-t/\tau_R))\,\mathrm{d}\tau_R$$

Equation (3) $\qquad\qquad\qquad\qquad$ Equation (4)

Kelvin model assuming a single retardation time which is numerically equal to the mean retardation time of the continuous Kelvin model.

The obvious feature of Fig. 14.9 is the flattened experimental curve which renders the point of inflexion much less incisive than is the curve calculated on the assumption of a single retardation time.

Suffice it to say, that eqns. (1) and (2) for creep and stress relaxation of generalised (but still finite) Kelvin and Maxwell models respectively, can be readily modified. All that is necessary is to make the corresponding concomitant notational changes as set out in Table 14.1.

The derivations of eqns. (3) and (4) are initially due to Wiechert[3] and Smekal[4] respectively and were subsequently reappraised by a number of workers.

G_∞ the integration constant in eqn. (4) is the value of the modulus after infinity time. It represents the elasticity of the one spring which is connected to a dashpot of infinite viscosity (see note to Fig. 14.8) in the infinite model just as in the finite generalised Maxwell model. The spring must be included to ensure that a finite stress remains after infinity time.

J_0 in eqn. (3) represents the instantaneous compliance, the integral term represents the viscoelastic deformation and t/η represents the flow (irreversible deformation) which occurs in creep.

The relaxation and retardation spectra $g(\tau_R)$ and $f(\tau_J)$ are appropriate for plots in which time is represented linearly. However, it is advantageous to use a logarithmic time scale for two reasons:

(1) the time dependence of the processes becomes more pronounced;
(2) extrapolation to long time behaviour becomes both easier and more exact.

The conversion of the linear spectra is easily carried out by manipulation of the integral terms of eqns. (3) and (4).

Since $d\tau = \tau\, d\ln\tau$, we need only multiply and divide the integral term by τ_R and τ_J respectively in order to convert the linear equations for G and J, viz.:

$$\int_0^\infty f(\tau_J)(1 - \exp(-t/\tau_J))\,d\tau_J = \int_0^\infty \tau_J f(\tau_J)(1 - \exp(-t/\tau_J))\frac{d\tau_J}{\tau_J}$$

Writing

$$\left\{\begin{array}{llllll} L(\ln\tau_J) & \text{for} & \tau_J f(\tau_J) & \text{and} & d\ln\tau_J & \text{for} \quad \dfrac{d\tau_J}{\tau_J} \\[3ex] H(\ln\tau_R) & \text{for} & \tau_R g(\tau_R) & \text{and} & d\ln\tau_R & \text{for}\,\dfrac{d\tau_R}{\tau_R} \end{array}\right\}$$

we then have the integrals as the convenient expressions

$$\int_{-\infty}^{+\infty} L(\ln \tau_J)(1 - \exp(-t\tau_J))\,d\ln \tau_J$$

and

$$\int_{-\infty}^{+\infty} H(\ln \tau_R)\exp(-t/\tau_R)\exp(-t/\tau_R)\,d\ln \tau_R$$

$L(\ln \tau_J)$ still has the dimensions of a compliance and represents the contribution to creep of those retardation processes which have characteristic retardation times between $\ln \tau_J$ and $(\ln \tau_J + d\ln \tau_J)$.

$H(\ln \tau_R)$ similarly still has the dimensions of a modulus and represents the contribution to stress relaxation of those relaxation processes which have characteristic relaxation times between $\ln \tau_R$ and $(\ln \tau_R + d\ln \tau_R)$.

Substituting these modifications of the integral terms into eqns. (3) and (4) we finally obtain

$$J = J_0 + \int_{-\infty}^{+\infty} L(\ln \tau_J)(1 - \exp(-t/\tau_J))\,d\ln \tau_J + \frac{t}{\eta} \tag{5}$$

and

$$G = G_\infty + \int_{-\infty}^{+\infty} H(\ln \tau_R)\exp(-t/\tau_R)\,d\ln \tau_R \tag{6}$$

The instantaneous compliance J_0 and the equilibrium modulus G_∞ are obtained by setting $t = 0$, when the exponential terms disappear.

It will be noted that the integrals involving logarithmic time go from $-\infty$ to $+\infty$ whilst the integrals applying to linear time go from zero to $+\infty$ for the obvious reason that $\ln 0 = -\infty$.

A knowledge of L and H allows one to calculate all the viscoelastic functions. Strictly speaking, since the two are related,[6] only one of them is necessary, but their relationship is cumbersome and one can carry out the necessary conversions by other means. This subject has been dealt with by a number of workers in standard texts.[7-10]

Two of the most useful equations relating the functions are:

$$H(\ln \tau) = -\frac{dG}{d\ln t} + \frac{d^2G}{(d\ln t)^2} \tag{7}$$

where $t = 2\tau_R$,[8] and analogously $L(\ln \tau_J)$ from J:

$$G = \frac{\sin m\pi}{m\pi J} \tag{8}$$

where m is the slope of the log J vs. log t plot.[9]

Equation (7) shows how the relaxation spectrum can be calculated from the relaxation modulus G; all that one requires is the first and second derivative of a plot of G vs. $\ln t$.

Equation (8) shows how the functions G and J can be calculated from one another.

An exceedingly concise and elegant way of arriving at eqns. (6) and (7) has been given in a paper by Turner[11] who uses as his starting point the principal feature of the theory of linear viscoelasticity as it has been developed over the last fifty years, namely the assumption that stress and strain are related through a linear differential equation

$$a_n \frac{\partial^n \sigma}{\partial t^n} + a_{n-1} \frac{\partial^{n-1} \sigma}{\partial t^{n-1}} + \cdots + a_0 \sigma = b_m \frac{\partial^m \gamma}{\partial t^m} + b_{m-1} \frac{\partial^{m-1} \gamma}{\partial t_{m-1}} + \cdots + b_0 \gamma$$

$$(9)$$

where $a_n \ldots a_0$ and $b_m \ldots b_0$ are material constants.

If this general equation is to be made rigorous, one would have to transpose it into tensor notation. This would liberate one from the restrictive assumption that all the components of the stress or strain tensor are isotropic, but would also lead to formidable manipulating difficulties. In using eqn. (9) then, as it stands, we are assuming:

(1)   that the stress–strain relationship is linear;
(2)   that the stresses and strains are isotropic.

Both these assumptions are basically unjustified and can give no more than approximations—in plastics materials this is particularly evident. But Turner does no more than use these as a *starting point* for a transformation from uni-dimensional to three-dimensional linear viscoelasticity and then proceeds to consider the further complications which arise from non-linear viscoelasticity in plastics.

If we accept eqn. (9) and also the above assumptions as serviceable approximations (however crude), then the equations for a Hookean body, a Maxwell element and a Voigt element follow immediately:

Hookean body: All constants a and b except a_0 and b_0 are zero.
   Equation (9) becomes

$$a_0 \sigma = b_0 \gamma \qquad (10)$$

Maxwell element: All constants a and b except a_0, a_1 and b_1 are zero.
   Equation (9) becomes

$$a_0 \sigma + a_1 \frac{\partial \sigma}{\partial t} = b_1 \frac{\partial \gamma}{\partial t} \qquad (11)$$

This is the spring and dashpot in series and applies to stress relaxation at constant strain.

Voigt element: All constants a and b except a_0, b_0 and b_1 are zero. Equation (9) becomes

$$a_0 \sigma = b_0 \gamma + b_1 \frac{\partial \gamma}{\partial t} \tag{12}$$

This is the spring and dashpot in parallel and applies to strain retardation at constant stress.

Neither the Maxwell nor the Voigt element and therefore neither eqns. (11) or (12) can represent both creep and stress relaxation simultaneously, but a combination of the equations—or a modification of eqn. (9) in which all constants a and b except a_0, a_1, b_0 and b_1 are zero—does so. The resulting eqn. (13) expresses this in mathematical terms:

$$a_1 \frac{\partial \sigma}{\partial t} + a_0 \sigma = b_1 \frac{\partial \gamma}{\partial t} + b_0 \gamma \tag{13}$$

The corresponding model is shown in Fig. 14.10. Equations (11), (12) and (13) have simple solutions involving the term $\exp(-t/\tau)$ where τ is a function of the constants a and b.

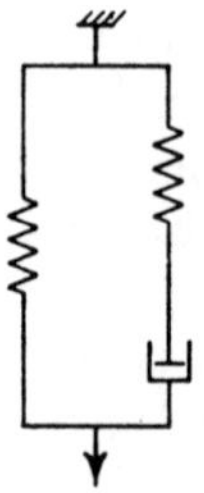

FIG. 14.10.   Model of eqn. (13).

If the strain is constant τ is τ_R, the relaxation time, and if the stress is constant τ is τ_J, the retardation time.

The approximate solution of these equations can be improved if the n and m terms in eqn. (9) are *not* assumed to be negligibly small. The alternative method of representation involves large numbers of Maxwell elements in parallel or Voigt elements in series. These are the generalised models which have been presented before, with a *series* of relaxation or retardation times respectively.

Equation (11), on solution, gives

$$\sigma = \sigma_0 \exp(-t/\tau_R)$$

which for a generalised model becomes

$$\sigma = \sum_{i=1}^{N} (\sigma_0)_i \exp(-t(\tau_R))$$

and

$$G(t) = \frac{\sigma}{\gamma_0} = \sum_{i=1}^{N} \frac{(\sigma_0)_i}{\gamma_0} \exp(-t(\tau_R)_i)$$

where $G(t)$ is the time dependent relaxation modulus, or

$$G(t) = \frac{\text{total (time-dependent) stress}}{\text{(constant) strain}}$$

Summing over the entire range of infinitely small increments of relaxation times, the summation is replaced by the integral

$$G(t) = \int_0^\infty g(\tau_R) \exp(-t/\tau_R) \, d\tau_R + G_\infty \tag{14}$$

where $g(\tau_R)$ is the distribution function of relaxation times (the relaxation spectrum) and G_∞ is the integration constant (the value of the modulus at infinity time).

This, then, is the treatment of Maxwell relaxation. The treatment of Voigt retardation is precisely analogous:

First we solve eqn. (12) to obtain

$$\gamma = \gamma_0(1 - \exp(-t/\tau_J))$$

which for the generalised model becomes

$$\gamma = \sum_{i=1}^{N} (\gamma_0)_i(1 - \exp(-t/\tau_J))$$

and

$$J(t) = \frac{\gamma}{\sigma_0} = \sum_{i=1}^{N} \frac{(\gamma_0)_i}{\sigma_0}(1 - \exp(-t/\tau_J))$$

where $J(t)$ is the time-dependent creep compliance, or

$$J(t) = \frac{\text{total (time-dependent) strain}}{\text{(constant) stress}}$$

Summing over the entire range of infinitely small increments of retardation times, the summation sign is replaced by the integral

$$J(t) = \int_0^\infty f(\tau_J)(1 - \exp(-t/\tau_J)) + J_0 + \frac{t}{\eta} \tag{15}$$

where $f(\tau_J)$ is the distribution function of retardation times (the retardation spectrum) and J_0 is the integration constant (the value of the instantaneous compliance at zero time).

The term t/η is the flow term which becomes important when t is very large (since η is *always* very large) and which is neglected in relatively short-time experiments.

There remains only one more modification to be made: it is obviously preferable to plot $G(t)$ or $J(t)$ vs. log time rather than vs. linear time; the transformation of eqns. (14) and (15) are readily accomplished in the manner already shown, giving respectively:

$$G(t) = \int_{-\infty}^{+\infty} H(\ln \tau_R) \exp(-t/\tau_R)\, d \ln \tau_R + G_\infty \tag{16}$$

and

$$J(t) = \int_{-\infty}^{+\infty} L(\ln \tau_J)(1 - \exp(-t/\tau_J))\, d \ln \tau_J + J_0 + \frac{t}{\eta} \tag{17}$$

Finally it should be mentioned that the *dynamic* parameters (storage and loss modulus, storage and loss compliance) can also be expressed in logarithmic distribution functions.[12]

The dynamic moduli and compliances are, of course, a function of frequency which replaces the simple time function and becomes its analogue when transposing from static to dynamic stress–strain relationships. The function of the form $\exp(-t/\tau)$ is replaced by the appropriate and characteristic Debye term of the form $1/(1 + \omega^2\tau^2)$. In the final result we obtain:

Storage modulus

$$G'(\omega) = \int_{-\infty}^{\infty} H(\ln \tau_R) \frac{\omega^2\tau_R^2}{1 + \omega^2\tau_R^2}\, \mathrm{d}\ln \tau_R + G_\infty$$

Loss modulus

$$G''(\omega) = \int_{-\infty}^{\infty} L(\ln \tau_J) \frac{\omega\tau_R}{1 + \omega^2\tau_R^2}\, \mathrm{d}\ln \tau_R$$

Storage compliance

$$J'(\omega) = \int_{-\infty}^{\infty} L(\ln \tau_J) \frac{1}{1 + \omega^2\tau_J^2}\, \mathrm{d}\ln \tau_J + J_0$$

Loss compliance

$$J''(\omega) = \int_{-\infty}^{\infty} L(\ln \tau_J) \frac{\omega\tau_J}{1 + \omega^2\tau_J^2}\, \mathrm{d}\ln \tau_J + \frac{1}{\omega\tau_J}$$

Why are eqns. (16) and (17) so important? In S. Turner's words their principal importance lies in this:

'It is characteristic of linear systems, i.e. those representable by a linear differential equation that if the response to a step input is known (e.g. from stress relaxation and creep experiments), then the response to any arbitrary input can be calculated from the superposition integral, a procedure often known as Boltzmann's Superposition Principle within the field of linear viscoelasticity.'[11]

The phenomenon of creep in thermoplastics has been very satisfactorily reviewed by Turner.[13] He points out that whilst stress–strain–time temperature relationships can be established in a number of ways, the creep experiment has two important advantages, namely:

(1) it is simple;
(2) it can be directly applied to service problems, especially with the more rigid plastics in which designers are becoming increasingly interested.

The importance of the creep function lies in the fact that extrapolation from a limited data range will give an acceptable account of creep behaviour over a much wider range of variables, even though extrapolation techniques

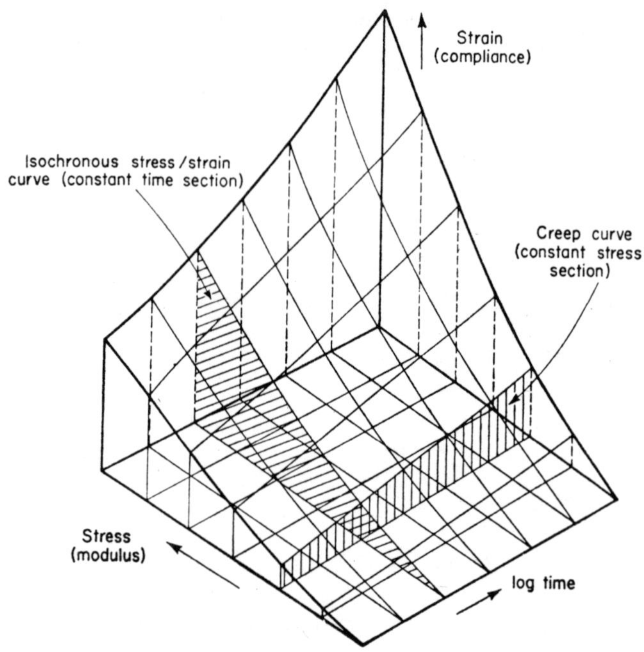

FIG. 14.11.   Stress–strain–time diagram.

are never entirely reliable. The viscosity term in the creep function (eqn. (6)) rarely occurs at small deformations, except in very low molecular weight polymers or at high temperatures.

By maintaining a constant temperature the stress–strain–time surface can be produced for a number of constant stress (creep) and constant time (isochronous) conditions as shown by sections in Fig. 14.11. The solid model so obtained can be reduced to a contour map in which each contour represents points of equal strain or stress relaxation. A series of contour maps at a number of temperatures completes the picture. Apparatus for creep measurements is reviewed by Dunn *et al.*[14] and a simple yet effective device is described by Scherr and Palm.[15]

The effects of inertia have been neglected so far. This is reasonably justified in plastics in which only static stresses are considered because in these no acceleration of volume elements occurs. But the sudden application of a stress could result in oscillating responses, especially in an elastomer. We are here entering the border region between static and dynamic stresses and the dynamic behaviour of polymer depends on both the inertia and the elastic restoring forces. Supposing that a polymer specimen has a sudden stress applied to it at the two ends. At the moment of stress application the central part of the mass experiences no stress at all and it will take some finite time before elastic waves become established throughout the specimen. The time necessary to establish this resonance state will depend on the distance of separation of the points of stress application as well as the velocity of wave propagation. At first however a

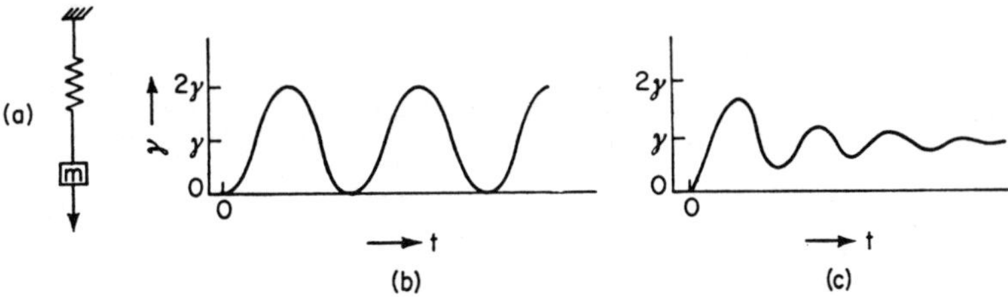

FIG. 14.12.   (a) Spring with inertial mass; (b) without damping; (c) with damping.

state of vibration is created which will be non-uniform throughout the specimen and which will also vary with time.

Let us first consider the simplest case of a spring, and lump the inertia together as a mass hanging from the bottom of the spring instead of distributing it uniformly along the spring (Fig. 14.12(a)). When a stress is now applied the strain is no longer given by Hooke's law $\sigma = G\gamma$ alone but includes an additional term of inertial acceleration, so that the equation becomes

$$\sigma = m\frac{d^2\gamma}{dt^2} + G\gamma \tag{18}$$

m being an inertial mass which is proportional to the density.

If a stress is suddenly applied at zero time, then the strain will not immediately settle into the equilibrium deformation $\gamma = \sigma/G$, but will oscillate between zero and 2γ as shown in Fig. 14.12(b). In actual fact, however, the oscillation is damped out by the internal viscosity of the material, so that the strain–time diagram takes the shape of Fig. 12(a).

Let us now consider a Kelvin element with an additional inertial mass (Fig. 14.13). Obviously, the basic equation used before,

$$\sigma = \eta\frac{d\gamma}{dt} + G\gamma$$

 Polymer Rheology

must be modified by addition of the term of inertial acceleration to give

$$\sigma = m\frac{\mathrm{d}^2\gamma}{\mathrm{d}t^2} + \eta\frac{\gamma}{\mathrm{d}t} + G\gamma \tag{19}$$

When the damping out of oscillations as shown in Fig. 14.12(c) is great, then the specimen attains its equilibrium deformation very rapidly and the inertial character of elasticity may be ignored. This is generally the case with plastics when a constant stress is applied gradually. Not so, however, in

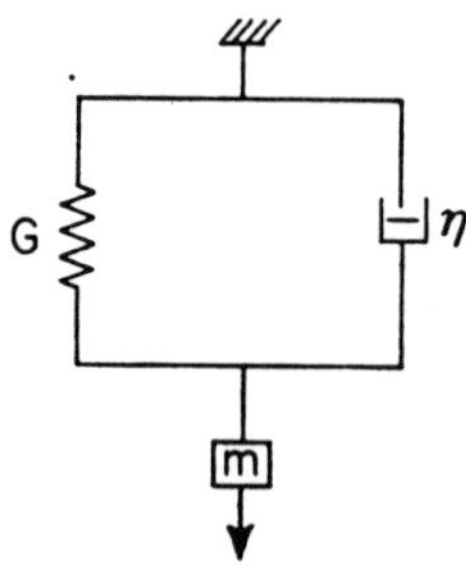

FIG. 14.13.

cases where the stress is applied very rapidly (impact conditions) or where it varies cyclically. When is a stress applied sufficiently rapidly for inertial effects to become important? This is obviously a function of the viscoelastic parameters of the material and on reflection it becomes obvious that relaxation times are decisively involved. Cyclic stresses will receive attention later. The mathematical solution of eqn. (19) is given in ref. 16.

The sudden application of a stress by mechanical means is impossible—the forces are applied at the surface and must therefore set up waves which travel at finite rates. A constant stress distribution is only obtained when the initial waves are damped out. If this takes an appreciable time one must be careful when interpreting stress responses in the initial stages of an experiment. The justification for introducing the subject of inertial and damped inertial elasticity which does not normally figure very importantly in static testing lies precisely in expressing this warning and explaining the reasons which underlie it.

Whilst mechanical stresses cannot be applied instantaneously and whilst they can only be transmitted from the surface, other types of stresses *can* be applied virtually instantaneously and uniformly throughout the entire volume of a test specimen. The most important type of stress which falls into the latter category is that which is due to the application of an electric field. This acts on such dipoles as may be present and distorts them like springs against the various opposing forces present. Mathematically, the differential equations which apply to mechanical systems apply equally to other force fields. Stambough[17] has drawn up a table of analogies between mechanical and electrical stress systems (see Table 14.2).

TABLE 14.2

Mechanical	*Electrical*
Load (gm mass)	Voltage
Strain	Current velocity
Behaviour of dashpot (viscosity η)	Behaviour of resistance (R)
Behaviour of spring (compliance J)	Behaviour of condenser (capacitance C)
Series connection of spring and dashpot (Maxwell)	Parallel connection of condenser and resistance
Parallel connection of spring and dashpot (Kelvin)	Series connection of condenser and resistance
Mechanical energy stored in springs	Electrical energy stored in capacitors
Mechanical energy dissipated in dashpots	Electrical energy dissipated in resistances
Calculation of G in a generalised Maxwell body (tedious)	Analysis of mechanical behaviour using electrical network systems (this is more easily variable to fit experimental curves)

The actual determination of relaxation spectra can be achieved from an evaluation of available relaxation and steady-state flow data.[18]

Starting from the relation between the decaying stress $\sigma(t)$ and the steady-state shear rate $\dot{\gamma}$

$$\sigma(t) = \dot{\gamma} \int_{-\infty}^{\infty} H(\tau)\tau \exp(-t/\tau)\,\mathrm{d}\ln\tau$$

where $H(\tau)$ is the distribution function of relaxation times on a logarithmic time scale; it is necessary to invert this function so as to make $H(\tau)$ the subject. This can be done by approximation methods as shown by Ferry and co-workers[19,20] and leads to a first approximation of

$$H(\tau)\Big|_{\tau=t} = -\frac{1}{\dot{\gamma}t}\frac{\mathrm{d}\sigma(t)}{\mathrm{d}\ln t}$$

The logarithmic time scale spectrum $H(\tau)$ can also be evaluated from steady-state flow data as shown by Faucher[21] who has extended the non-Newtonian flow theory of Eyring and Ree.[22] Faucher, assuming a continuous distribution of flow units, rewrites the Eyring–Ree equation

$$\sigma(\dot{\gamma}) = \int_0^\infty G(\tau)\,\text{arc sinh}\,(\dot{\gamma}\tau)\,d\tau$$

where $G(\tau)$ is the ordinary relaxation spectrum as defined before, namely

$$\tau G(\tau) = H(\tau)$$

Differentiating the Eyring–Ree equation, one obtains

$$\frac{d\sigma}{d\dot{\gamma}} = \int_0^\infty \frac{\tau G(\tau)\,dr}{\sqrt{1 + \dot{\gamma}^2\tau^2}}$$

which can be transposed into two parts, viz.:

$$\frac{d\sigma}{d\dot{\gamma}} = \int_0^{\tau = (1/\dot{\gamma})} \frac{\tau G(\tau)\,d\tau}{\sqrt{1 + \dot{\gamma}^2\tau^2}} + \int_{\tau = (1/\dot{\gamma})}^\infty \frac{\tau G(\tau)\,d\tau}{\sqrt{1 + \dot{\gamma}^2\tau^2}}$$

For small shear rates, i.e. $\dot{\gamma} \leq 1$ the first integral represents the main contribution, and, neglecting the second, one obtains

$$G(\tau)\bigg|_{\tau = (1/\dot{\gamma})} \cong -\dot{\gamma}^3 \frac{d^2\dot{\gamma}}{d\dot{\gamma}^2}$$

or

$$H(\tau)\bigg|_{\tau = (1/\dot{\gamma})} \cong -\dot{\gamma}^2 \frac{d^2\dot{\gamma}}{d\dot{\gamma}^2}$$

On the other hand, if $\dot{\gamma} > 1$, the second integral provides the main contribution, which, as Faucher also shows, leads to

$$G(\tau)\bigg|_{\tau = (1/\dot{\gamma})} \cong \dot{\gamma}^3 \frac{d^2\sigma}{d\dot{\gamma}^2} + \dot{\gamma}^2 \frac{d\sigma}{d\dot{\gamma}}$$

or

$$H(\tau)\bigg|_{\tau = (1/\dot{\gamma})} \cong \dot{\gamma}^2 \frac{d^2\sigma}{d\dot{\gamma}^2} + \dot{\gamma} \frac{d\sigma}{d\dot{\gamma}}$$

Clearly, we have here the 1st and 2nd derivatives of the $\sigma/\dot{\gamma}$ curve which can be determined by graphical differentiation.

Experimentally, it is necessary to follow the stress-relaxation of a melt (say in a cone/plate viscometer) after attainment, maintenance and cessation of a steady-state shear.

In their work on polystyrene and PMMA, Ajroldi *et al.*[18] found that the equation

$$H(\tau)\Big|_{\tau=(1/\dot{\gamma})} \cong -\dot{\gamma}^2 \frac{\mathrm{d}^2\sigma}{\mathrm{d}\dot{\gamma}^2}$$

gave good agreement between experimental and calculated spectra *even for* $\dot{\gamma} > 1$ and that the alternative equation

$$H(\tau)\Big|_{\tau=(1/\dot{\gamma})} \cong \dot{\gamma}^2 \frac{\mathrm{d}^2\sigma}{\mathrm{d}\dot{\gamma}^2} + \dot{\gamma}\frac{\mathrm{d}\sigma}{\mathrm{d}\dot{\gamma}}$$

gave poor results in any case, evidently because the second integral of the equation

$$\frac{\mathrm{d}\sigma}{\mathrm{d}\dot{\gamma}} = \int_0^{\tau=(1/\dot{\gamma})} \cdots + \int_{\tau=(1/\dot{\gamma})}^{\infty} \cdots$$

approaches zero and equation

$$H(\tau)\Big|_{\tau=(1/\dot{\gamma})} \cong -\dot{\gamma}^2 \frac{\mathrm{d}^2\sigma}{\mathrm{d}\dot{\gamma}^2}$$

can be applied to *any* range of times or shear rates.

On the other hand, the Eyring–Ree equation is

$$\eta_a = \sum \frac{x_n}{\alpha_n} \beta_n \frac{\text{arc sinh}\,(\beta_n\dot{\gamma})}{\beta_n\dot{\gamma}}$$

where β_n is proportional to the τ of the nth flow unit, x_n is the frictional area of the nth flow unit, and α_n is the ratio (shear volume)/(2 × kinetic energy) of the nth flow unit. If one assumes β_n to be identical with the relaxation time τ and also assumes a continuous distribution of τ one obtains

$$\eta_a = \int_0^{\infty} \tau G(\tau) \frac{\text{arc sinh}\,(\dot{\gamma})}{\dot{\gamma}\tau}\,\mathrm{d}\tau$$

If we plot (arc sinh x/x) vs. log x it is seen that the point of maximum slope is at $\tau\dot{\gamma} = 2.9$. Now,

$$\frac{\text{arc sinh}\,\dot{\gamma}\tau}{\dot{\gamma}\tau} \begin{cases} = 0 & \text{when } \dot{\gamma}\tau < 2.9 \\ = 1 & \text{when } \dot{\gamma}\tau > 2.9 \end{cases}$$

and by substituting for the inverse hyperbolic sine function one obtains

$$\eta_a \cong \int_0^{\tau = (2 \cdot 9/\dot{\gamma})} \tau C(\tau)\,d\tau$$

from which, on differentiation

$$H(\tau)\Big|_{\tau = (2 \cdot 9/\dot{\gamma})} \cong -\frac{\dot{\gamma}^2}{2 \cdot 9}\frac{d\eta_a}{d\dot{\gamma}}$$

Now this equation can be directly applied to flow data and affords a second means for obtaining $H(\tau)$ which agrees quite well with

$$H(\tau)\Big|_{\tau = (1/\dot{\gamma})} \cong -\dot{\gamma}^2\frac{d^2\sigma}{d\dot{\gamma}^2}$$

both of which are really equivalent by inspection.

One of the most interesting aspects which emerges from this, by the way, is the fact that a relaxation spectrum equation originally derived from solid state viscoelastic considerations yields valid results when applied directly to the liquid (melt) state of polymeric materials, a further powerful corroboration of the concept of the integrity of a generalised fluid state.

For the conclusion of this account nothing could be more suitable than examples which illustrate how a knowledge of moduli can be used to solve design problems. The two examples given are taken from Du Pont technical literature.

Example 1

A cantilever 100 mm long is to hold 4·5 kp under continuous load for one year. The maximum temperature of the ambient air is 40°C and the deflection of the beam must not exceed 3·2 mm. If a rectangular beam of acetal resin is to be used, the modulus has been found to be 10·550 kp cm^{-2}. What are the required dimensions of the beam?

Solution: The standard formula for a stressed beam is

$$\gamma = \frac{\sigma l^3}{3GI}$$

where γ = deflection = 3·2 mm (max.), σ = load = 4·5 kp, l = length of beam = 100 mm, G = modulus = 10·550 kp cm, and I = moment of inertia. Hence $I = 0.445\,\text{cm}^4$.

For a rectangular cross-section

$$I = \frac{bd^3}{12}$$

where b is the width and d is the depth of the beam normal to the load.
If b is 6 mm, then

$$I = \frac{0 \cdot 25 d^3}{12} = 0 \cdot 289$$

whence $d = 20 \cdot 6$ mm. Similarly, it is possible to plot d as a function of b and the customer can take his pick.

Example 2

Determine the bulge to be expected in an aerosol bottle with flat base and cylindrical walls which is to remain stored on a shelf for one year with an internal pressure of 7 kp cm^{-2}. The thickness of the base is 4 mm, the radius 20 mm. Assuming that the equation for the deflection at the centre of a plate with supported edges is

$$\gamma = \frac{3Pr^4(5 - 4\mu - \mu^2)}{16G_y d^3}$$

where $P =$ pressure $= 7$ kp cm^{-2}, $R =$ radius $= 20$ mm, $\mu =$ Poisson's ratio $= 0 \cdot 38$ for the acetal resin intended for use, $G_y =$ apparent modulus at one year $= 11 \cdot 250$ kp cm^{-2} at 40°C, and $d =$ thickness $= 4$ mm. Hence $\gamma \cong 1$ mm.

An actual bottle showed a bulge of 8·64 mm under these conditions. When similar calculations were made for a circular plate fixed at the edges, so that the internal diameter cannot be reduced, then the estimated bulge is 0·228 mm. The actual figure is intermediate because the bottle walls behave in a manner intermediate between the two calculated cases. A bottle designed with an inward bulge of more than 1 mm will therefore not be likely to deform under pressure sufficiently to give a convex bottom which will rock.

REFERENCES

1. R. S. LENK, *Plastics Rheology*, p. 205, Maclaren, London (1968).
2. R. S. LENK, *Plastics Rheology*, p. 209, Maclaren, London (1968).
3. E. WIECHERT, *Ann. Physik*, **50**, 335, 546 (1893).

4. A. SMEKAL, *Z. Phys. Chem.*, **B44,** 286 (1939).
5. T. ALFREY, *Mechanical Behaviour of High Polymers*, High Polymer Series, Wiley Interscience, New York (1948).
6. T. ALFREY and P. DOTY, *J. Appl. Phys.*, **16,** 700 (1945).
7. E. PASSAGLIA and J. R. KNOX, in *Engineering Design for Plastics*, p. 162, E. Baer (editor), Reinhold, New York (1964).
8. R. D. ANDREWS, *Ind. Eng. Chem.*, **44**(4), 707 (1952).
9. H. LEADERMAN, in Eirich: *Rheology*, Vol. III, Acad. Press (1960).
10. J. D. FERRY, *Viscoelastic Properties of Polymers*, John Wiley, New York (1961).
11. S. TURNER, *Trans. & J. Plast. Inst.*, **34**(11), 127 (June 1966).
12. A. V. TOBOLSKY and H. EYRING, *J. Chem. Phys.*, **11,** 125 (1943) and others subsequently.
13. S. TURNER, *Brit. Plastics*, **37**(6), 322 (June 1964).
14. C. M. R. DUNN, W. H. MILLS and S. TURNER, *Brit. Plastics* (July 1964).
15. H. J. SCHERR and W. E. PALM, *J. Appl. Poly. Sci.*, **7,** 1273 (1963).
16. R. S. LENK, *Plastics Rheology*, p. 200, Maclaren, London (1968).
17. R. B. STAMBOUGH, *Ind. Eng. Chem.*, **44,** 1590 (1952).
18. G. AJROLDI, C. GARBUGLIO and G. PEZZIN, *J. Poly. Sci.*, A-2, **5,** 289 (1967).
19. F. W. SCHREMP, J. D. FERRY and W. W. EVANS, *J. Appl. Phys.*, **22,** 711 (1951).
20. J. D. FERRY and M. L. WILLIAMS, *J. Colloid Sci.*, **7,** 347 (1952).
21. J. A. FAUCHER, *J. Appl. Phys.*, **32,** 2336 (1961).
22. H. EYRING and T. REE, *J. Appl. Phys.*, **26,** 793 (1955).

15

Deformation in the Solid State—Large Strains

Just as we have a generalised flow curve for the liquid state (see earlier), so there exists a generalised deformation curve for the solid state which shows comparable features. Such a stress–strain curve is not *directly* obtainable from tensile measurements since the stress is expressed in force per unit area of the *original* cross-section of the specimen and the cross-section changes continuously. This results in a curve as shown in Fig. 15.1(b). In order to correct the 'conventional' stress of Fig. 15.1(b), it is necessary to record the change in cross-sectional area during the experiment and express the stress as the force per unit area of the *actual* cross-section for any given strain. This is the *true stress* and the resulting true stress–strain curve is shown in Fig. 15.1(a).

Only the linear region of small scale deformations was treated in the modulus considerations in Chapter 14. Modulus, like viscosity, has its proportionality limit set by the position of point 'A' and to determine apparent moduli as the slope of the line drawn from the origin to a point beyond 'A' is, strictly speaking, just as meaningless as to determine the apparent viscosity from a flow curve by the analogous procedure. However, since the proportionality limit is quite low in some materials, engineers sometimes go slightly beyond point A' so long as the apparent modulus is not less than 85% of the initial modulus, and use the corresponding deformation in design calculations.

Whilst small scale deformations are characterised by the modulus which is constant up to the proportionality limit, the modulus changes at larger deformations and it is obviously essential to know the previous stress history of a component under test. No mouldings, extrusions, or calendered products are ever free from built-in stresses unless they are carefully annealed and processing conditions can therefore cause great differences in products made from the same material. Assuming, however, that the stress–strain history has been obliterated by annealing and that it has thus been returned to its ground state, then it should be

193

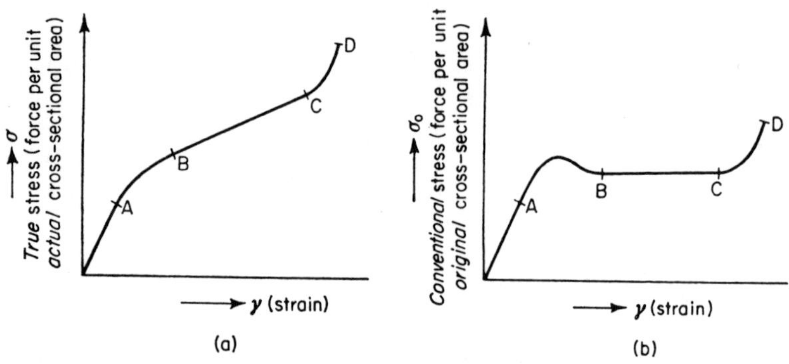

FIG. 15.1.   Stress–strain diagrams.

possible to lay down the conditions for strain softening, necking, cold-drawing and strain hardening, as has been shown by Vincent.[1]

Let the conventional stress σ_0 be defined by L/A_o, where L is the load and A_o is the original cross-sectional area; let the true stress $\sigma = L/A$, where A is the actual cross-sectional area at the time of applying load L.

Let l be the length of a tensile specimen under load L, l_o the original length and $l/l_o = R$, the 'elongation ratio'.

If σ_0 and σ are both plotted as a function of strain defined by

$$f(\gamma) = \frac{l - l_o}{l_0} = R - 1$$

then two stress–strain curves are obtained as shown in Fig. 15.2.

If σ_0 falls below the arrowed maximum then the specimen begins to neck. Assuming incompressibility and constant geometrical shape, the

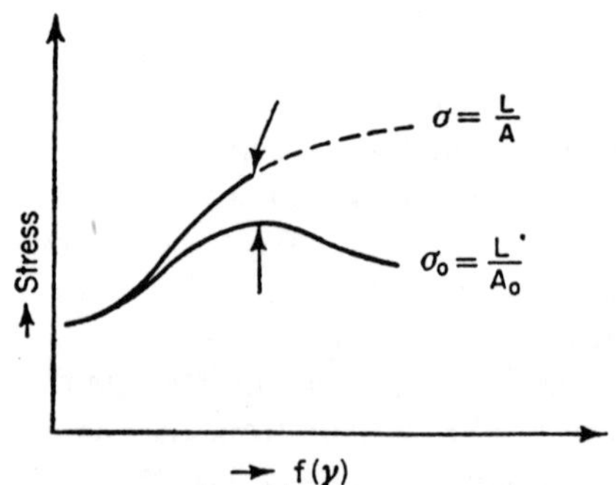

FIG. 15.2.   True stress and 'conventional' stress.

volume remains constant during the experiment and also remains definable by

$$Al = A_o l_o$$

This can be written

$$\frac{A_o}{A} = \frac{l}{l_o}$$

which is identical with the 'elongation ratio' R.

Hence $A = A_o/R$, and substituting for A we get

$$\sigma = \frac{L}{A} = \frac{LR}{A_o} = \sigma_0 R$$

so that σ, the true stress, is readily accessible up to the point where necking starts. Necking is a shape discontinuity characterised by a change of the specimen boundary surfaces from the condition of parallelism with the direction of the force (Fig. 15.3).

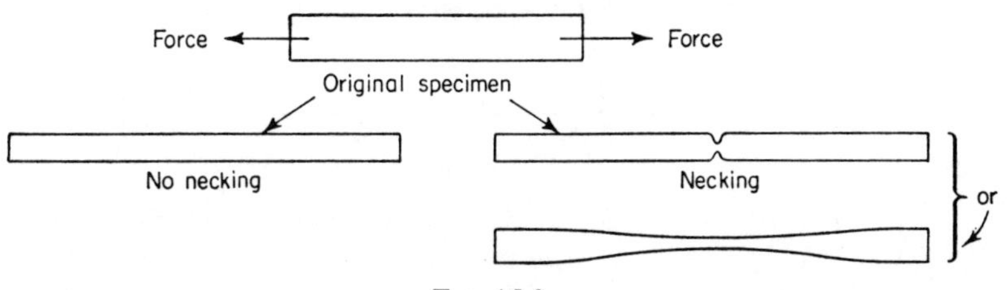

Fig. 15.3.

Once necking has started, the relationship $\sigma = R\sigma_0$ can no longer be used, because the prismatic or cylindrical shape assumed in deriving the relationship is no longer in existence. If the curve for σ in Fig. 15.2 were to be extended, one would have to obtain the true stress from the force and the dimensions of the thinnest part of the neck (broken continuation in Fig. 15.2).

Up to the point of necking (arrowed in Fig. 15.2) the relationship $\sigma_0 = \sigma/R$ holds. The point of necking is defined by the maximum of the σ_0 curve, where the slope of the tangent $d\sigma_0/dR$ is equal to zero. We can therefore differentiate σ_0 with respect to R and set to zero:

$$\frac{d\sigma_0}{dR} = \frac{1}{R}\frac{d\sigma}{dR} - \frac{\sigma}{R^2} = 0$$

whence

$$\frac{\mathrm{d}\sigma}{\mathrm{d}R} = \frac{\sigma}{R}$$

and this we also know to be equal to σ_0 (see before).

Therefore, if the true stress σ is plotted vs. R, the point where necking starts is that point on the curve where the slope of the plot is equal to σ_0 (the conventional stress). This point may be found by drawing a tangent to the curve from the origin ($\sigma = 0$, $R = 0$) and this is known as 'Considère's construction' (Fig. 15.4).

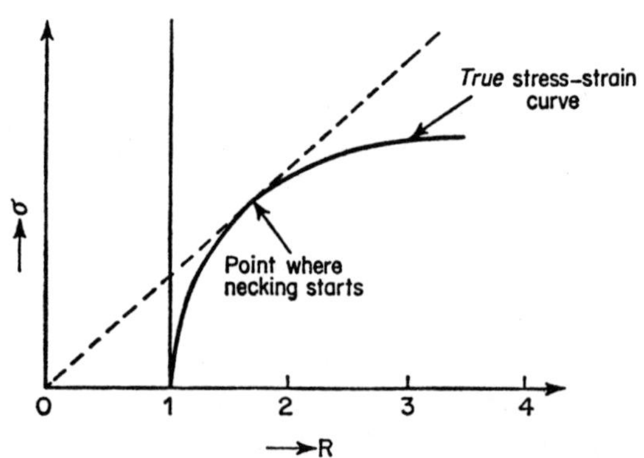

FIG. 15.4.   Considère's construction (see also Fig. 15.5).

The actual place where the neck forms is a place of weakness in the specimen which is always present because of the impossibility of making the specimen perfectly uniform along its entire length. The stress will therefore be greater at one spot than anywhere else. In the initial Hookean region of the stress–strain curve where the curve rises steeply this non-uniformity does not yet cause a neck to appear because the slight extra strain can be supported in the vulnerable spot without producing an excessive stress. But as the modulus decreases beyond the proportionality limit, additional stresses produce progressively greater strains and this will tend to make the vulnerable spot thinner than the rest of the specimen. Eventually the system becomes unstable and a neck is formed. Strain softening, having started at the point of departure from Hookean proportionality, culminates in producing maximum instability in the system at the point given by Considère's construction.

When a neck has been formed one of two things can happen on further application of stress:

(a) *Either* the neck becomes progressively thinner and eventually the specimen breaks;

(b) *Or* the neck may stabilise itself to a constant cross-sectional area as the shoulders travel along the specimen until they have 'consumed' all available undrawn material. During this process the nominal load remains constant. This is called *cold-drawing*.

The neck is finally stabilised after cold-drawing is complete. At that stage the flow units will all be oriented in the direction of stress and the system is once again stable. The stress required to produce further strain increases becomes increasingly great, the curve once again sweeps upward and does so increasingly steeply. At this stage the applied force, having overcome the secondary cohesive forces of the original specimen earlier on, comes up against primary forces (molecular bonds, bond angles, hydrogen bonds, tightened entanglements). At the moment when instability would once again start, the high stress cannot be sustained by a vulnerable spot in the same way as it could initially, because no reserve mechanism is left for this to happen, and rupture occurs. The point where the neck regains maximum stability can be determined by an extension of Considère's construction, as Vincent[1] has pointed out (Fig. 15.5).

The region between C and D is clearly a region of 'solid flow' during which maximum instability is maintained, whilst between D and E gradual

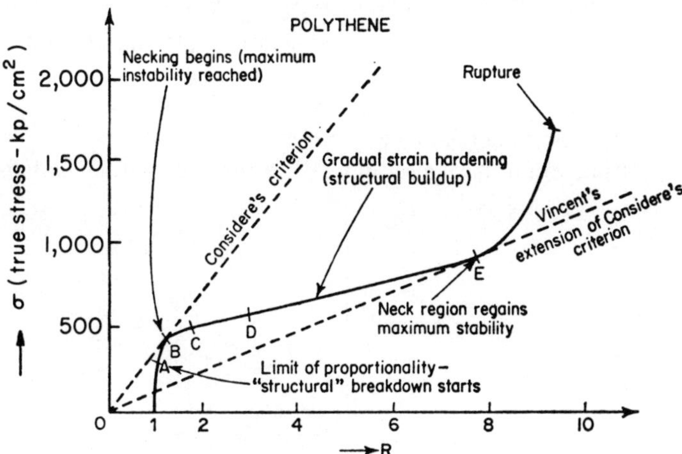

FIG. 15.5.　Considère's construction, including Vincent's extension.

strain hardening (structural build-up) takes place—the exact reverse of the gradual strain softening (structural breakdown) which had previously taken place between A and B. Point B lies in what corresponds to the pseudoplastic region and point E lies in what corresponds to the dilatant region of the generalised liquid flow curve.

We shall now leave the true stress curve and confine ourselves to conventional stress curves which are directly and therefore most readily determinable.

Not all thermoplastic polymers are able to stabilise the neck and whether the neck becomes thinner and breaks or stabilises itself depends not only on the polymer and other compounding ingredients, but also on the molecular weight of any one polymer itself. If the flow units are too small to form entanglements or can only form few and loose entanglements which easily slip apart in the 'solid flow' region, then rupture will occur at relatively low stresses because the stresses required for strain hardening cannot be supported. The amount of strain hardening decreases with the molecular weight of the flow units as can be seen qualitatively even from the conventional stress–strain curves in Fig. 15.6, which were obtained for polythenes of different melt flow indexes.

The highest molecular weight material (lowest MFI) shows the full extent of cold drawing, medium molecular weight material shows much less and low molecular weight material shows none at all.

It is true that thermoplastics, when tested under arbitrary conditions do not necessarily show the full development of the generalised stress–strain curve. The curve may stop short because of rupture:

(a)   whilst it is still within the Hookean proportionality limit;
(b)   in the region of increasing strain softening;
(c)   at the point where necking begins;
(d)   during necking (as in Fig. 16.6(c));
(e)   during strain hardening but before the neck is fully stabilised.

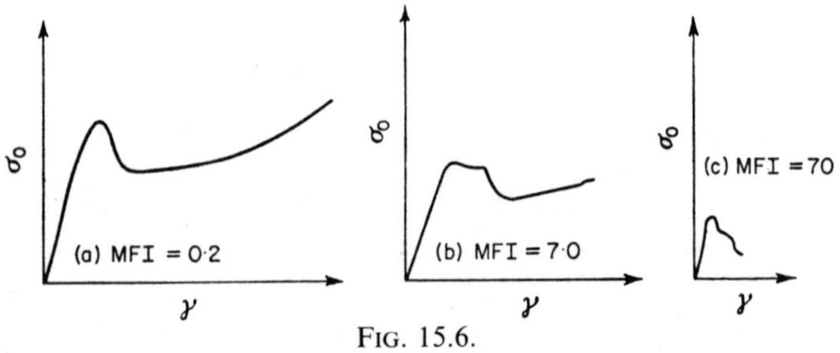

Fig. 15.6.

In typical thermoset materials it is never possible to exceed the proportionality limit sufficiently to obtain necking because, owing to the density of cross-linkages, viscous shear flow cannot get under way to such as extent as to dominate the stress response. But in lightly cross-linked thermosets (which are untypical plastics) and in many rubbers elastoviscous responses of the Voigt model type (see Chapter 14) may involve retarded dashpot flow. With filled phenolic and aminoplastic moulded specimens failure generally occurs within the proportionality limit and involves little deformation up to the rather high stresses necessary to produce rupture.

In thermoplastics, on the other hand, the stress–strain curve can take any form derived from the generalised curve; it should, however, be added that curves which fall progressively shorter of full development through premature rupture of the specimen are also progressively depressed towards the strain axis and in the extreme case the maximum itself will degenerate into an apparently insignificant knee. The actual curve obtained will depend on the severity of the experimental conditions (Fig. 15.7). In (j) no flow is observed, the response appears to be ideally elastic. In (h) some flow (shear) is indicated by derivation from linearity just before rupture, whilst in (g) dominance of flow deformation is just reached. Up to this point a material is said to fail in a *brittle* manner. From (f) to (c) permanent deformation involves progressively greater deformation and smaller stress at failure. Such materials are said to fail in a *tough* manner. The case of (c) deserves further comment. Drawn polymers such as nylon or polyester fibre, monofilament or film have had a previous stress history when they are being tested. What appears to be the starting point of the test (dotted coordinates) is not the true starting point at all. If the curvature of the stress–strain curve of, say, a drawn nylon monofilament shows a curvature 'the wrong way round' and fails in a brittle manner, this is entirely due to the fact that, after stress hardening, shear flow has ceased to be the dominant response and tension has reasserted its dominance. In (b) and finally in (a) the divide between tensile elastic and viscous flow has become rather smudged, the viscous forces tend to dominate and the deformation at failure may not be of a high order—the rubbery state has been reached.

Clearly, if a material fails in a brittle manner it fails in tension, and if it fails in a tough manner it fails in shear. The probability of brittle failure therefore increases with the severity of the test.

Tests can be made more severe in a number of ways:

(1) The free volume can be decreased so that the viscous shear-deformational response to an applied stress becomes increasingly difficult.

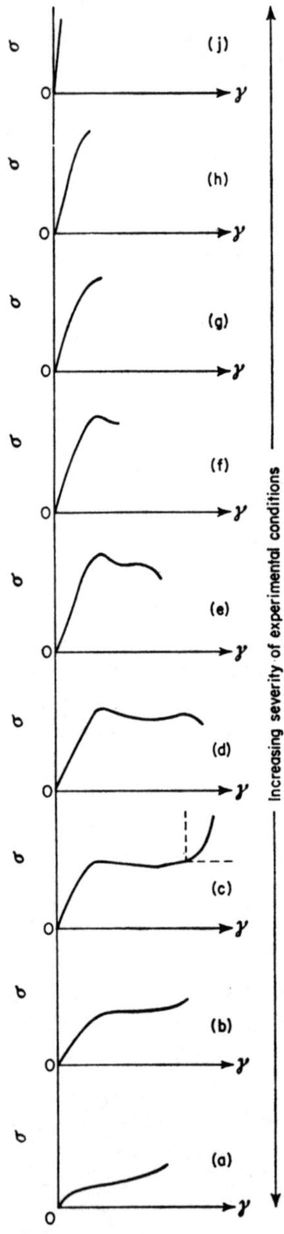

FIG. 15.7.   Shapes of stress–strain curves as a function of the severity of testing conditions.

This is simply achieved by reducing the temperature, or by reducing the amount of plasticising additives (including low molecular weight fractions) in the compound.

(2) The time available for the flow units to get the viscous deformational process under way and so to dissipate the imparted stress energy may be reduced so that the only response then open to the flow units is a tensile response. This is fairly readily done by increasing the rate of test, in the extreme case up to impact conditions.

(3) The applied stress can be chosen so that its tensile component is increased at the expense of the shear component. Indeed, if an exclusive triaxial tensile stress could be applied to a specimen (which is practically impossible), then the specimen would fail in a brittle manner whatever the state of the other variables. Triaxiality can be approached by notching the specimen increasingly sharply, as will be seen later.

(4) The specimen may be given a previous stress history which has already traversed the region of flow instability so that the specimen has regained stability to a greater or lesser extent. Alternatively the specimen may contain stored energy in the form of frozen-in stresses which will make a contribution equivalent not only to the true total stress but which will also accelerate the rate of stress application when additional stresses are applied. This, clearly, is a function of processing conditions and is absent in properly annealed samples in which the stresses have been fully dissipated by allowing corresponding strains to develop.

(5) The free volume could also conceivably be reduced by the application of very high pressures. It is likely that a tensile specimen tested whilst immersed in a hydraulic liquid under high pressure will fail in a brittle manner when the same specimen will fail in a tough manner at atmospheric pressure but at otherwise equal conditions.

(6) The cross-linkage of polymer chains, whether achieved chemically (as in the case of unsaturated polyesters) or by irradiation, will produce a network. The cross-link density of this network will determine the degree to which solid flow under stress is restricted. The greater the flow restriction, the less stress can be dissipated and the greater will be the relative severity of a tensile test, so that brittle failure becomes increasingly probable.

(7) The chemical nature of the polymer chains themselves can be modified either directly or by modification of the monomer, say by inclusion of comonomer to the monomer feed prior to polymerisation. This will promote tough rather than brittle failure. For example:

(i) The chlorosulphonation of polythene to 'Hypalon' increases free volume and destroys crystallinity. It also gives subsequent cross-linking possibilities with the result that a material is obtained which is a rubber at ordinary temperatures and which is 'tougher' than the original polythene.

(ii) Copolymerisation, e.g. of vinylidene chloride with butyl acrylate or of ethylene with vinyl acetate. This also destroys—at least partially—the crystallinity which is characteristic for poly(vinylidene chloride) and polythene respectively. The result in the case of the vinyl polymer is a corresponding increase in free volume and the conversion of an otherwise brittle, rather insoluble and thermally intractable polymer to a copolymer which is tough, tends to be elastomeric, is soluble in suitable solvents and—most important—which is thermally processable. This is known as 'internal plasticisation'. Many similar examples could be given. External plasticisation—as in flexible PVC—has the same effect, although it is physically rather than chemically induced.

Apart from the obvious effect of deliberate notching one may consider other situations where defects are most probably the root cause of brittle failure in otherwise tough materials:

(1) The fact that polymers are, after all, molecular chains, implies that the inter-chain regions and especially chain ends constitute regions of weakness compared to the chains themselves. The regions of weakness at chain synapses may in fact be regarded as micro-defects on a molecular scale. These may be exaggerated in environments which are capable of further interfering with the cohesive forces at molecular interfaces and synapses, so that actual micro-cracks or crazes are developed during 'environmental stress cracking' (ESC). Polymers which consist pre-dominantly of crystalline material will be less prone to ESC than those with substantial amounts of amorphous phase and this is borne out when the more highly crystalline linear (high density) polythenes are compared to the less crystalline (low density) polythenes. The latter are highly susceptible to surface active agents whilst the former are virtually immune to ESC in the presence of surface active agents. Again, polycarbonates (which are essentially amorphous) are highly susceptible to ESC by vapours of hydrocarbon solvents. It does not follow that *all* amorphous polymers are susceptible to ESC. PVC, for example, is remarkably resistant to ESC. But it would be fair to say that ESC would be likely to occur in amorphous rather than crystalline material if it occurs at all. If the length of flow units becomes so small, however, that each unit consists of a whole molecule the chain length of which is sufficiently small for that whole molecule to form a discrete unit of a multi-molecular crystal, then the synapses will cease to be

points of attack for ESC, because there will then be little if any amorphous phase left. This is the case in *very* low molecular weight polythenes ('polythene waxes') and, in the extreme case, in paraffin wax. The increase in crystallinity—as the molecular weight reduces to that of paraffin wax—also reduces the probability of solid flow (viscous deformation) and the material will again fail in a brittle rather than in a tough manner. Despite the fact that *intermediate* molecular weight polythenes have more amorphous phase than very low or very high molecular weight polythenes, the expected increase in toughness is more than counterbalanced by their relatively high susceptibility to ESC and consequent embrittlement through micro-defects.

(2) The total energy required to produce brittle failure is really made up of two components: a crack *initiation* energy and a crack *propagation* energy. A material will always fail in the manner which corresponds to the lowest energy necessary to produce failure. If the energy necessary to produce brittle failure is higher than the energy necessary to produce tough failure (i.e. failure by permanent viscous flow deformation), then the specimen will fail in a tough manner, and vice versa. This provides a key for 'toughening' otherwise brittle materials. Crack initiation can be reduced by addition of plasticisers; crack propagation can be reduced by arresting the crack after it has formed. The former is achieved, for example, by addition of plasticiser to rigid PVC, the latter by the inclusion of a highly elastic

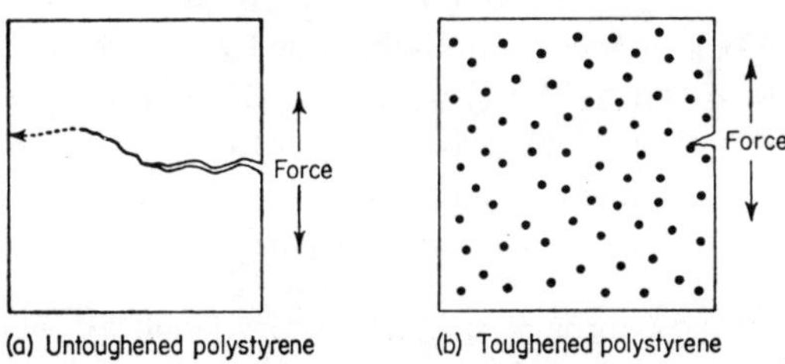

FIG. 15.8.

disperse phase as for example in toughened polystyrene (Fig. 15.8). In Fig. 15.8(a) the force is providing sufficient energy to initiate a crack and is propagating that crack across the specimen which is about to fail in a brittle manner. In Fig. 15.8(*b*) the same energy has merely initiated the crack but finely dispersed rubber particles are absorbing and dissipating the remaining energy, so that the specimen cannot fail in a brittle manner under the same conditions. The brittle strength has been raised above the tough strength and when failure does occur it will therefore be tough failure. The efficiency of toughening polystyrene in this way depends on:

(i) the probability of the propagating crack meeting a crack-arresting rubber particle; that is to say, it depends on the amount of rubber present and on the fineness of dispersion;

(ii) the intimacy of contact and the degree of adhesion of the interfaces between the phases so that the approaching crack is not deflected away from the energy-absorbing rubber particles by physical approach barriers such as voids;

(iii) the mass of each dispersed particle, since the energy absorbing capacity is necessarily a function of the mass of the particle. There is an optimum dispersion of such a fineness that the probability of a propagating crack being intercepted is high and the energy absorbing capacity for arresting the crack on interception is adequate. This optimum is experimentally determinable and it has been proved that 'over-fine' dispersion reduces the effective toughening obtainable from a constant weight proportion of disperse phase.

Unfortunately polystyrene loses its optical clarity by the inclusion of a disperse phase. The drilling of articles, of course, constitutes notching and one would very much like to toughen the impact-brittle drilled acrylic windscreen of a motorcycle which invariably cracks from the drill-hole when the machine is dropped. Unfortunately the inclusion of dispersed rubber is not feasible—although this might well achieve the object of overcoming shock brittleness in the notched windscreen—for the obvious reason that the excellent clarity which, in addition to its good weathering properties, led to the use of acrylic sheet for this application, would be lost.

(3) The inclusion of fillers generally promotes embrittlement because it *also* constitutes notching of a kind. On the other hand:

(i) if fillers have compensating properties (such as energy absorption potential, as in dispersed rubbers, or the spreading of stresses over larger volumes, as in fibrous fillers),

(ii) or if they have specific chemical or physical interactions with the polymer (as is sometimes the case with special clays, silicas and carbons) which may outweigh the intrinsic notch embrittlement due to their inclusion, then these fillers may promote toughness in the net result where embrittlement would otherwise have been the dominant effect.

It is now necessary to give a number of definitions which will enable us to make a more detailed examination of stress–strain curves:

Modulus: the slope of the tangent to the curve.

Yield point: that point on the conventional stress–strain curve where necking starts. This is the maximum of the curve where the modulus (tangent to the curve) is zero.

Yield strength: the stress per unit area at yield.

Brittle strength: the stress per unit area at break if the material breaks before it yields, i.e. if it fails in a brittle manner.

Ultimate strength: the stress per unit area at rupture (identical with the brittle strength in brittle materials, but different from the yield strength in tough materials).

Energy at yield: the area under the stress–strain curve ($\int \sigma \, d\gamma$) up to the yield point.

Energy at break: the total area under the stress–strain curve ($\int \sigma \, d\gamma$) from its origin up to rupture.

These are all parameters of the stress–strain curve. In addition there are the complementary parameters of *elongation at yield* and *elongation at break* which define themselves and which complete the list.

It is important to note that while the initial modulus is of fundamental significance (analogous to the 1st Newtonian viscosity in liquids), it is only a function of stiffness (rigidity, resistance to flow deformation) and cannot be a function of strength. This is illustrated in Fig. 15.9. Material 'a' has the highest initial modulus. It is the stiffest material. But whilst 'b' is much less stiff than 'a' it is obviously stronger because the stress at break is greater. Both 'a' and 'b' are brittle, but 'c' which is tough is also stronger than both 'a' and 'b', intermediate in stiffness (initial modulus) between 'a' and 'b'.

Obviously, the initial modulus (or for that matter, any modulus taken from the curve) has no connection with either the strength or the type of failure of the material. This is why static modulus determinations which are

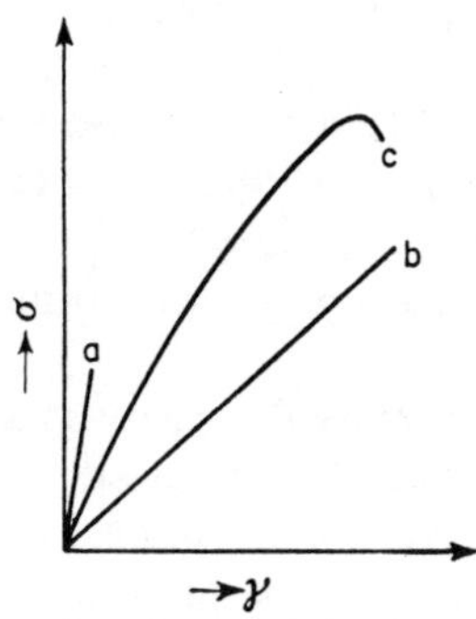

Fig. 15.9. Illustration showing that modulus is not a *strength* parameter.

so useful in characterising thermal transitions and give so much structural information are of rather limited interest to the *design engineer* who is concerned with the stresses which a component can stand before it becomes unserviceable and who may perhaps also wish to know what deformation is involved should the material be tough.

A design engineer is only interested in *one* strength parameter. If the material fails in a brittle manner the brittle strength is the only available strength parameter anyway. But it is pointless to give the second (ultimate) strength in a material which is tough and has already yielded. Nobody will be interested in the ultimate strength of a PVC raincoat after it has already been pulled out of shape to such an extent that the owner cannot wear it any more. But the stress which it will stand before it yields obviously constitutes useful information. The only exception to this is for drawn fibre, film and monofilament. Here the previous stress history (including the yield point) is ignored and the ultimate strength is quoted—but the drawn material can obviously be regarded as a qualitatively different species from the undrawn material.

In order to determine the strength of a product one must test it to destruction—to rupture in a brittle material and to yield in a tough material. This means that if one were to guarantee the strength of a component, there would be none to sell. The alternative is to test randomly drawn samples under simulated use (or abuse) conditions, apply a suitable safety factor and state the confidence in satisfactory performance in probability terms. The reader is referred to standard texts on quality control.

The design engineer has an additional problem: whilst data on strength parameters of materials prepared according to standard specifications abound, these may give an entirely misleading picture when they are applied to the actual component because:

(1) The component will be different in size and shape.
(2) It could conceivably have greater stiffness designed into it by suitable changes in its geometry.
(3) It may have internal corners constituting notching to varying degrees of severity.
(4) The level of residual stresses may be very different from those which a material test specimen moulded under arbitrary standard conditions may have. Furthermore, the distribution of residual stresses is almost always non-uniform.
(5) The service temperature conditions may be different.

(6)	The stress levels which the component should withstand without failure may be very different.

(7)	If cyclic stresses are involved their frequency may be very different.

(8)	The stress system (to some extent idealised in standard tests) may be different in either type or complexity.

As a result it is futile to base a product design on materials tests unless all the variables are fully covered and this is such a stupendous task that it could not be undertaken even for a single polymer type—remembering that each type comes in many grades of molecular weights, molecular weight distribution, fillers, plasticisers or other modifying additives—let alone for the whole spectrum of plastics. The best that can be done is to draw intelligent inferences from carefully chosen materials tests, select the possible materials, make the component and test it under actual service conditions in the hope that some correlation of ranking will exist between materials and product tests.

If this sounds like a counsel of despair, this is not altogether so—it is merely a warning against possible pitfalls and premature conclusions. In particular, no single test can characterise the strength of a plastic. Having accepted that, what *can* be done?

If the severity of a test can be changed by changing one of several experimental conditions while keeping all the others constant, then it is clear that identical stress–strain curves can be obtained in a variety of ways. This has some important implications: an impact test can only tell us qualitatively whether a material is shock brittle or not, and if the material *is* shock brittle then the impact test will also give the energy at break. Even that energy is rather inaccurate, however, since it makes no allowance for kinetic energy, nor for the fact that the clamping jaws absorb energy. Nevertheless, components are very commonly subjected to impact stresses during their functional life. In the short moment of impact it is obviously difficult to obtain a stress–strain curve. But if the same degree of severity of test can be achieved by reducing the temperature, then the test need not be carried out under impact conditions at all; instead a curve obtained over a time interval which can be fully scanned for stress–strain relationships and which will give a result which is fully valid for predicting the behaviour of the material or component under impact conditions can be used. All that needs to be known additionally is what decrease in temperature is equivalent to a corresponding increase in the rate of stress application. In the same way notch embrittlement can be related to the embrittlement caused by an equivalent temperature reduction or by an increase in the rate of stress application. The *brittle point* can be expressed as the temperature at which the application of a given stress at a given rate (other variables also being constant) will produce a stress–strain curve which just terminates at the maximum, that is to say, at the tough/brittle transition, as in Fig. 15.7(g). It can also be expressed in terms of stress application rate, degree of notching, etc., and thus acts as a watershed for the macroscopic type of

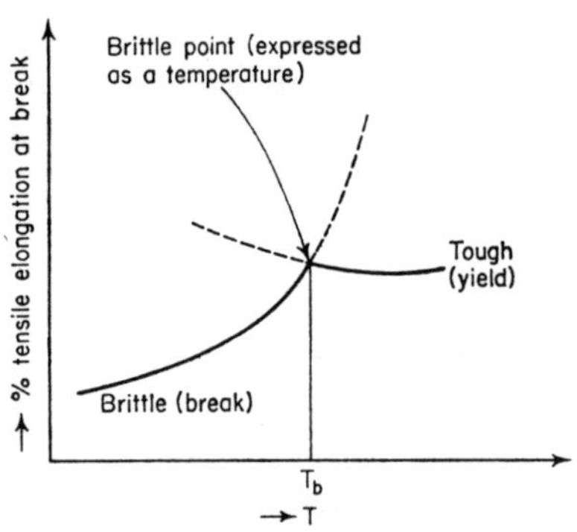

FIG. 15.10.   Discontinuity at the 'brittle point' (deformation vs. temperature).

failure. Indeed, the major problem in the study of the strength of plastics is to measure, understand and explain the trends of brittle strength whilst changing the temperature and other experimental variables one at a time.

A good way of exploring the failure characteristics of a material consists in plotting the strength parameters of stress–strain curves, such as elongation at yield and break, stress at yield and break, and energy at yield and break against the temperature or against the rate of stress application. Both yield and break will follow separate curves which cross over at the brittle point T_b (Fig. 15.10). At *that* point the probability of tough and brittle failure are exactly equal. Below T_b the probability of brittle failure will increase as the two curves (tough curve now broken) diverge. Above T_b the probability of tough failure will increase analogously. Around T_b there will always be some specimens out of a suitably large number tested which fail in the untypical manner at much higher stresses and strains. The absolute trend of the curves is immaterial in this respect, the only thing that matters is their degree of divergence. Thus, while the elongation to *yield* in amorphous polymers (e.g. PMMA—polymethyl methacrylate) decreases along its observed course, that of crystalline polymers (e.g. PP—polypropylene) increases through most of its course (Fig. 15.11). But this does not affect the principal argument of tough/brittle transition.

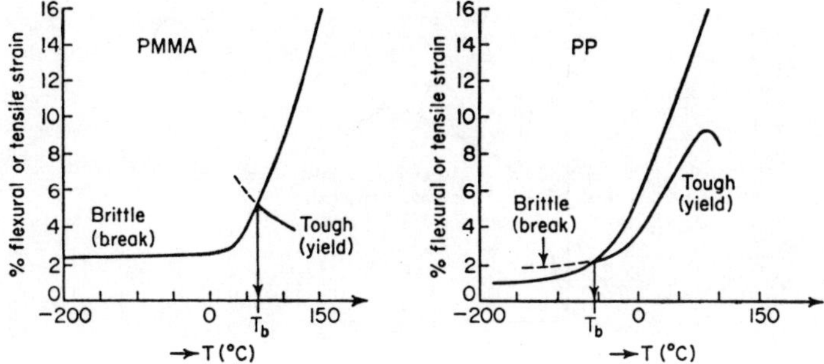

FIG. 15.11. Discontinuity at the 'brittle point' (deformation vs. temperature—typical thermopiastics).

The stress at break and yield also show a similar bifurcation which indicates the brittle point (Fig. 15.12) and so does the energy at yield and break (Fig. 15.13).

An exactly similar picture emerges when the *straining rate* is changed instead of the temperature. The tough/brittle transition will then be expressed in reciprocal seconds and the symbol $\dot{\gamma}_b$ for the transition will take the place of the symbol T_b.

Some materials have more than one apparent yield point (from the point of view of serviceability). In such a case (Fig. 15.14) the *lower* yield point is the relevant one for design considerations.

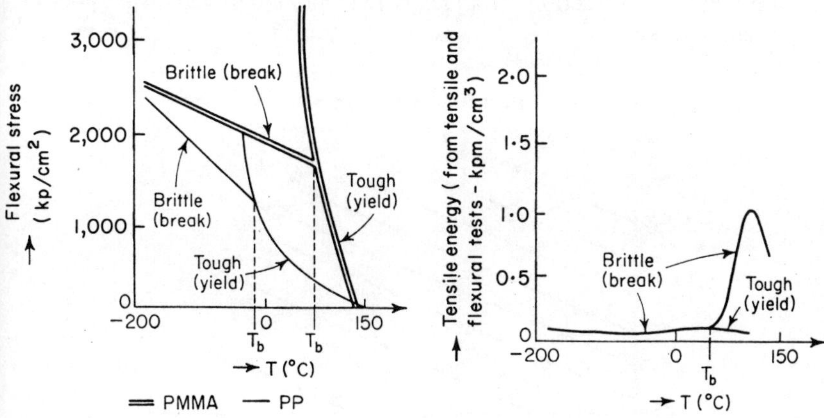

FIG. 15.12. Discontinuity at the 'brittle point' (stress vs. temperature—typical thermoplastics).

FIG. 15.13. Discontinuity at the 'brittle point' (energy vs. temperature—typical thermoplastics).

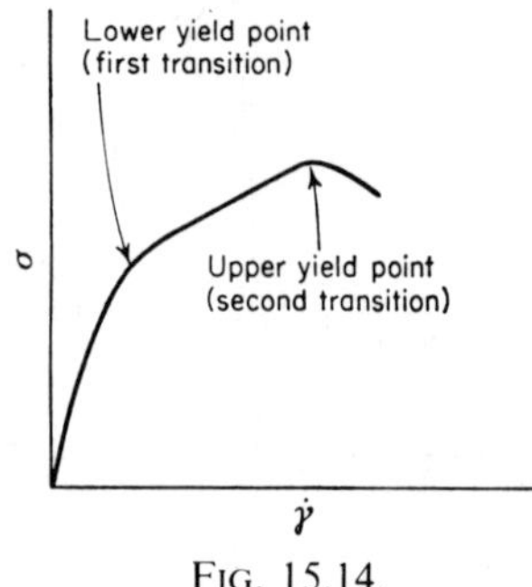

Fig. 15.14.

If the tensile straining rate $\dot{\gamma}$ in log form is plotted vs. temperature,[2] then a family of parallel straight lines is obtained each of which refers to a constant yield stress (Fig. 15.15).

Results on rigid PVC in tension at seven different straining rates and six different temperatures produced curves from which the relationship was deduced for the material.

In laboratory tests specimens are usually subjected to 'simple' tests such as tension. In service the type of stress which puts the component at risk may be more complex. Obviously, one cannot test all materials for all possible stress systems, so one has to study the relationship between test results from 'simple' stresses with those obtained from complex stresses. The word 'simple' has been put into inverted commas because, as will be seen presently, considerable complexity may be hidden under the cloak of apparent simplicity.

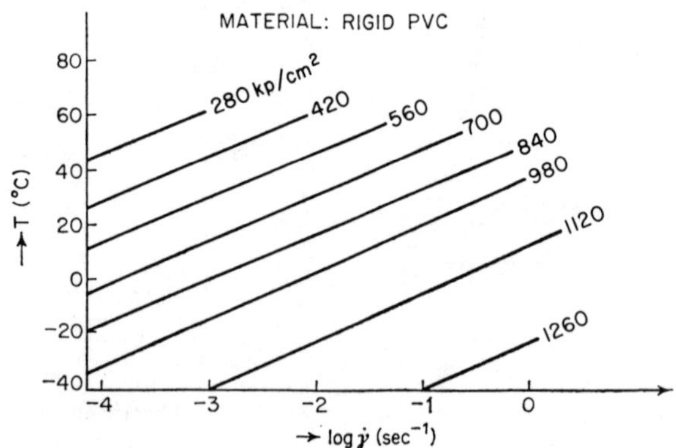

Fig. 15.15.   Straining rate vs. temperature at different loads (PVC).

The question which must be answered is: What type of stress systems are likely to promote tough and brittle failure respectively? The stress at any point in a loaded solid can be reduced to three-component tensile or compressive stresses. To simplify matters, let us assume that one of these three components is zero, so that we are left with two tensor components, σ_1 and σ_2. If a positive sign denotes tension and a negative sign compression, then it is clear that a two-component stress system can be devised in five ways:

(1) uniaxial compression $(-\sigma, 0)$;
(2) uniaxial tension $(+\sigma, 0)$;
(3) biaxial tension $(+\sigma, +\sigma)$;
(4) biaxial compression $(-\sigma, -\sigma)$;
(5) shear $(+\sigma/2, -\sigma/2)$.

We can ignore *biaxial compression* because it is a somewhat contrived rather than a naturally occurring stress system. The other four, one would imagine, can be simulated quite easily in the laboratory: *Uniaxial compression* by compression of a specimen between two parallel plates; *Uniaxial tension* in a tensile test; *Biaxial tension* in a clamped membrane which is stressed at one point; *Shear* by applying a uniaxial tensile stress at an angle of 45° (Fig. 15.16[3]). Note that a shear stress is equivalent to a combination of a tensile stress and a numerically equal compressive stress at right angles to each other. The BS shear test for plastics which specifies the punching out of a disc from a sheet specimen falls far short of the application of 'pure shear'. However, we know that in a uniaxial tensile test

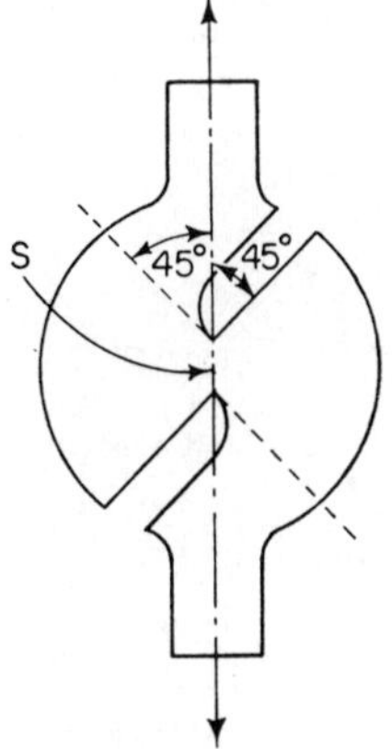

FIG. 15.16.   Shear test specimen suggested by Goldenberg *et al.*[3]

the specimen is simultaneously compressed in a lateral plane. Conversely, the skin of a specimen under uniaxial compression is simultaneously strained in tension.

Photoelastic patterns (strain lines in polarised light) showed clearly that the stress concentration was greatest in the region S (Fig. 15.16) and that the strain lines intersected the axis of locus S at an angle of 45°. Reasonably pure shear was thus ensured and the shear stress was, moreover, approximately constant over the cross-sectional area at S.

However, we know that in a uniaxial tensile test the specimen is simultaneously compressed in a lateral plane. Conversely, the skin of a specimen under uniaxial compression is simultaneously strained in tension.

It is clear that:

either: shear is a composite system consisting of a uniaxial tensile and compressive component;

or: uniaxial tension (compression) is itself a complex system consisting of a shear and a biaxial tensile (compressive) component.

This can be graphically and mathematically expressed as follows:[4]

In Fig. 15.17(a) uniaxial tensile components were taken as the primary coordinates, whilst in 15.17(b) the coordinates have been rotated clockwise through 45° and the shear and biaxial tensile components are considered to be the primary coordinates. In the first case shear appears as a composite stress made up of uniaxial tension and compression whilst in the second uniaxial tension (compression) appears as a composite stress made up of shear and biaxial tension (compression). Either of these views is equally tenable. In the first case

$$(+\sigma, -\sigma) = (\sigma, 0) + (\sigma, -\sigma)$$

shear   uniaxial tension   uniaxial compression

and in the second

$$(+\sigma, 0) = (+\sigma/2, -\sigma/2) + (+\sigma/2, +\sigma/2)$$

uniaxial tension   shear   biaxial tension

The actual *magnitude* of the stress (whether σ or $\sigma/2$) is unimportant because we are only concerned with a qualitative resolution at this stage.

Which of the two representations in Fig. 15.17 has physical reality? The answer to this is provided when a tensile specimen of rigid PVC yields. Markings appear in the direction of shear and this is at an angle of *less* than 90° to the tensile stress axis. If these markings were at exactly right angles to the tensile stress axis, then the shear component would be zero. The angled pattern of parallel shear lines was first observed in mild steel under tension and is known as the 'Lüder bands'.

When a tensile specimen fails in a tough manner, it always fails in *shear*. Now, shear is a flow phenomenon—therefore a solid that yields is in reality a liquid of sorts. Moreover, once a shear force can cause solid flow (yield) the internal friction will generate heat. This in turn pushes the stress–strain

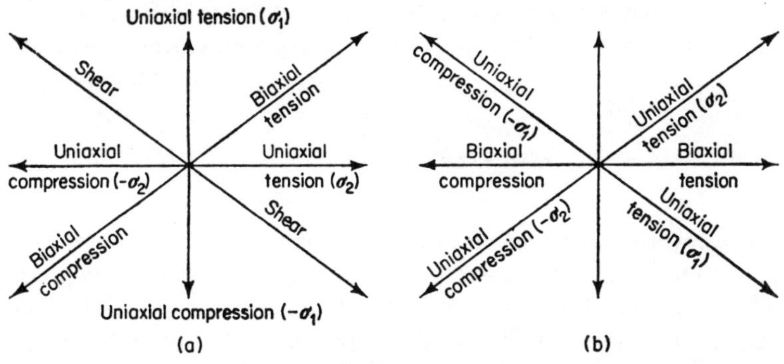

Fig. 15.17.

curve in the same direction as if the test specimen temperature had been raised, with the result that failure must necessarily be tough.

If a tensile test, is, however, carried out very rapidly (tensile impact conditions), that is to say, over a time interval which is shorter than the relaxation time of the plastic and shorter than the time necessary for the establishment of resonating waves, then the flow mechanism cannot get under way, no Lüder bands will appear and the specimen fails in a brittle manner with a fracture surface at right angles to the tensile stress axis.

If the shear component is great enough, then tough failure will always occur. The critical proportion of shear in the stress system which just produces tough failure varies from plastic to plastic and is of course also dependent on other variables.

Of course, one must consider the problem in all three dimensions. Notching will not only increase the straining rate, it will also affect the proportion of shear to tensile stress in a stress system: the stress at the base of a notch will tend to convert a uniaxial tensile stress $(\sigma_1, 0, 0)$ into a triaxial tensile stress $(\sigma_1, \sigma_2, \sigma_3)$,[5] which is due to the fact that the struts created by notching will oppose contraction at right angles to the applied stress. If we accept that it is *shear stress* which controls the yielding of the specimen, then we can see why notches increase the probability of brittleness:

(*a*) In a uniaxial stress system $(\sigma_1, 0, 0)$ the maximum tensile stress conceivable is σ and the maximum shear stress is $\sigma/2$, so that the ratio of the two is given by:

$$\frac{\text{tensile stress}}{\text{shear stress}} = \frac{\sigma}{\sigma/2} = 2$$

(*b*) On the other hand, in a *notched* specimen where the uniaxial tensile stress $(\sigma_1, 0, 0)$ has become converted to a triaxial tensile stress $(\sigma_1, \sigma_2, \sigma_3)$ where $\sigma_1 > \sigma_2 > \sigma_3$, the maximum conceivable tensile stress is σ_1 and the maximum shear stress is $(\sigma_1 - \sigma_3)/2$. In this case the ratio of the two is given by:

$$\frac{\text{tensile stress}}{\text{shear stress}} = \frac{2\sigma_1}{\sigma_1 - \sigma_3} \geq 2$$

In other words, notching has caused a reduction in the share of the shear stress within the total stress system, and since shear stress tends to produce tough failure the probability of brittle failure is increased. Obviously, therefore, notching should be avoided in plastics design, or at least reduced in severity by generous radiusing at the notch base.

Where a plastic is already brittle, notching cannot cause further embrittlement although it will reduce the strength and the energy-to-break. But near the tough/brittle transition even a slight notch can swing the balance sufficiently to make a normally tough unnotched material fail in a brittle manner (e.g. rigid PVC at room temperature). If the brittle strength is *very much* greater than the tough strength, however, then even severe notching will not be sufficient to induce brittleness and all that happens is that the material will yield in the notch region. In so doing it will increase the notch radius and make the probability of brittle failure still more remote. This is the case, for instance, with low density polythene at room temperature even at deformation rates corresponding to free-fall impact.

One more complication in viscoelastic responses must be dealt with which has already been hinted at. One usually considers these responses to be associated with stress and strain rate tensors, but this presupposes that the tensors have sufficient time to resolve themselves into components, whatever the appropriate coordinate system may be.[6] If, however, this is not the case, then the relaxation processes (including stress distribution by stress-wave propagation) cannot get under way and the stress will remain anisotropic. Indeed, we may assign, for any time interval insufficient for resonating stress waves to become established along the length of a tensile test specimen, an anisotropic stress contribution τ_a to the system, *in*

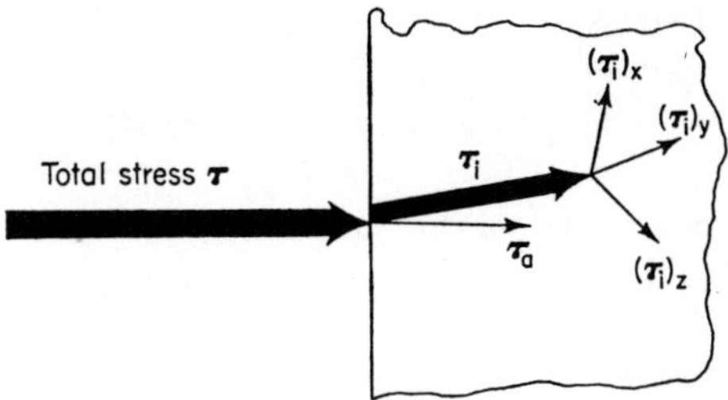

Fig. 15.18.

addition to the *isotropic* stress contribution τ_i which later is resolvable into components (Fig. 15.18): whilst

$$\tau = \tau_a + \tau_i = \tau_a + [(\tau_i)_x + (\tau_i)_y + (\tau_i)_z]$$

$$\frac{\tau_a}{\tau_i} = f(t)$$

where $f(t)$ tends to infinity as time tends to zero, and $f(t)$ tends to zero as time tends to infinity.

In order to clarify the relative proportion of anisotropic stress in the stress system it is necessary to evaluate τ_a/τ_i as a function of time. This cannot readily be done directly, but one can get *some* information by making use of the fact that brittle failure is associated with the absence of relaxation mechanism and also with the presence of notches which serve to increase the triaxiality of the stress. To put it another way: if notches make it sufficiently more difficult for the material to resolve a stress input into isotropic components, then the unresolved (anisotropic) portion becomes more important. Eventually the anisotropic portion predominates or becomes at any rate, sufficiently large to reduce the dissipation of stress significantly, so that, as a net result, the probability of brittle failure is increased.

It was seen that the complexity of nonlinear viscoelastic response to large deforming stresses causes enormous difficulties in assessing plastics in a manner which is reliable and meaningful for design purposes. The following variables are involved:

(1) Temperature
(2) Strain and straining rate
(3) Stress and stressing rate
(4) Time (frequency)

(5)   The type of stress
(6)   The proportion of anisotropy in the total stress
(7)   The previous stress history (e.g. moulding stresses)
(8)   Notching
(9)   The presence of modifying additives (fillers, plasticisers, degradation products, comonomer built into the chain, cross-linking agents, etc.)
(10) Molecular weight
(11) Molecular weight distribution
(12) The morphology of the polymer (crystallinity, size of crystallites, orientation)
(13) Environmental stresses
(14) Pressure

Faced with this formidable list one can nevertheless devise a system which gives useful design information without spending a lifetime on the evaluation of just one grade of one material. According to Turner[6] this should be based on the following experiments:

(1) Two or three creep tests at different stresses at 20 °C, followed by a study of recovery after creep.

(2) An isochronous stress–strain curve at 20 °C at, say, 100 s on the same standard material as was used for the creep test.

One obtains a three-dimensional stress–strain–time response surface of a solid model which can be represented in planar fashion by means of contour lines with any two out of the three variables constituting the base coordinates. The isochronous stress–strain curve represents the section through the solid model in a plane at right angles to the time axis.

(3) A study of the effects of moulding stresses (and their removal by annealing) and of environmental effects such as humidity, ultraviolet light, etc., on the material by means of creep and isochronous stress–strain tests.

(4) A more detailed study of the effects of *temperature* on

  (i)   the isochronous stress–strain curve,
  (ii) creep curves at various stresses,
  (iii) the storage and loss modulus of the complex dynamic modulus at varying frequencies.

To these four main experiments the following information might be added:

(5) Quotation of at least *some* shift factors caused by increasing the test

severity in terms of their temperature equivalent, especially those for notching, time (frequency), type of stress, and stress history.

Before leaving the subject of large strain deformation in the solid state it should be pointed out that plastic deformation is always accompanied by the generation of substantial amounts of frictional heat. Thus, Marshall and Thompson[7] have observed temperature increases of about 80 °C when drawing amorphous poly(ethylene terephthalate). Similar effects in crystalline polymers such as polyethylene and polyamides are common knowledge.

Whilst mechanical energy is stored during elastic deformation, an energy loss occurs during plastic deformation through generation of frictional heat. If the heat is not removed instantaneously and quantitatively, then the specimen under test fails to remain at constant temperature, whatever effort is made to control the experimental conditions. With a change in temperature the mechanical properties of the test specimen will also change, the extent of these changes being material-specific. Spectacular changes in mechanical properties within a narrow temperature range are always likely to occur in the region of the glass transition.

Rigbi[8] has analysed the relationship between deformation and the heat generated during the process. He found that:

(i) the creep coefficient $(\partial \gamma/\partial t)_T$ (at constant temperature) is smaller than $(\partial \gamma/\partial t)_\sigma$, the creep coefficient at constant stress;

(ii) the relaxation coefficient $(\partial \sigma/\partial t)_T$ (at constant temperature) is smaller than $(\partial \sigma/\mathrm{d}t)_\gamma$, the relaxation coefficient at constant strain.

If this appears strange at first sight it must be pointed out that the stress is directly influenced by the temperature change caused by relaxation at constant strain. This is due to the fact that the specimen does not tend to stretch or shrink so much after the appearance of that temperature change.

If heat exchange is prevented by total isolation of the system, then an *adiabatic* modulus can be formulated which differs from the *isothermal* modulus in that the former is a function of the stress and the heat capacity C_v of the material. This also applies to adiabatic and isothermal creep and relaxation coefficients.

REFERENCES

1. P. VINCENT, in *Physics of Plastics*, Ch. 1, P. D. Ritchie (editor), Iliffe Books Ltd (1965).

2. P. Vincent, *Plastics*, **26**, 141 (Nov. 1961).
3. N. Goldenberg, M. Arcan and E. Nicolau, in Proc. Int. Symp. Plastics Testg. & Standardisation, Philadelphia (1958).
4. P. Vincent, *Plastics*, **27**, 115 (Jan. 1962).
5. E. R. Parker, *Brittle Behaviour of Engineering Structures*, Wiley, New York (1957).
6. S. Turner, *Trans & J. Plast. Inst.*, **33**, 106 (1965).
7. I. Marshall and A. B. Thompson, *Proc. Roy. Soc.*, A 221, 541 (1954).
8. Z. Rigbi, Res. Report TDM 69-9, Technion—Israel Institute of Technology (Haifa), Dept. of Mechanics (Aug. 1969).

16

Deformation in the Solid State—Cyclic Strains

A dynamic test subjects a volume element to a stress which varies with time, but it is more particularly regarded as one in which the applied stress varies periodically. The commonest variation is sinusoidal; it lends itself readily to mathematical treatment. Relaxation and retardation times, complex moduli, and storage and loss moduli have been mentioned earlier on. Both the model theory and the phenomenological theory of linear viscoelastic behaviour may be applied to dynamic stresses and the Debye equations for the storage and loss modulus were given. The importance of sinusoidal stresses and strains are well appreciated in applied mechanics; they play a decisive part in the failure of components and assemblies through fatigue. They have been used for chemical and structural analysis in the visible, infra-red, ultra-violet and X-ray regions; they are the mechanism of acoustic transmission; they have played a dominant part in electrical conduction and insulation. Indeed, the fundamental mathematics has been worked out from electrical theory and the conclusions reached were found to be fully applicable to sinusoidal stresses and strains in general, irrespective of the size of the volume elements or the nature of the deformation which may be involved.

Just as an elastic modulus and loss modulus (viscosity) can be determined on liquids, so the storage and loss modulus can be determined in solids. The two moduli represent the components of the complex dynamic modulus. The electrical elasticity modulus ε' is known as the 'permittivity' or 'dielectric constant' whilst the electrical loss modulus ε'' is known as the 'dielectric loss'. Both are functions of the applied field frequency and of the dipole relaxation time of the material under the test conditions. The two moduli are given by the Debye equations:[1,2]

$$\varepsilon' = \varepsilon_\alpha + \frac{\varepsilon_s - \varepsilon_\infty}{1 + \omega^2 \tau^2} \tag{1}$$

$$\varepsilon'' = \frac{(\varepsilon_s - \varepsilon_\alpha)\omega\tau}{1 + \omega^2 \tau^2} \tag{2}$$

Polymer Rheology

where ε_∞ = the dielectric constant at infinity frequency, ε_s = the static dielectric constant (zero frequency), ω = frequency in radians, and τ = relaxation time.

Equations (1) and (2) can be written:

$$\frac{\varepsilon' - \varepsilon_\infty}{\varepsilon_s - \varepsilon_\infty} = \frac{1}{1 + \omega^2 \tau^2} \tag{3}$$

and

$$\frac{\varepsilon''}{\varepsilon_s - \varepsilon_\infty} = \frac{\omega\tau}{1 + \omega^2\tau^2} \tag{4}$$

If the left-hand sides of eqns. (3) and (4) are plotted vs. $\log \omega\tau$, then the symmetrical curves as shown in Fig. 16.1 are obtained .

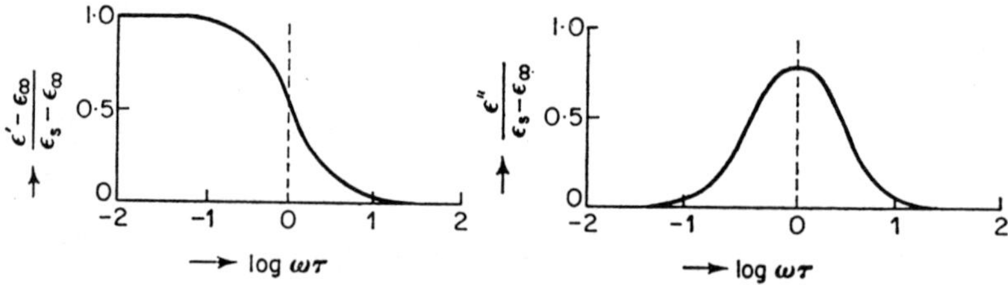

FIG. 16.1.   Electrical storage and loss modulus.

The loss peak for ε'' is reached when $\log \omega\tau = 0$, or when $\omega\tau = 1$, that is to say, when the frequency is equal to the reciprocal of the relaxation time.

The largest value for the dielectric constant ε' and loss ε'' can be obtained directly from eqns. (1) and (2) by putting $\omega\tau = 1$, when

$$\varepsilon'_{max} = \frac{\varepsilon_s + \varepsilon_\infty}{2} \tag{5}$$

$$\varepsilon''_{max} = \frac{\varepsilon_s - \varepsilon_\infty}{2} \tag{6}$$

A neat method of checking whether the Debye equations are rigorously obeyed is due to Cole and Cole. It consists of plotting ε'' vs. ε' at constant temperature with varying frequency:

$$\varepsilon' - \frac{\varepsilon_s - \varepsilon_\infty^2}{2} + (\varepsilon'')^2 = \frac{\varepsilon_s - \varepsilon_\infty^2}{2} \tag{7}$$

Thus, by plotting ε'' vs. ε', a semicircle should be obtained with radius

$(\varepsilon_s - \varepsilon_\alpha)/2$. Its centre must be on the ε' axis (abscissa) at a distance $(\varepsilon_s + \varepsilon_\alpha)/2$ from the origin. The points of intersection of the semicircle with the abscissa are those for which ε' is numerically equal to ε_s and ε_∞. Assuming that $\varepsilon_s = 10$, $\varepsilon_\alpha = 2$ and $\tau = 10^{-10}$ s, the graph obtained would be as shown in Fig. 16.2.

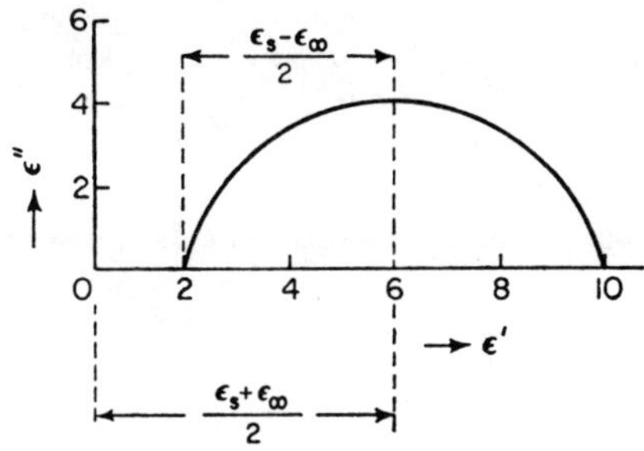

Fig. 16.2.   Cole–Cole plot.

Note that ε'' and ε' are completely defined if ε_s and ε_∞ are known and that the Cole–Cole plot is independent of relaxation time which does not enter into the relationship directly.

For the total area under the $\varepsilon''/\ln \omega$ curve a simple expression can be derived:

Using the identity

$$\int_0^\infty \frac{d\omega}{1 + \omega^2\tau^2} = \frac{\pi}{2\tau}$$

we obtain from eqn. (2)

$$\int_0^\infty \varepsilon'' \frac{d\omega}{\omega} = \frac{\pi}{2}(\varepsilon_s - \varepsilon_\infty) \tag{8}$$

This enables us to calculate $(\varepsilon_s - \varepsilon_\infty)$ when ε'' is known over the entire frequency range.

Experimental results are often expressed in the form of the ratio $\varepsilon''/\varepsilon'$ as a function of frequency. This ratio, also known as $\tan \delta$ (see later) follows from the Debye equations

$$\tan \delta = \frac{\varepsilon''}{\varepsilon'} = \frac{(\varepsilon_s - \varepsilon_\infty)\dfrac{\omega\tau}{1 + \omega^2\tau^2}}{\varepsilon_\infty + \dfrac{\varepsilon_s - \varepsilon_\infty}{1 + \omega^2\tau^2}} = \frac{(\varepsilon_s - \varepsilon_\infty)\omega\tau}{\varepsilon_s + \varepsilon_\infty\omega^2\tau^2} \tag{9}$$

On differentiating eqn. (9) with respect to ω and setting d tan $\delta/d\omega$ to zero, we see that the maximum value of tan δ is obtained at a critical frequency given by

$$\omega_c = \frac{1}{\tau}\sqrt{\frac{\varepsilon_s}{\varepsilon_\alpha}} \tag{10}$$

and is equal to

$$(\tan \delta)_{max} = \frac{\varepsilon_s - \varepsilon_\alpha}{2\varepsilon_s\varepsilon_\alpha} \tag{11}$$

The corresponding values ε_c' and ε_c'' at which tan δ is at a maximum are obtained from the Debye equations substituting ω_c (as per eqn. (10)) for ω:

$$\dagger\varepsilon_c' = \frac{2\varepsilon_s\varepsilon_\alpha}{\varepsilon_s + \varepsilon_\alpha} \tag{12}$$

$$\dagger\varepsilon_c'' = \frac{\varepsilon_s - \varepsilon_\alpha}{\varepsilon_s + \varepsilon_\alpha}\sqrt{\varepsilon_s\varepsilon_\alpha} \tag{13}$$

In practice experimental curves for ε' and ε'' vs. log ω deviate considerably from the values predicted by the Debye equations. The ε' curve is flatter and extends over a wider frequency range and the ε'' curve is broader and has a lower maximum than that predicted by eqn. (6). Generally, however, the curves still remain symmetrical (Fig. 16.3).

As a consequence, the Cole–Cole plot also deviates from the semi-circular. Cole and Cole showed that the experimentally obtained ε'' vs. ε' plots still tend to be circular, but that the centre of the circle lies under the abscissa. The reason for this is that the idealised model is too simple. It is just not true that for each molecule the magnitude of the interacting forces, the magnitude of the applied stress and the temperature are constant throughout the bulk of the test specimen. Each dipole has its own relaxation

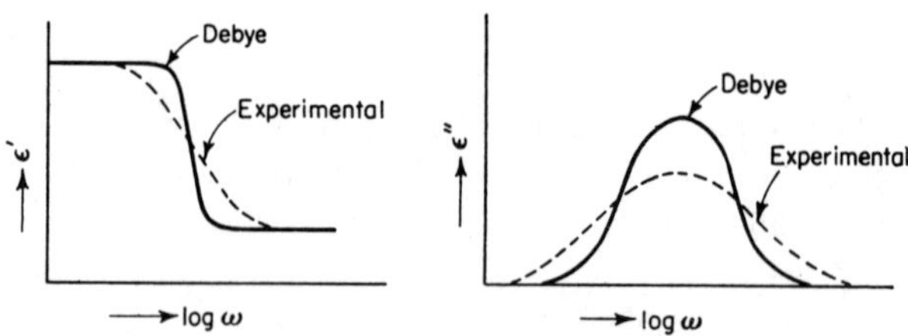

FIG. 16.3.   Storage and loss moduli (see also Fig. 16.1).

$\dagger$ Note that ε_c' and ε_c'' are different from ε_{max}' and ε_{max}''.

time and is also affected by neighbouring dipoles. This is true for relatively simple molecules, but even more so for complex ones. In measurements of moduli one therefore obtains an average value, a most probable value, around which other values will cluster.

If $G(\tau)$ is the distribution function of relaxation times, then $G(\tau)\,d\tau$ represents the fraction of volume elements associated, at a given instant, with relaxation times which range from τ to $(\tau + d\tau)$. $G(\tau)$ is 'normalised' such that the total area under the distribution curve is equal to unity:

$$\int_0^\infty G(\tau)\,d\tau = 1 \tag{14}$$

The modified Debye equations are

$$\varepsilon' = \varepsilon_\infty + (\varepsilon_s - \varepsilon_\infty)\int_0^\infty \frac{G(\tau)\,d\tau}{1 + \omega^2\tau^2} \tag{15}$$

and

$$\varepsilon'' = (\varepsilon_s - \varepsilon_\infty)\int_0^\infty \frac{\omega\tau\,G(\tau)\,d\tau}{1 + \omega^2\tau^2} \tag{16}$$

Using eqn. (16) it can be shown that eqn. (8) for the total area under the $\varepsilon''/\ln\omega$ curve is also valid for a material which has a *distribution* of relaxation times:

$$\int_{-\infty}^\infty \varepsilon''\,d\ln\omega = (\varepsilon_s - \varepsilon_\infty)\frac{\pi}{2} \tag{17}$$

Cole and Cole showed that for a distribution of relaxation times the equation for the complex modulus ε^*, namely

$$\varepsilon^* = \varepsilon_\infty + \frac{\varepsilon_s - \varepsilon_\infty}{1 + i\omega\tau} \tag{18}$$

must be modified to

$$\varepsilon^* = \varepsilon_\infty + \frac{\varepsilon_s - \varepsilon_\infty}{1 + (i\omega\tau_0)^{1-h}} \tag{19}$$

where τ_0 and h are constants.

If there is only *one* relaxation time and not a distribution, then $h = 0$ and eqn. (19) reduces to eqn. (18). The parameter h is a function of the subtending angle ϕ of the circle sector in the Cole–Cole plot (Fig. 16.4), namely

$$\phi = \pi - 2\alpha$$

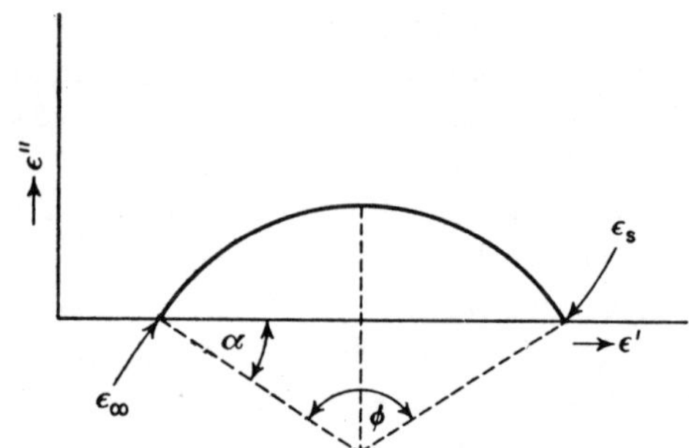

FIG. 16.4.   Modified Cole–Cole plot.

where

$$\alpha = \frac{\pi h}{2}$$

so that

$$\phi = \pi(1 - h)$$

Unfortunately for structural analysis the distribution of relaxation times in plastics is rather scattered and h is large (the limiting value is unity). Only a shallow segment of the Cole–Cole circle shows above the ε' horizon and this is easily missed. Moreover, there may be a *number of species* of dipoles and the distribution of their relaxation times may therefore overlap. The resulting smudging of the Cole–Cole segments tends to make the clear identification of individual segments exceedingly problematical (Fig. 16·5). The continuous line is the experimental observation, whilst the broken lines represent hypothetical distributions of relaxation times of which only the arrowed one stands out, and that one is obviously *not* coincident with a true loss maximum. Misinterpretation is therefore all too easy.

What is unfortunate for the structural analyst is, however, a great

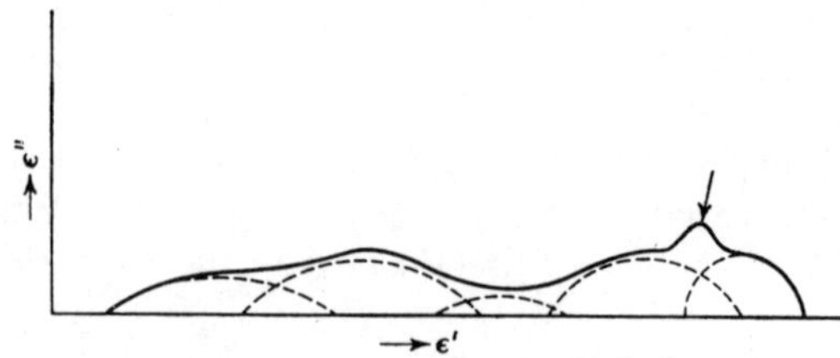

FIG. 16.5.   Cole–Cole plot for material with multiple maxima.

advantage for plastics generally. If h is very large indeed, say, close to its limiting value of unity, then the Cole–Cole segment degenerates into a tangential point and the exponential term in the denominator of eqn. (19) becomes unity. This means that the imaginary (loss) portion of the complex modulus (ε'') disappears and $\varepsilon^* \equiv \varepsilon_s$. h is close to unity in plastics and therefore they are not very lossy electrically, in fact they are excellent insulators.

However, excellence is a relative term. Looking at plastics in isolation, it soon becomes obvious that these materials vary greatly amongst themselves. Clearly, polar polymers will have loss peaks at frequencies which are close to the mean of the reciprocal relaxation times of the dipoles involved. There are not many commercially important polymers which are *not* polar by nature—the list is virtually complete with polyolefines, polystyrene, PTFE and polyvinylidene chloride. Amongst the more highly polar polymers are the polyamides and PVC, whilst the polyesters, acrylics, phenolics, aminoplastics and epoxides have weaker dipoles. Oxidative degradation in polyolefines can increase the vanishingly small loss factor of polythene by two or three decades. This affords a means of assessing the amount of antioxidant present in the compound. Polar impurities (especially catalyst residues in stereospecific polyolefines) greatly increase the magnitude of dielectric losses. This is particularly serious in long distance cable insulation and it is probably true to say that only the availability of highly sophisticated ultra-low-loss polythenes of power factor 0·000 02 to 0·0001 have made the land–sea cable from Britain to Australia possible. Even small losses are serious because the low thermal conductivity of plastics makes it difficult for the heat which is generated in the process to be dissipated. As the heat builds up, so the relaxation times decrease, the material may become more lossy and eventually thermal breakdown may occur. In certain cases losses can be put to good use: PVC can be high-frequency welded, phenolics and aminoplastics can be preheated by applying a dielectric field of suitable frequency for a few seconds. Preheating reduces moulding cycles, makes for better flow, lower residual stresses, reduced mould wear and more uniform mouldings of thermosetting moulding materials such as PF and the aminoplastics.

It is now appropriate to examine briefly the mechanism which causes the loss modulus to arise.

Polarity exists in certain chemical bonds, especially carbon to halogen and carbon to nitrogen, giving rise to large dipole moments. Dipole moments arising from carbon to hydroxyl and carbon to oxygen bonds are also clearly detectable. On the other hand, the polarity arising from carbon to hydrocarbon substituent groups (alkyl, aryl, cycloaromatic) is small. The higher the polarity, the greater the susceptibility of a dipole to align itself to conform to an applied electric field. This field will then further distort the dipole. Indeed, it may even induce the formation of dipoles where dipoles do not exist in the ground state.

If the electric field is static, then all the energy of polarisation is stored in the distorted dipole which acts like a spring and reverts to its rest position when the field is removed. But if the field alternates in polarity, then the dipole will tend to move with the field and in so doing will work against the frictional resistance (internal viscosity) of its environment. Heat is generated and energy is lost. The 'brake' on the oscillating dipoles will become more and more noticeable as the field frequency is gradually increased and is a direct function of the phase difference between the field and the lag in the response. At a certain frequency the phase lag reaches a maximum; on further increasing the frequency, the dipoles will become less and less sensitive to field changes because they are less and less able to respond within the decreasing time interval available between changes of polarity. The frequency at which the lag (loss) is greatest is around the inverse relaxation time of the dipole. The dielectric constant (storage) is highest at zero frequency, that is to say, when the field is static, and least at very high frequencies after the lossy region has been traversed. Maxwell has observed that the dielectric constant at very high frequencies in nonpolar liquids is very nearly equal to the square of the optical refractive index:

$$\varepsilon'_\alpha \cong n^2$$

Electrically the storage modulus (dielectric constant) ε' is defined as the ratio of the capacitance C of the dielectric and the capacitance of the empty space C_0 of the same dimensions; the loss modulus is a measure of the resistance current going through the dielectric and is given by

$$\varepsilon'' = \frac{1}{\omega R C_0} = 36\pi 10^{11} \frac{K}{\omega}$$

where R = resistance of the dielectric in ohms, K = specific conductance in ohm^{-1} cm^{-2}, and C_0 = vacuum capacitance in Farads (ohm^{-1} s).

In a simple capacitor the capacitative current I_c and the resistive current I_R are given by

$$I_c = i\omega\varepsilon' C_0 E \qquad \text{and} \qquad I_R = \omega\varepsilon'' C_0 E$$

where $E = E_0 \cos \omega t$ and E_0 = the amplitude of the alternating voltage E.

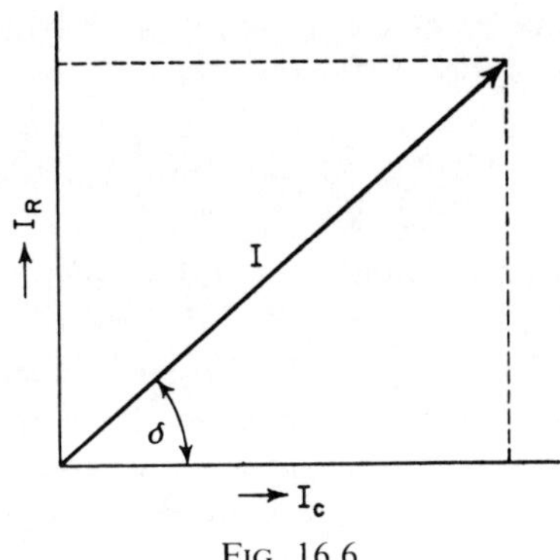

FIG. 16.6.

The relationship between total current I, capacitative and resistive current is given in Fig. 16.6.

It is clear that I_R/I_c is numerically equal to $\varepsilon''/\varepsilon'$ and when plotted as in Fig. 16.6 the ratio can be regarded as the tangent of an angle δ. The significance of the angle δ becomes apparent when the stress and the strain response are plotted together as a function of time (frequency). It is clear that the stress leads the strain by a constant phase difference which can be expressed in terms of radians (Fig. 16.7) and the phase angle δ. This remains constant since the strain follows the stress at the same frequency.

The study of the electrorheology of plastics is a fascinating subject but since there is no space for a detailed treatment which would do it justice the reader is strongly urged to consult refs. 3–5.

Briefly, the following features should be noted:

(1) The dielectric constant is a function of the electronic polarisability α_e

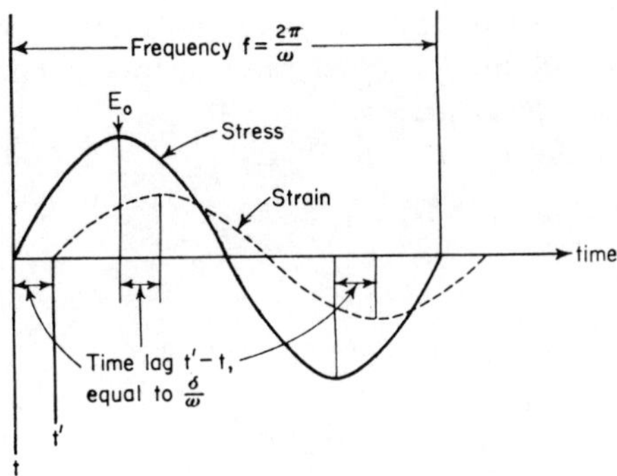

FIG. 16.7.

and density ρ in nonpolar materials and is given by the Lorenz–Lorentz equation through the optical refractive index n:

$$\frac{\varepsilon' - 1}{\varepsilon' + 2} \cong \frac{n^2 - 1}{n^2 + 2} = \frac{4\pi N}{3M} \alpha_e \rho$$

where $N =$ the Avogadro Number, $6 \cdot 023 \times 10^{23}$, and $M =$ molecular weight.

It follows that the electronic polarisability is inversely proportional to the density. This means that in crystalline materials where the density changes abruptly at the melting point T_m, and in amorphous materials where the density changes at different rates with temperature around T_g, the transitions are indicated by dielectric constant measurements as a function of temperature, and that these transitions are frequency-independent. It is a useful exercise to consider these transitions with the free volume concept well in mind.

(2) The very small loss ε'' of nonpolar polymers is still unexplained. Extensive milling on hot rolls say, of polythene in the presence of air cause substantial losses because of the appearance of carbonyl groups, but it is difficult to ascribe the small basic loss to this or similar forces or to impurities alone.

(3) In polar materials there is much greater scope for electrical investigation of dipole structures from which chemical and morphological details may be inferred. Here we have a permanent dipole moment the orientation of which is measured by the *orientational* polarisability α_μ:

$$\alpha_\mu = \frac{\mu^2}{3kT}$$

where k is the Boltzmann constant.

This is additional to the *electronic* polarisation α_e which is *always* present. The contribution of α_μ can raise the total polarisation to very high levels. For highly polar materials we have

$$\varepsilon_s - \varepsilon_\infty = \frac{32\pi N\rho}{3M} \frac{\mu^2}{3kT}$$

This means that:

(a)   if μ is large, $\varepsilon_s - \varepsilon_\infty$ may be very large;
(b) $(\varepsilon_s - \varepsilon_\infty)$ varies inversely with temperature. Hence the value of ε_s tends to decrease markedly as the temperature increases.

Why does the orientational polarisability (and hence ε_s) depend so strongly on temperature in polars, and why is the temperature dependence of the inverse type? The thermal agitation interferes with the orientating effect of the field—at high temperatures the dipoles already move vigorously because they are thermally energised and the relatively gentle field influence cannot therefore make as great a contribution to ground-state-departure as at lower temperatures where the thermal agitation is much less pronounced.

(4) The orientational polarisability depends strongly on the ability of the dipoles to turn in a field. In a liquid this can readily manifest itself, but less so in an amorphous solid and even less so in a solid in which the dipoles are firmly locked in a crystallite. This means that, although ε_s tends to reduce with increasing temperature, the unlocking effect when crystals melt at T_m will reverse the downward trend of $d\varepsilon_s/dT$ and overwhelm it. The same argument also applies qualitatively to the glass transition in amorphous polymers.

Polar side groups, if present, can make a distinct contribution to the overall polarisability which thus does not have to be restricted to the main chain. There can therefore exist additional glass transitions which refer specifically to dipoles other than main-chain dipoles. These can be readily picked out over the temperature spectrum of dielectric loss.

(5) The dielectric relaxation times which characterise the loss maxima can be of the order of 10^{-10} s in the liquid state, but at low temperatures they can be very long and will therefore only show up at low frequencies. In order to obtain the overall picture one must therefore traverse many decades of frequencies. This can fairly readily be done electrically and makes dielectric measurements a powerful research tool.

(6) The activation energy E for dipole orientation processes can be obtained from a plot of $\log \tau$ vs. $1/T$ which, in conformity with the Arrhenius equation gives a slope of $2\cdot303E/R$. The τ value used is that obtained from a Cole–Cole plot or an ε'' vs. $\log \omega$ plot for each of a series of temperatures. The measured activation energy in polars usually lies between 5 and 50 kcal/mole. A significant point is that the activation energy for some polar polymer crystals is not very different from that found for dipole orientation in systems which consist of randomly distributed (amorphous) chain molecules. This proves that the *whole* chain need not move to permit the required degree of polar mobility. The higher the activation energy, the greater is the change of the relaxation time with temperature.

(7) As plasticiser is added, so the relaxation time becomes shorter. This is entirely expected from the free volume concept: the dipoles are freer and the

loss maximum occurs at higher frequencies. This shifting in the loss maximum by means of plasticisation is sometimes very useful in driving the loss maximum from certain frequency ranges in which one wants to use an insulating polymer but is prevented from doing so by excessive dielectric loss. The same 'plasticising' effect can also be achieved by judicious copolymerisation. But the first step in such 'tailoring' is an exact knowledge of the lossy regions (see earlier).

The dielectric moduli have been very fully investigated for a number of polar polymers, including poly(ethylene terephthalate)[6] and polymethyl methacrylate. On the basis of these measurements solid models and contour maps of the dielectric constant/temperature/frequency and dielectric loss/temperature/frequency relationships have been constructed.

A number of attempts have been made to relate polymer structure with the dynamic–electrical or dynamic–mechanical properties.[6–9]

Deutsch *et al.*, for instance, point out that the crystallites of poly(ethylene terephthalate) can have units arranged in a symmetrical *trans*-form and in an asymmetrical *cis*-form as shown in Fig. 16.8.

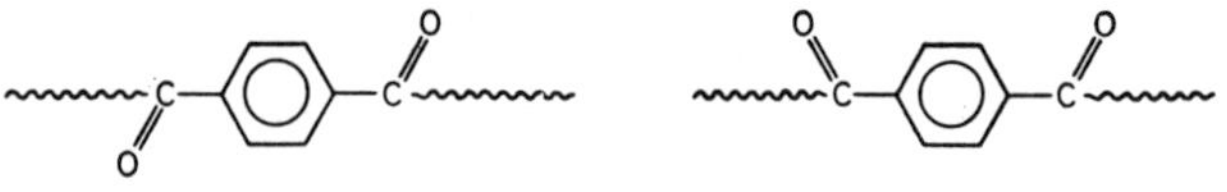

Trans   Cis

Fig. 16.8.

The polyester has a dielectric loss peak around 80 °C which is also near its mechanical glass transition at which temperature crystallisation begins to occur on reducing the temperature. It is therefore reasonable to suppose that the dielectric loss is associated with a transition in the amorphous phase which becomes impossible in the crystalline phase. That the packing of chains is similar in the two phases is indicated by the fact that the density difference is only about 10%. The benzene rings are probably in parallel planes locally and the oxygens are coplanar with the benzene rings. In the crystalline phase the large dipole moment of the *cis*-form is frozen (so is the *trans*-form, but in the *trans*-form the dipole moments are in opposite directions and therefore cancel out). In the amorphous phase, on the other hand, some distortion in response to an electric field is likely to occur and at the absorption peak the *cis–trans* transition, which is associated with the freedom of movement due to T_g, occurs frequently enough for the repeating unit to have a measurable average dipole moment. The relaxation time of this dipole moment must be related to the average time of the transition process.

Optical and electrical loss spectra are fairly familiar, but it has been demonstrated for a number of polymers that *mechanical* loss peaks are also characteristic. On the whole, mechanical spectra are of rather limited use in structural research. Firstly, the available frequency range is incomparably smaller than for optical and electrical spectra; and secondly the applied stress is rather indiscriminate in the response it excites. A gross mechanical stress will not only affect mechanical flow units, but also such electrical or magnetic dipoles as may be present, as well as the chemical bond geometries. Since the latter three involve much more subtle changes their effect will be swamped by the mechanical macro-effect. The subtle changes of the nonmechanical responses can only be isolated by applying correspondingly subtle and highly specific modes of excitation. Mechanical and electrical excitations are qualitatively the same thing, but only in the sense that a tornado is qualitatively the same as a gentle zephyr. A zephyr will rustle the leaves of a tree, a tornado will undoubtedly do the same, but this will not be very remarkable considering the violent vibrations and swaying of twigs, branches and even the trunk itself. On the whole, each mechanical loss region (in polar compounds at least) also corresponds to an electrical loss region and both tend to occur in the same temperature and frequency regions. This is not surprising, since both are associated with flow units which become immobile as the temperature drops (or as the frequency increases). Whether an electrical loss peak occurs in a mechanical loss region depends, of course, on whether the flow unit contains dipoles or not.

Polymers containing water may be regarded as two-phase systems. The water may act as a plasticiser (as in polyamides), but some water may also be present in polymers which are intrinsically hydrophobic, as in polystyrene and PVC, especially if hygroscopic additives are present. It will therefore contribute to the comparatively high dielectric loss in emulsion polymers when these are compared with mass, solution or suspension polymers. In the case of polyamides the increase in conductance is so great that one must look for a second effect. This is provided if one assumes that the polyamides also act as polyelectrolytes.

In order to assess the electrical suitability of a plasticiser one must carry out the measurements at an isoviscous point, that is to say, at the various temperatures at which the flexibility of all compositions is the same. One will then prefer the plasticiser which has the fewest number of dipoles per unit volume of plasticised compound whilst yet having an acceptable cold flex temperature.

We must now leave the border region of interplay of mechanical and electrical loss spectra and consider the mechanical loss spectra more fully.

We have already seen that mechanical spectra can contribute to an understanding of polymer structure and morphology but it has been pointed out that the scope is limited because of the limitation to conveniently realisable frequency ranges. However, those frequency ranges that *are* available are also those which are important for the mechanical and acoustic performance of plastics engineering components.

If a simple polymer is cyclically stressed at a given temperature and frequency it will respond in one of two ways: *either* it will behave as an elastic rubber with a modulus of about 10^7 cgs units; *or* it will behave as a glass with a modulus of about 10^{10} cgs units, with little deformation occurring.

If the time between segmental movements is short compared to the time of the stress period, then the material will comply readily. This is the case for the rubbery condition. The reverse will apply to the glassy condition.

Obviously, if the segmental periodic time decreases with increasing temperature, a transition will occur beyond which the material will behave as a rubber and this transition will occur within the narrow limits around T_g. But if the periodic time of stress is decreased (frequency increased), then the transition will occur at a higher temperature. Therefore, as Karas points out,[10] it is possible to make a rubbery material behave as a glass by striking it very quickly: an ordinary impact blow takes 1–2 ms and under such a blow the material will behave as a rubber; but if the same rubber is struck at 1 μs it may shatter like a glass. Again we have time/temperature superposition and if we operate in the linear viscoelastic region then this will apply even more rigorously than in the nonlinear region of viscoelastic behaviour which was the subject matter of Chapter 15.

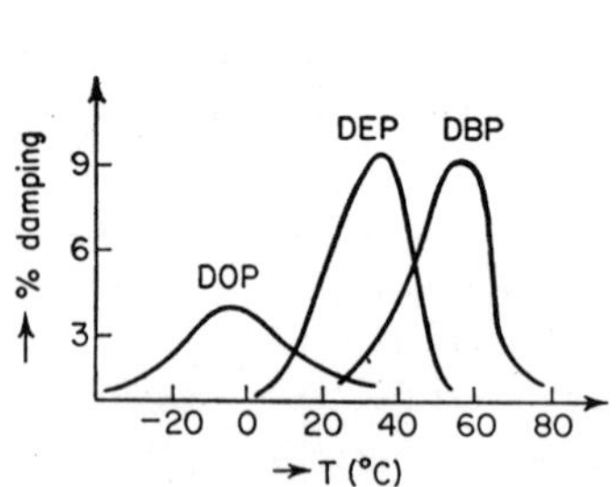

FIG. 16.9. Absorption maxima: PVC with different plasticisers (after Nielsen[13]).

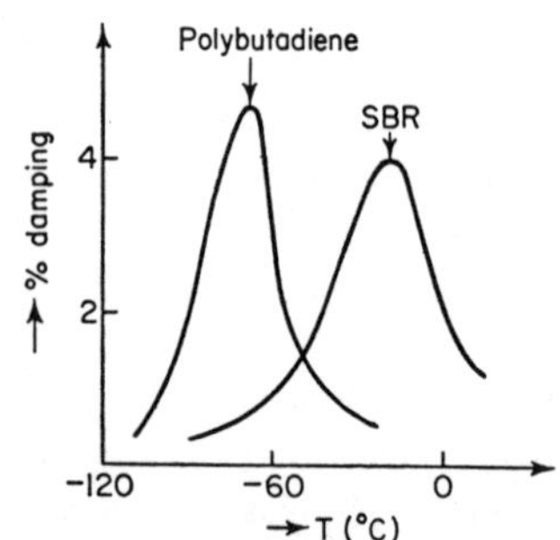

FIG. 16.10. Absorption maxima: damping for rubber-modified polystyrene containing 5% added rubber.

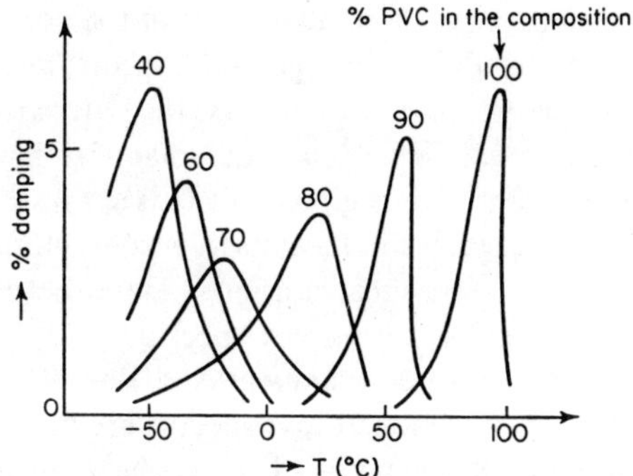

Fig. 16.11.   Absorption maxima as a function of temperature: PVC with different amounts of the same plasticiser (DOP) (after Wolf[12]).

If additives are introduced, several possible effects may occur, depending on the nature of the additive. The addition of varying amounts of plasticiser to PVC will alter the position of the transition region. Alteration in the type of plasticiser or rubber additive will change the form and position of the transition (Figs. 16.9 to 16.12).

With rubber-modified polystyrene the addition of rubber may produce a new molecular species with its own damping peak in the same way as do solvated complexes in simple molecules. Whether it does or not will depend on the type and quantity of rubber used.

A change in molecular weight will not generally alter the dynamic properties, except at quite low molecular weights. It should be clear why this

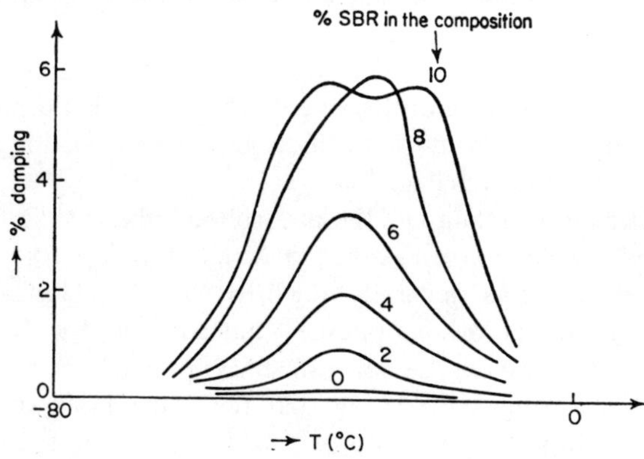

Fig. 16.12.   Absorption maxima as a function of composition; polystyrene with different amounts of the same added rubber (SBR).

must be so. But a change in the regularity or in the configuration of side groups on a chain will result in a change of the spectrum. Atactic polymers differ greatly from stereoregular ones because the latter are generally highly crystalline. Where crystallinity occurs the material has effectively two phases and separate transition regions will be seen in the spectra. If the phases are incompatible, each will have its own characteristic transition. If they are mutually soluble the spectra approximate those of a copolymer. Partial compatibility results in complex spectra.

An interesting situation arises when a crystalline polymer is cold drawn and when the dynamic–mechanical loss spectra are compared before and after draw and subsequent annealing. This has been investigated by Prof. Keller's school of polymer physics at the University of Bristol. An example of this work is a monograph by Stachurski and Ward.[16] They prepared specially oriented sheet from low-density polyethylene by cold-drawing, calendering and annealing and found that the anisotropy of one of the two relaxation processes was largely determined by molecular orientation. The second relaxation process had a morphological basis, apparently the mechanical loss associated with it being related to interlamellar shear stresses. The structure of the lamellae is therefore determined by the anisotropy of the relaxation. This applies to polyethylene of both low and high density.

When relaxation maxima appear in a spectrum they are usually denoted by Greek letters in alphabetical order with decreasing temperature. If additional peaks are found later, or if it is seen that neighbouring peaks are linked by closely related relaxation processes, then primes are added to the Greek letters. The peaks were intensively studied with the aim of assigning relaxation processes to the appropriate molecular and morphological structure as well as to the configuration of the polymer mass. It was found that the α-relaxation of polyethylene is related to the morphology of the crystal lamellae—specifically to the motion of chain folds and reorientation of chains within the lamellae.[11]

The β-relaxation relates to the amorphous phase.[12,13] Its activation energy is of the same order as that involved in a glass transition.

The γ-relaxation is generally regarded as characterising the glass transition of the amorphous phase,[14] but some authors think that the crystalline phase also exerts some influence upon it.[11,15] This may not be altogether surprising, considering that the two phases are not sharply separated but linked by a region of partial crystallinity.

Stachurski and Ward[16] succeeded in identifying one of the relaxation processes with interlamellar shear and they showed that there exists a

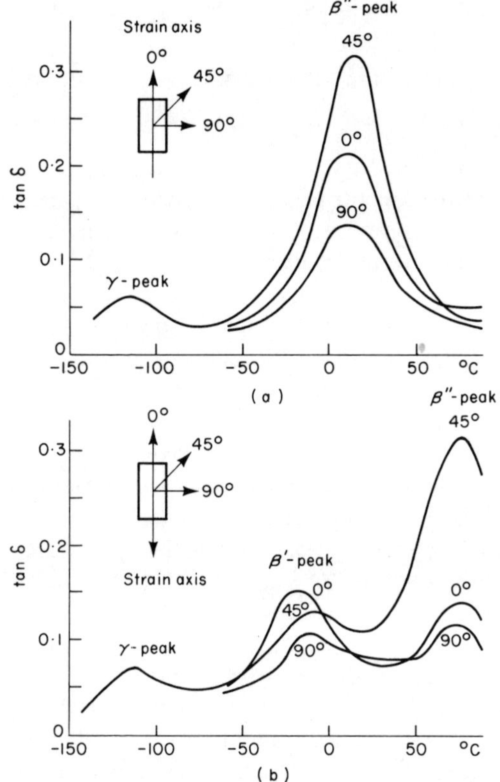

Fig. 16.13. Modification of absorption maxima due to changes in orientation with respect to the direction of test. (a) Cold drawn specimen; (b) annealed specimen.

second relaxation process which is linked to another feature of anisotropy. The latter was confirmed by examining the dynamic–mechanical loss spectra of cold drawn film. The samples were subjected to a sinusoidal oscillation in torsion such that the oscillation direction was parallel, normal to, and at a $45°$ angle to the drawing axis. The results are shown in Fig. 16.13.

The high β''-peaks of the cold drawn specimen near $0°C$ indicate considerable anisotropy (Fig. 16.13(a)). The annealed film specimens in (b) point to a second relaxation process characterised by a β'-peak. The order of magnitude of the peaks for specimens which were cut at different angles to the draw axis show variations in anisotropy—compare Fig. 16.13(a) and (b):

(a)   (β''-peak): $\tan \delta (45°) > \tan \delta (0°) > \tan \delta (90°)$

(b)   (β')-peak: $\tan \delta (0°) > \tan \delta (45°) > \tan \delta (90°)$

Stachurski and Ward discussed the physical significance of the relaxation characteristics of the spectra based upon further experiments which also included the storage component of the complex dynamic modulus over a frequency range of several decades.

These methods, and others, show how dynamic mechanical spectra can serve to provide a valuable insight into the mechanism of relaxation processes in polymers in which the processing conditions and/or after-treatment have produced structural anisotropy.

It is also of considerable practical and theoretical interest to investigate the changes in dynamic–mechanical behaviour when the specimen is simultaneously subjected to uniaxial tension at constant draw rate.

It has already been stated that any effective structural rearrangement of volume elements in a polymer mass is immediately indicated by shifts in the dynamic modulus and in the loss spectrum. Stachurski and Ward[16] have examined such rearrangements during the drawing of polyethylene film, whilst Bodner and his colleagues at the Haifa Technion[17–21] have developed a torsion pendulum which makes it possible to observe rearrangements which occur in the course of a uniaxial tensile experiment. The specimen is subjected to a simultaneous oscillation of very small amplitude during elongation. The materials investigated were epoxy and polyester resins filled with quartz sand and with glass fibre. It was the object of the experiments to gain an insight into the processes which occur at the resin/filler interface and which result in eventual fracture as the strain increases. It was of particular interest to ascertain whether fracture occurs as a result of progressive dehesion of resin and filler with increasing strain, as a result of fibre break or cracking of the resinous matrix, or by uniform failure of the composite without any preceding appearance of interphase weaknesses. This was further examined as a function of straining rate and thus led to an elucidation of the time-dependence of the material properties.

It was seen that the time functions of the quartz-filled epoxy resins were identical with those of the unfilled epoxy resin, a result which had been predicted by Hashin[22] earlier. The only apparent effect of including the filler was a several-fold increase in the modulus. (A similar result had been reported by Schwarzl[23] for filled natural rubber which was tested at constant frequency.)

Since the dynamic modulus of an epoxy resin containing 38 volume–per cent of quartz filler is three times greater than that of the unfilled resin, one would expect that dehesion of filler and resin would cause great changes in the dynamic modulus. By the same token, the loss modulus would increase substantially. An examination of the quartz-filled epoxy resin, however, showed no such thing. This shows clearly that the eventual break in the quartz–epoxy composite was *not* preceded by dehesion at the quartz–resin interface and subsequent growth of voids.

An altogether different picture emerged when filled polyester resins were tested: in this material the quartz filler begins to part from the resin even at quite low tensile stresses, with a simultaneous increase in the loss modulus, whilst the unfilled polyester resin showed no significant change in loss modulus from beginning to break.

Evidently, a number of factors have a decisive influence on the behaviour of a resin/filler composite:

(i)   The interfacial forces between resin and filler.

(ii)   Geometrical factors, such as filler shape and particle size. It should be borne in mind that differences in shape (e.g. spherical particles as against glass fibre) greatly influence the interfacial forces, if only because of the enormous increase in the surface/volume ratio.

(iii)   Changes in the interphase forces under the influence of an applied tensile stress. These may, additionally, be accompanied by time-dependent effects which can be studied by varying the straining rate.

(iv)   Changes in the interphase forces as a function of temperature. These cannot be studied at all easily, since 'isothermal' conditions are not *truly* isothermal when one considers the localised temperature at the interface on a molecular level.

(v)   The degree and direction of orientation (if any) of non-spherical fillers in the composite—especially of fibrous fillers—and the angle of orientation relative to the strain axis.[20]

It is clear that dynamic–mechanical methods can be rather useful for gaining an insight into the mechanism by which epoxy and polyester composites break under stress. It would be of interest to extend the method to certain thermoplastics, for example, to glass-filled nylon and to unfilled crystalline polymers. In the latter one might obtain information on the action of the crystalline upon the amorphous phase, and on the role which the intervening regions of partial crystallinity play during tension and in particular during the cold drawing process.

The work on dynamic–mechanical spectra involves frequencies below

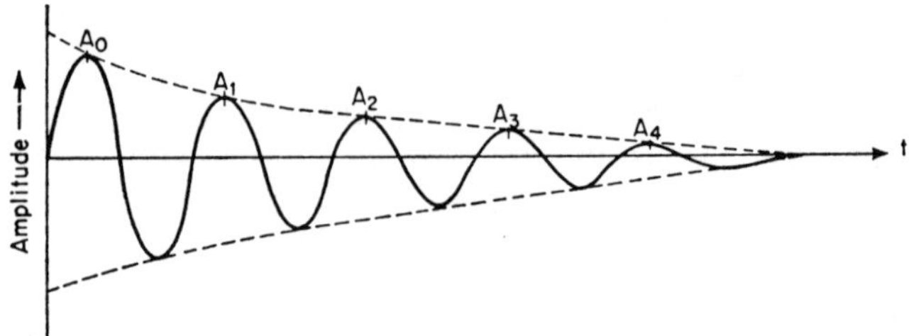

FIG. 16.14.   Damping.

10^4 cycles per s, mostly less than 2000 cycles per s. At such frequencies simple apparatus is available. Although forced vibrations give perfectly acceptable results, the simplest apparatus relies on free oscillation techniques. A sample is slightly strained and on releasing the stress the period of oscillation and the decay in amplitude of stress is measured. The rate at which the oscillations will die away depends on 'damping', i.e. the loss due to internal friction (internal viscosity, irreversible deformation, absorbed energy). It is therefore the precise analogue of the dielectric loss ε'' (see Fig. 16.14).

The following instruments are used:

(a)   Torsion pendulum.
(b)   Vibrating reed
(c)   Ball rebound apparatus
(d)   Rolling ball loss spectrometer.

The theory and method of operation of these instruments is described in 'Plastics Rheology'[24] and in refs. 25–28.

We have seen how dynamic spectra can be used to elucidate polymer structure, to evaluate the suitability of polymers for dielectric insulation and to supply mechanical data which are related to the performance of plastics under cyclic stresses.

Since the vibrations, rotations and other motions of chemical bonds are also highly frequency-selective it is not surprising that they show up as distinct spectra. Since many types of motions involving chemical bonds are possible, even in relatively simple molecules, infra-red spectra are very complex, but the peaks have been accurately charted and they are fully indicative of the chemical nature of materials. Infra-red spectroscopy is a big subject and is dealt with in a number of standard reference works. No detailed account of optical spectra can be given here.

Electron spin and nuclear magnetic resonance spectra have a more recent history but their discussion lies also beyond the scope of this book.

Acoustic spectra can supply useful information on sound damping and one may expect more work to be done in this field in the future.

It is clear that dynamic spectra are of absorbing interest in every sense of the word.

REFERENCES

1. C. J. F. BÖTTCHER, *Theory of Electric Polarisation*, Ch. X, Elsevier, Amsterdam (1952). (This contains a derivation of the Debye equations.)
2. H. FRÖHLICH, *Theory of Dielectrics*, Clarendon Press, Oxford (1949).
3. J. D. HOFFMAN, The mechanical and electrical properties of polymers, *I.R.E. (Component Parts)*, **CP-4**, 2, 42–69 (June 1957).
4. G. P. MIKHAILOV and B. I. SAZHIN, Macromolecular dielectrics, *Russian Chem. Reviews* (trans. into English), **29**(7), 410 (July 1960).
5. K. N. MATHES, in *Engineering Design for Plastics*, Ch. VII, E. Baer (editor), Reinhold, New York (1964).
6. W. REDDISH, *Trans. Farad. Soc.*, **46**(6), 330 (June 1950).
7. K. DEUTSCH, E. A. W. HOFF and W. REDDISH, *J. Poly. Sci.*, **13**(72), 565 (1951).
8. F. WÜRSTLIN and H. THURN, in *Die Physik der Hochpolymeren*, Vol. VI, H. A. Stewart (editor), Springer Verlag, Berlin (1956).
9. C. MUSSA, *Trans. & J. Plast. Inst.*, **31**(96), 146 (Dec. 1963).
10. G. C. KARAS, Principles of Dynamic Testing, *Brit. Plastics*, p. 59 (Feb. 1964).
11. J. D. HOFFMAN, G. WILLIAMS and E. PASSAGLIA, *J. Poly. Sci.*, C-14, 173 (1966).
12. K. SCHMIEDER and K. A. WOLF, *Kolloid-Z.*, **134**, 144 (1953).
13. L. E. NIELSEN, *J. Poly. Sci.*, **42**, 357 (1960).
14. A. H. WILLBOURN, *Trans. Farad. Soc.*, **54**, 717 (1958).
15. K. M. SINNOT, *J. Appl. Phys.*, **37**, 3385 (1966).
16. Z. H. STACHURSKI and I. M. WARD, paper presented at the British Society of Rheology autumn meeting, Shrivenham (Sept. 1968).
17. S. R. BODNER, Materials Mechanics Laboratories Report (MMLR), **12**, Technion—Israel Institute of Technology (Oct. 1967).
18. S. R. BODNER and J. M. LIFSHITZ, MMLR, **8**, Technion—Israel Institute of Technology (June 1967).
19. J. M. LIFSHITZ and A. ROTEM, MMLR, **14**, Technion—Israel Institute of Technology (June 1967).
20. J. M. LIFSHITZ, MMLR, **15**, Technion—Israel Institute of Technology (Jan. 1969).
21. J. M. LIFSHITZ and A. ROTEM, MMLR, **17**, Technion—Israel Institute of Technology (July 1969).
22. Z. HASHIN, *J. Appl. Mech.*, **32**, 630 (1965).
23. F. R. SCHWARZL, in *Mechanics of Solid Propellants* by S. Fringen *et al.*
24. R. S. LENK, *Plastics Rheology*, pp. 194–9, Maclaren, London (1968).
25. K. H. ILLERS and E. JENCKEL, *Kolloid-Z.*, **160**(2), 98 (Oct. 1958).
26. T. RAPHAEL and C. D. ARMENIADES, *SPE Trans.*, **4**, 2 (April 1964).
27. I. CHEETHAM, *Trans. Proc. Rubber Ind.*, **40**(4), T156 (1965).
28. F. WÜRSTLIN, K. SCHMIEDER, K. A. WOLF and A. J. STAVERMAN, *Kolloid-Z.*, **134**(2/3) (1953).

<h1 style="text-align:center">17</h1>

Critical Strain

J. POHRT

*Materials Testing Department (Polymers),
BASF, D-6700 Ludwigshafen, Federal Republic of Germany*

INTRODUCTION

The properties of a frozen polymer melt can only be properly understood and described after considering the rheology of the glassy state. In the following we shall consider flow processes during mechanical loading as well as the effect of the processing history on the mechanical properties of plastics. It will be seen that a combination of the results obtained affords an insight which has important theoretical and practical implications.

FLOW LIMITS DURING LONG-TERM LOADING

Experiments under long-term loading conditions are generally represented by creep curves which reflect the visible macroscopic effects of flow processes in the glassy state (Fig. 17.1). The flow limit is conventionally represented by the break point. On joining the break points in Fig. 17.1 one obtains a break line which is a function of both stress and time, and which crosses lines representing lower and lower stresses and strains.

It may be assumed that break is preceded by a phase involving the formation of some defect which may be experimentally observed provided that there is some identifiable manifestation of damage in the test specimen. Some relevant data of this kind are available from dynamic tests on metals and these are illustrated in Fig. 17.2.

It has been found that in thermoplastics such initial damage is normally microscopic or submicroscopic and that it occurs under static as well as under dynamic loading conditions. This is illustrated in Fig. 17.3 which gives the results of long-term creep experiments. The line of damage runs

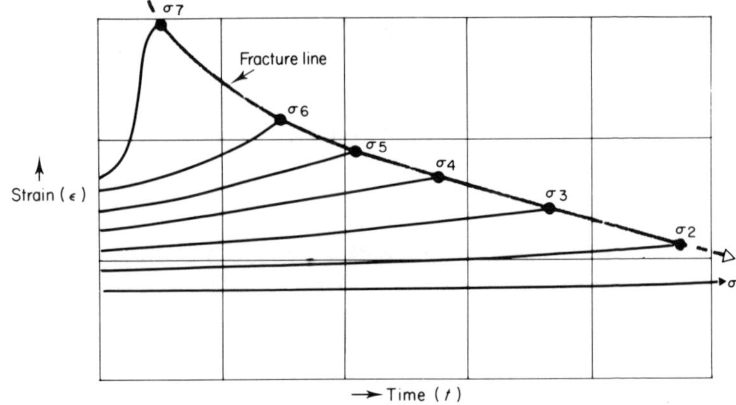

FIG. 17.1.   Typical creep behaviour under constant load, where $\varepsilon = f(t)$ and the stress σ increases from σ_1 through σ_7 (● indicates the point of rupture).

below the line of failure and indicates the appearance of identifiable defects known as 'initial stress defects'; these approach an ultimate value of 0·7 % of the failure strain and are referred to as the 'critical strain', ε_c, on a time scale extrapolated to infinity.

In the particular example just given the initial stress defect occurred in air; the action of aggressive media such as material-specific solvents or

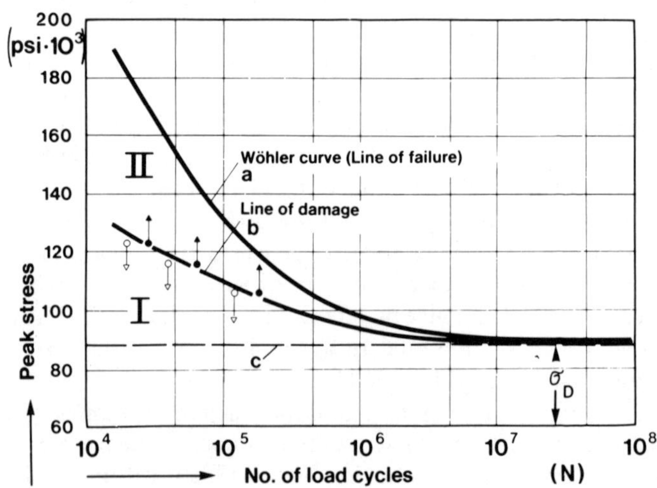

FIG. 17.2.   Fatigue testing of steel.

wetting agents have received far more attention, while air has not generally been considered an aggressive medium in its own right. However, the formation of initial stress defects in air was considered by Cheetham and Dietz, Sauer and Hsiao, Maxwell and Rahm, and others during the 1950s;[1-5] in the course of this they referred, for the first time, to apparent crack formations which they called 'crazes'. These were later described by

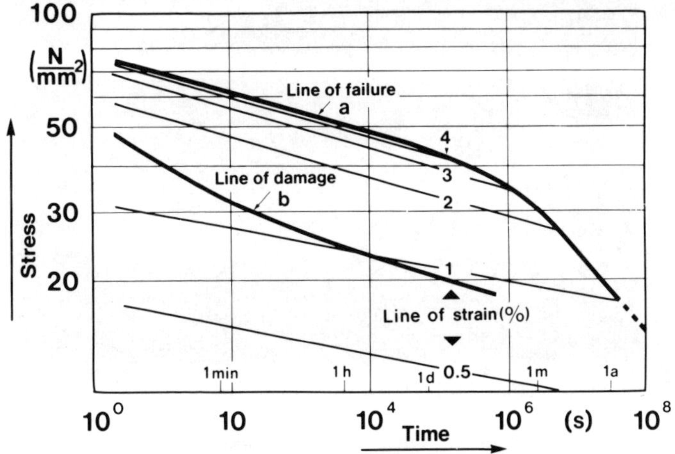

FIG. 17.3.   Creep of PMMA at room temperature.

Kambour by means of extensive illustrations of partially stretched materials with a fibrillar structure. Kambour distinguished between crazes and actual failure manifestations such as cracks.[6-11]

Craze formation became a topic of polymer physics in its own right;[12-16] it proved to be particularly timely and relevant for the understanding of energy absorption during failure initiation in rubber modified materials. Whilst the appearance of crazes had first been noted in (and is still commonly assigned to) amorphous materials, the writer has succeeded in producing craze-like formations in partially crystalline polymers as well (Fig. 17.4).

Returning to the formation of initial stress defects and their consideration as flow phenomena, it is necessary to draw special attention to the extensive work of Menges[17-22] because of its particular value in determining failure limits in terms of critical strain; the data are also particularly important in engineering design. We agree with the view that a correlation of creep behaviour based upon damage in terms of initial stress

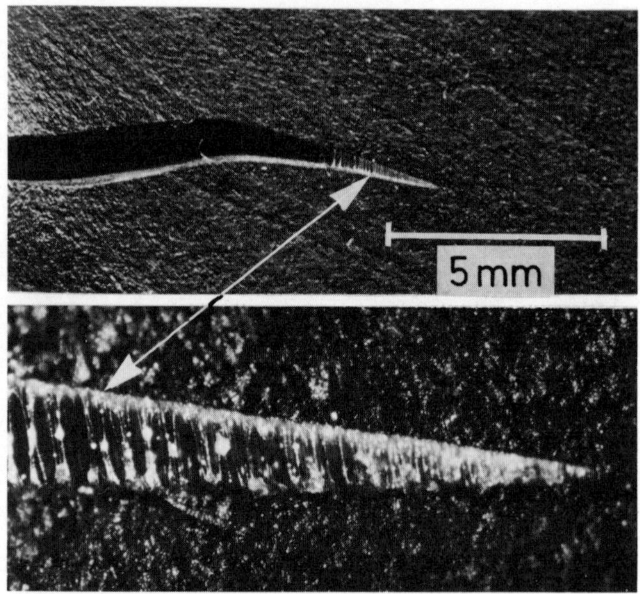

FIG. 17.4.   Macrocrazes in high-density polyethylene caused by insertion of an oversize pin and 7 days immersion in a surface active (wetting) agent at 70 °C. *Top:* the end of a crack. *Bottom:* Tip of the same crack showing macrocrazes at an approximate further seven-fold magnification.

defects is most useful, since it enables one to dispense with long-term experiments which would have to be continued right up to the moment of actual rupture.

FUNDAMENTAL EXPERIMENTAL REQUIREMENTS

Before establishing a standard test method for the determination of the formation of initial stress defects there are a number of *a priori* factors which need to be discussed.

Criterion of Damage

If 'mechanical damage' is to be the criterion for the limiting stress which can be applied it is necessary that some method be devised for its measurement. To do no more than try to distinguish between cracks and crazes in pinpointing the onset of failure by visual methods alone would lead to considerable confusion, not least in all those cases which involve coloured or opaque materials. When trying to determine a limiting value in actual practice a mechanical assessment has been found to be satisfactory; this involves the measurement of the (residual) strength which remains after exposing test specimens to a precisely known stress or strain history.

In a physical sense there exists no instantaneous limit—only a transition region—and it thus becomes necessary to define a substitute limit.

Loading Stresses and Loading Strains

When considering the stress history one must distinguish carefully between subjecting a specimen to a constant load and applying a constant strain.

In the first case the initially absorbed energy increases because the externally applied load continually supplies additional energy so that the total work results in the eventual formation of defects and reaches its highest value at the end of the experiment. In the second case the initial energy is continually reduced due to relaxation.

Under constant load the line of damage (Fig. 17.2) will be exceeded and the line of failure will inevitably be approached and likewise eventually exceeded; on the other hand, no break will occur at constant strain if the remaining energy after the creation of the initial defect proves to be insufficient for the propagation of that defect, i.e. if it falls below a certain critical value.

The type of test that is selected will depend, not least, upon the service conditions for which the component is intended.

Multiaxial Strain

In actual practice, components are almost invariably strained in more than one direction. The energy involved in the formation and propagation of initial defects will depend on the test direction and will thus be found to be both above and below the value expected for an ideally isotropic specimen; indeed, the difference between the observed maximum and minimum values will be a measure of the degree of anisotropy which largely results from the strain history of the specimen.

The testing strain or the testing load may, in turn, be multiaxial, and this will further distort the strain ratio with respect to the axes involved. Tough failure will then tend to change to brittle failure. This change implies an increase in the severity of the test conditions—this has already been discussed at some length in Chapter 15. The net result is that anisotropy and multiaxiality become additional factors to be considered when setting up standard testing methods.

CHOOSING A TEST METHOD

Available Methods

The multiplicity of methods for assessing the extent of the damage caused by an applied stress is due to the fact that a remarkable variety of individually held and cherished theories exist. However, any given method is relevant only in specified circumstances. Moreover, most of the methods tend to ignore crack formation in air (or water) as a basis of comparison and confine themselves to working in an aggressive environment, which admittedly makes matters simple and straightforward in the first instance. One such method is the well known 'Bell Telephone Test' (ASTM 1693) for polyethylene. The relatively short time of exposure to produce failure on immersion in a wetting agent and the low cost, as well as the fact that numerous determinations can be carried out simultaneously are certainly highly commendable features. On the other hand, the test does not attempt to control the uncertainty regarding the stress which produces the deformation. Long-term experiments have therefore been necessary on occasions, even though this involved a substantially greater expenditure on test equipment, as well as a great deal of experimental time.

Long-term creep experiments have the advantage that they may be applied to any material, but they do present great difficulties in interpretation if critical strain rather than rupture is to be taken as the assessment criterion. This applies particularly to the establishment of the limit when long-term failure in air is investigated.

When applying a flexural strain one often states the contact length with the surface fibre rather than the cross-sectional area; the test rig adequately defines the edge stress of the surface fibre, but this is not the case in the Bell Telephone Test. Flexural tests are certainly advantageous whenever the behaviour of a material is to be determined near the edges where an aggressive medium penetrates into the sample. On the other hand, it is not technically possible to produce high deformations, so that the test is effectively restricted to stiff plastics.

While tensile and flexural tests are essentially (or at least nominally) uniaxial, the Bell Telephone Test involves forces which act in more than one direction. The multiaxial principle is also implicit in the ball impression method according to DIN 53 449 which we shall now consider in detail.

The Ball Impression Method

Fundamental Principles

This method was chosen because it enables one to reach a number of conclusions which extend the scope of conventional testing methods.[23–30] It is based upon forcing oversize balls into standard holes of 3 mm diameter (Fig. 17.5). This causes the development of compressive and tensile stresses which reach a maximum at the ball equator. The corresponding strain can be stepwise adjusted by using balls of increasing oversize. As damage is caused, so its extent manifests itself as a reduction in the residual strength.

FIG. 17.5.   Ball impression testing (preliminary steps): A hole of 3 mm diameter has been drilled (top); the selected oversize ball is placed in position (centre); the oversize ball has been impressed (bottom).

This, in turn, is a function of the time interval between impression and the actual destructive test itself. On an empirical basis it has been agreed that the lower limit of permanent damage is reached when a subsequent tensile or flexural test shows a 5% reduction in the original strength of a drilled specimen in the absence of an impressed ball. The corresponding ball oversize then represents the critical strain.

Attempts to express this in terms of uniaxial strain were initiated quite some time ago and involved the use of specific aggressive media. This made it possible to define the limit for eventual rupture within a very narrow margin. In these experiments we compared the critical strain of isotropic test specimens in uniaxial tension at constant strain with the critical ball oversize observed after impression of a specimen which had been prepared in identical manner and which had been wetted with the appropriate aggressive medium. It was found that the critical strain in uniaxial tension (%) was approximately equal to the tenfold ball oversize (mm) for bars of 6 mm width. It became possible to attain an accuracy of about 1% strain for every 0·10 mm ball oversize relative to the equator of the maximum zone of deformation round the ball, on the assumption that we are operating within the linear viscoelastic region. (We must, in any case, confine ourselves to that region in order to avoid unnecessary complications; however, one's conscience can rest easy, bearing in mind that critical strain is a low-strain property.)

Examples for the Determination of Critical Strain in Air
Critical strain in tough failure. In determining the critical strain, that is to say, the maximum permissible strain which just avoids (or just causes) permanent damage to be inflicted, it was found that in the case of polycarbonate this was equivalent to approximately 1·8 % in test specimens without heat treatment (A) and to approximately 1·7 % after annealing for 6h at 110 °C (B) (see Fig. 17.6). All the specimens had been ball-impressed 24 h before testing and had been kept in air. The results were obtained using the criterion of a 5 % reduction in flexural strength relative to that of control specimens without impressed balls (Fig. 17.6). At the time of the experiments no visible signs of defects could be observed which might have led one to suspect that the polymer mass had suffered any permanent damage. It was only seven years later that similar specimens which had been prepared at the time and which had been stored away in air began to show visible signs of stress cracking. Their appearance is pictorially represented in Fig. 17.6 by means of cut-outs of three samples each of which are placed alongside the flexural strength curves for specimens of the corresponding ball oversize. The same photographs are seen again in Fig. 17.7 at much greater magnification.

Since the elastic modulus increased after annealing one might ask whether the greater damage might not have been the consequence of the increased stress produced by impressing balls of the same oversize compared with unannealed specimens and whether it was not therefore due to more work being expended in creating the defects. However, we were able to note during the initial experiments that there still remained a distinct difference in the flexural strength curves of annealed and unannealed polycarbonate, even after compensating for the higher deformation stress by using balls of smaller oversize. This means that the increased damage necessary to reach the critical strain reflects a (morphological) change to a higher degree of order and a reduction in the number of voids or, in thermodynamic terms, to a reduction in entropy. The ability to recognise changes in the morphology of specimens which show themselves macroscopically as cracks only years later and to predict them on the basis of short-term mechanical tests is a major feature of the sensitivity of the ball

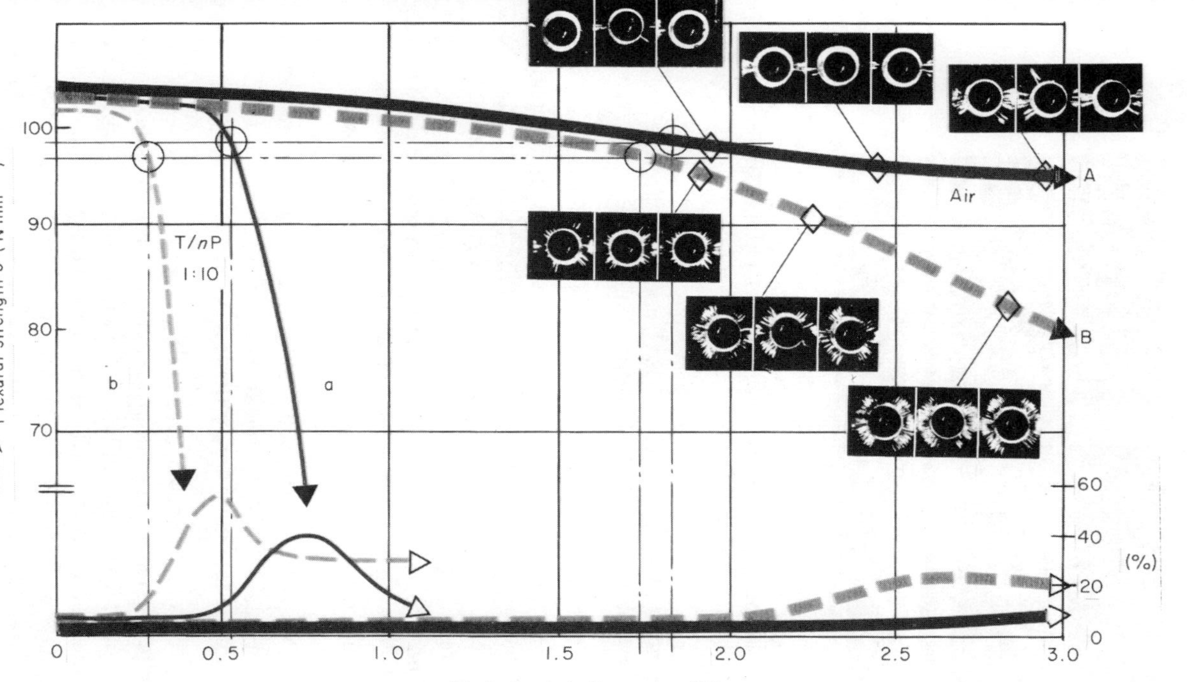

FIG. 17.6.   Strength of polycarbonate after the impression of oversize balls into unannealed specimens (A, a) and specimens which had been heated at 110 °C for 6 h (B, b). Specimens A and B had been kept in air for 24 h after ball impression before being tested and specimens a and b had also been immersed in toluene/n-propanol (T/nP 1 : 10 v/v) for 60 min. Photographic cut-outs of 3 specimens in each case indicate the damage which has become visually apparent at the appropriate strain due to the impression of oversize balls after keeping in air for 7 years. The lower curves indicate the coefficient of variation V. The crack formation limits according to DIN 53 449 which denote a 5 % strength decrease are shown as circles. All tests were carried out at 23 °C and 50 % relative humidity.

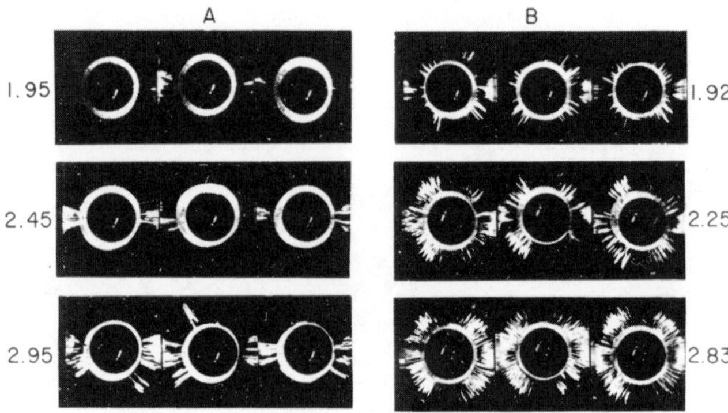

FIG. 17.7. Enlargement of the photographic cut-outs of Fig. 17.6 showing cracks in specimens of polycarbonate containing impressed oversize balls after 7 years in air. A—unannealed; B—annealed (6 h at 110 °C) before ball impression. The figures represent strain due to ball oversize, ε (%).

impression method. This sensitivity is based upon two decisive factors which distinguish the method from all other systems:

(a)   The development of stress cracking is concentrated in one location, the hole periphery; this is also the locus of maximum loading during the strength tests which follow ball impression. The hole constitutes a notch and even the slightest damage will manifest itself right here.

(b)   The concentration of the subsequent testing stress around the hole is a consequence of (i) the reduction in cross-sectional area and (ii) the notch sensitivity of the hole. Both these factors ensure that rupture does not occur just anywhere along the specimen length as a result of statistically distributed bulk defects, but that it occurs exclusively as a result of the damage which is caused by the impression of the ball prior to the test proper.

Together with curves (A) and (B) (in air) (Fig. 17.6) we see curves which refer to the use of an aggressive medium specific to polycarbonate, in fact the same aggressive medium which has also been used for comparing the results of the uniaxial tensile test with those obtained in the ball oversize test. When comparing curve (B) (annealed specimens) with curve (A) (unannealed specimens), the critical strain reduction in an aggressive medium stands out even more clearly than a similar comparison of annealed and unannealed specimens in air. Tests in an aggressive medium thus make the morphological changes caused by ball impression even more striking than similar tests in air.

Critical strain with ensuing brittle failure. It is somewhat more difficult to interpret the submicroscopic changes in a material such as styrene homopolymer if the initial damage upon which these changes are based produce a notch effect which is no greater than the inhomogeneities (the 'internal notches') of the polymer *per se*. Twenty-four hours after the initial deformation the critical strain in air is still above 1 % (Fig. 17.8(a) and (b)); the crack formation in the lower region is not, as yet, sufficiently pronounced. Four weeks later, however, the critical strain was found to be 0·1 % and 0·3 % for annealed samples. These figures are thought to be near enough identical with the ultimate values. This illustrates the importance of the time interval between impression of the balls and the final test. It is seen, just as in the case of polycarbonate earlier on, that the use of an aggressive medium instantly shows up the differences in the mechanical properties of annealed and unannealed specimens, and that this occurs at a lower level of critical strain than in air.

In the case of polystyrene, however, annealing produces an increase in critical strain; this is explained by the relaxation of energy-elastic stresses in the internal structure.

The determination of critical strain in air by the ball impression method requires a specimen in the shape of a rectangular bar. This should be as narrow as possible in order to identify the effect upon the cross-sectional area on either side of the hole during the formation of submicroscopic defects when the zone of stress concentration is subjected to a subsequent tensile or flexural test. We have been using bars which are preferably 6 mm wide, leaving 1·5 mm on either side after drilling the hole.

The tests in air do require a certain expertise and familiarity with the ball impression method because the method is exceedingly sensitive and responds to any variations in the condition of the test specimens.

The results for the materials which have been selected here were the subject of earlier publications; they are comparable to those obtained by Kambour. Our results were based upon the determination of critical strain of stress-free polycarbonate and of annealed polystyrene bars, while Kambour's data derive from creep experiments on the same polymers which had been carried out at a later date. These are shown in Table 17.1. These results should only be regarded as typical guideline data since critical strain cannot, by itself, have a unique value for any given polymer. Not only does there exist a time function with which it is associated, but critical

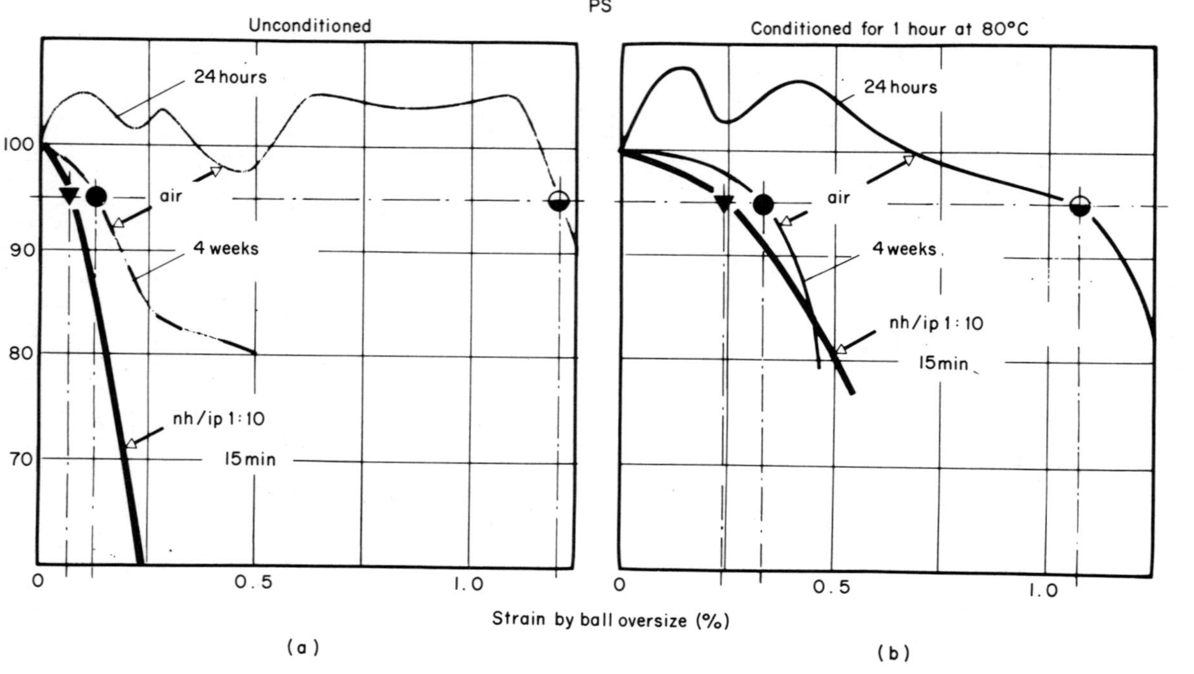

FIG. 17.8. Comparison of the residual strength of standard polystyrene. Flexural tests after 24 h and 4 weeks storage with impressed oversize balls in air at 23 °C of unannealed specimens (a), and after annealing at 80 °C for 24 h (b). Also shown are the strength curves after 15 min immersion in *n*-heptane/isopropanol (1 : 10 v/v) at 23 °C followed by subsequent storage in air for 24 h when the specimens were subjected to the flexural test. The flexural strength σ (N mm^{-2}) was converted to % relative to that of specimens which had not been subjected to straining by means of oversize balls. Crack formation limits after DIN 53 449 denoted by a 5% loss of strength and indicated by ▼ ● ◐.

strain is also highly dependent upon the manufacturing details which characterise each commercial grade and trade mark itself. (In that sense, the concept of critical strain may be said to resemble the concept of relaxation time.) Polycarbonate, in particular, can be made to attain even higher critical strains as a result of suitable polymer modification.

TABLE 17.1

	Critical strain (%)	
	Ball impression method	*Creep experiments after Kambour*
Polycarbonate	1·80	1·80
Polystyrene	0·30	0·35

In Chapter 18 we shall see how aggressive media and specimens containing oversize balls can be used for the deliberate induction of cracks as a function of time and for pinpointing the factors which cause variations in critical strain (and hence in performance) after the materials have been processed into actual components.

REFERENCES

1. R. G. CHEETHAM and A. G. DIETZ, Transact., ASME (1951).
2. C. C. HSIAO and J. SAUER, *J. Appl. Phys.*, **21,** 1071 (1950).
3. J. A. SAUER and C. C. HSIAO, Transact., ASME (1952).
4. B. MAXWELL and L. F. RAHM, *Ind. Eng. Chem.*, **41,** 1988 (1949).
5. J. A. SCHMITT and H. KESKKULA, *J. Appl. Poly. Sci.*, **3,** 132 (1960).
6. R. P. KAMBOUR et al., *J. Poly. Sci.*, Part A, **3,** 1713 (1965).
7. R. P. KAMBOUR et al., *J. Poly. Sci.*, Part A-2, **4,** 327 (1966).
8. R. P. KAMBOUR et al., *Macromolecules*, **1,** 393 (1968).
9. R. P. KAMBOUR et al., *Macromolecules*, **5,** 335 (1972).
10. R. P. KAMBOUR et al., *J. Poly. Sci.*, **11,** 1879 (1973).
11. R. P. KAMBOUR et al., *J. Poly. Sci., Macromolecular Reviews*, **7,** 1 (1973).
12. M. MATSUO et al., *Poly. Eng. Sci.*, **9,** 5, 197 (1969).
13. M. MATSUO et al., *Poly. Eng. Sci.*, **10,** 9, 253 (1970).
14. I. P. BERRY, *J. Poly. Sci.*, Part A, **3,** 2027 (1965).
15. S. S. STERNSTEIN and F. A. MYERS, *J. Macromol. Sci.* (Physics), 8, 539 (1973).
16. C. B. BUCKNALL, *Toughened Plastics*, Applied Science Publishers, London (1977).
17. G. MENGES et al., *Kunststoffe*, **57,** 11, 885 (1967).
18. G. MENGES et al., *Materialprüfung*, **14,** 5, 141 (1972).

19. G. MENGES *et al.*, *Kunststoffe*, **63**, 2, 95 (1973).
20. G. MENGES *et al.*, *Kunststoffe*, **63**, 3, 173 (1973).
21. G. MENGES *et al.*, *Kunststoffe*, **65**, 6, 368 (1975).
22. G. MENGES *et al.*, *Kunststoffe*, **66**, 11, 735 (1976).
23. E. BUCHHOLZ and J. POHRT, *Kunststoffe*, **54**, 10, 635 (1964).
24. J. POHRT, *Kunststoffe*, **59**, 5, 299 (1969).
25. J. POHRT, *J. Macromol. Sci.* (Physics), **5**, 6, 299 (1971).
26. J. POHRT, *Gummi Asbest Kunststoffe*, **24**, 6, 700 (1971).
27. J. POHRT, in *Konstruieren mit Kunststoffen*, Part 2, p. 1057, G. Schreyer (ed.), C. Hanser, Munchen (1972).
28. J. POHRT, *Kunststoffe*, **63**, 3, 163 (1973).
29. J. POHRT, *Gummi Asbest Kunststoffe*, **29**, 6, 384 (1976).
30. J. POHRT, *Kunststoffe*, **66**, 8, 481 (1976).

18

Critical Strain: The Effect of Processing History and Associated Factors

J. POHRT

*Materials Testing Department (Polymers),
BASF, D-6700 Ludwigshafen, Federal Republic of Germany*

INTRODUCTION

Since the mechanical properties of plastics components are greatly influenced by the processing history it is scarcely surprising that the lower limit of irreversible damage which is caused by a stress just sufficient to produce a critical strain is also highly dependent upon the previous processing history. The two examples given in Chapter 17 show the results obtained when testing in air, as well as those obtained after immersion in a specific aggressive medium. Tests in aggressive media provide an instantaneous indication of the changes induced during melt processing, and these changes may be expressed in terms of critical strain. This implies that the critical strain is not a unique value, but one which depends on a number of factors. Seen in this light a test in an aggressive medium is not merely the visible consequence of the presence of internal stresses, but also a strength determination in its own right. Having once obtained standard reference data on the behaviour of an isotropic specimen in an aggressive medium it immediately becomes possible to assess the degree of anisotropy and the presence of inhomogeneities generally. No longer do we look upon critical strain as a material-specific parameter based upon some hypothetical intrinsic structure; instead we begin to analyse defects and localised weaknesses in Griffiths' sense. In fact, we start to determine flaws in the bulk structure as a function of the variables involved in melt processing and as a function of the variables involved in the after-processing of glassy polymers.

ENTROPY-ELASTIC AND ENERGY-ELASTIC INTRINSIC (EIGEN-) STRESSES

When an extrudate has frozen-off the orientation imparted during melt flow produces mechanical weaknesses in the cross-flow direction. The strength properties are determined by the anisotropy, and not by the entropy-elastic stress (which is a function of the orientation).

The entropy-elastic stress is of a low order of magnitude. It only manifests itself as a restitutive force at elevated temperatures and the retraction associated with it can itself be used to determine the approximate degree of orientation in thermoplastics.

The energy-elastic eigenstress, on the other hand, is caused by a volume decrease in the bulk interior so that a force is generated which tends to pull the frozen-off surface layers inwards. This force contributes to the total stress and thus diminishes the external load required to produce failure in a strength test.

Styrene Polymers

If styrene homopolymer is heated to a temperature which is not high enough to produce measurable shrinkage then only the energy-elastic eigenstresses are affected. This is illustrated in Fig. 18.1 in which compression moulded and injection moulded bars which had been prepared at three different temperatures are compared both before and after annealing. The aggressive medium used was n-heptane/isopropanol (nh/ip) 1:10. The formation of cracks before annealing is clearly seen right down to the lowest ball oversizes (left-hand side), but the compression moulded bars (A) are not affected in this way: their orientation is zero, which is confirmed by zero contraction after exposure to a suitably elevated temperature.

No change is seen in the compression moulded bars after annealing (Fig. 18.1, top right); this proves that the bars were free from eigenstresses. The injection moulded bars, on the other hand, show crack formation limits (in terms of ball oversize) after annealing which increase with increasing orientation as indicated by a corresponding shrinkage. In other words: before annealing the strength increases from D to A because the eigenstress decreases. After annealing the position is reversed because the orientation increases from A to D.

Those bars which were subjected to a subsequent flexural test were strained along the orientation direction. This is a serious drawback since it is impossible, in these circumstances, to obtain any meaningful assessment of those cracks which are preferentially propagated along the direction

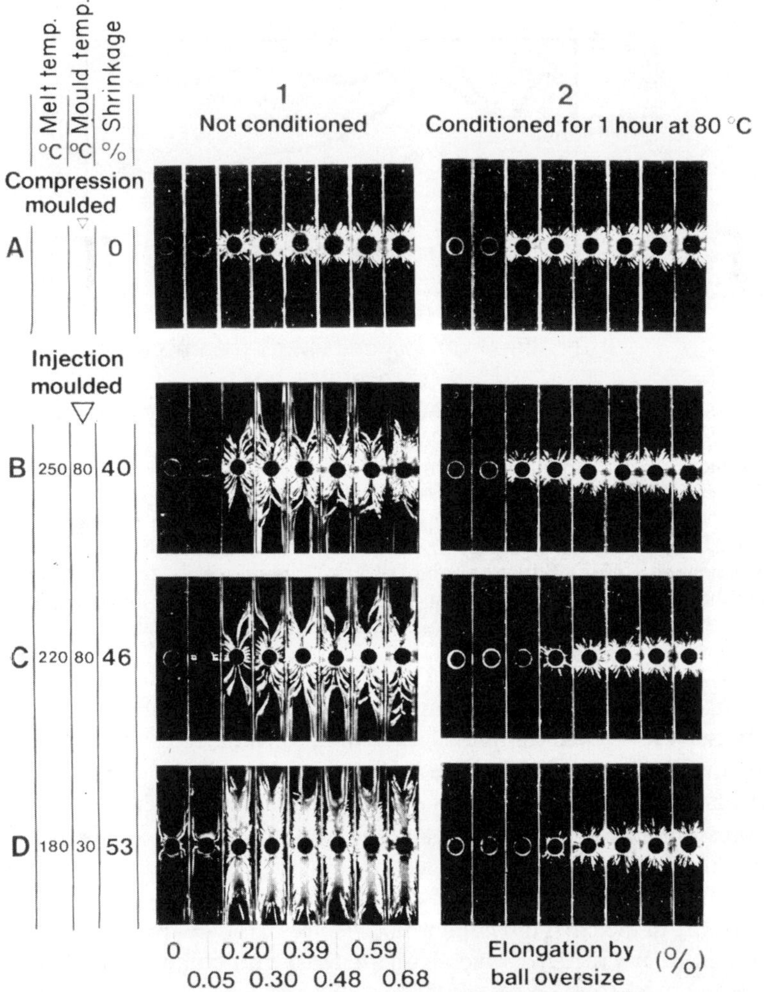

Fig. 18.1. Crack formation limits of general purpose polystyrene after 60 min exposure to *n*-heptane/isopropanol 1:1 (v:v) at strains caused by balls of between 0 and 0·068 mm oversize (equivalent to 0 and 0·68 % strain) in eight increasing oversize steps. The degree of orientation increases from A through D and is expressed in terms of maximum shrinkage from 0 to 53 %. In group 1 (unannealed) the eigenstresses become more and more dominant as the melt temperature decreases from 250 to 180 °C, with a corresponding *decrease* in the crack formation limits. In group 2 the eigenstresses have been removed by annealing and only the effect of orientation remains. Hence the *increase* in the crack formation limits as the temperature decreases from 250 to 180 °C.

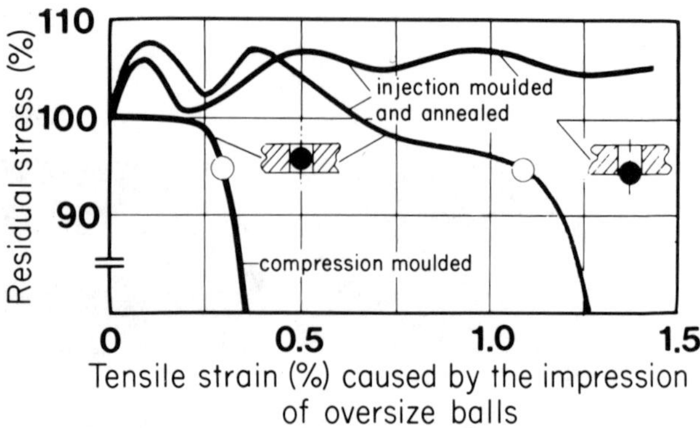

FIG. 18.2.   Residual strength in flexure of general purpose polystyrene 24 h after the impression of oversize balls: Comparison of injection moulded and oriented specimens with unoriented specimens obtained by compression moulding. (Reference standard: Flexural strength of unimpressed (zero oversize) specimens = 100.) The injection moulded specimens have had the eigenstresses (prior to oversize ball impression) removed by annealing at 80°C for 1 h. The specimens also afford a strength comparison after impression with the ball equator near the surface and in the specimen centre. The crack formation limit, represented by a 5% decrease in flexural strength, is denoted by a circle.

of orientation. Figure 18.2 shows the residual strength of a styrene homopolymer 24 h after ball impression and storage in air.

We have compared compression and injection moulded bars with the ball equator at the specimen surface and at the centre. The highest degree of orientation at the surface also produced the greatest strength after ball impression if one considers only the strain which is applied in the longitudinal direction and provided that the stress in the final flexural test is applied in the same direction.

Flexure in the cross direction would reduce the crack formation limit to (or even below) the level which is characteristic for compression moulded bars.

The fact that testing in the direction of orientation produces greater strength was reported some time ago by Fischer who examined high-impact polystyrene on the basis of creep experiments, using a mixture of olive oil/oleic acid 1:1 as a 'developing' medium (Fig. 18.3).

Henceforth it will certainly be necessary to carry out tests in the directions of both maximum and minimum strength. For this specimens appropriately cut from injection moulded rectangular plaques are eminently suitable. (An analogous procedure is also used in stress-optical measurements—see Chapter 21.)

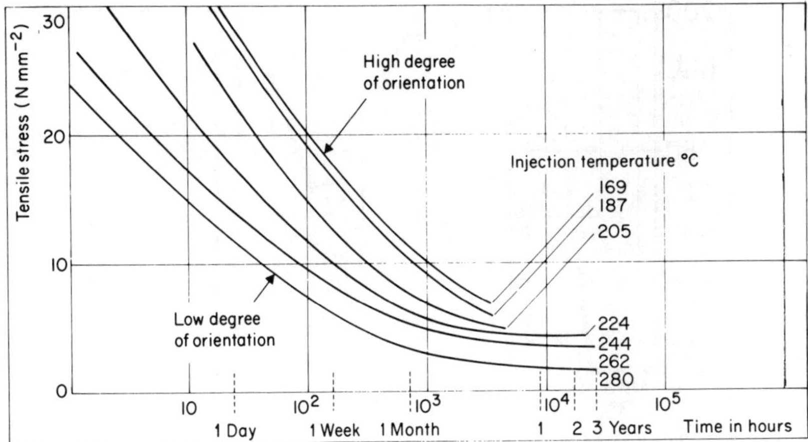

FIG. 18.3. Creep strength of toughened polystyrene of different degrees of orientation in 1:1 mixture of olive oil and oleic acid. The tensile stress at zero time corresponds to that of an arbitrary stress at constant load. It represents the magnitude of the initial stress which will ultimately lead to rupture.

Polyethylene

When considering the effect of orientation the literature concerns itself predominantly with amorphous polymers. However, anisotropy (which is structurally significant, but of no significance with respect to stress) also presents a problem in partially crystalline thermoplastics.

Tensile impact tests on 10×50 mm bars in which the notch is a 3 mm hole have shown that in the case of a partially crystalline type, injection conditions which produce higher orientation cause the impact strength to be correspondingly lower (Fig. 18.4). The reduction in impact strength is caused by directional crystallisation which, in turn, reduces the ability of the material to deform in shear; this clearly implies a reduction in tensile strain. Since the deformation energy is a function of the product of stress and strain any reduction in tensile strain would need to be compensated by a corresponding increase in tensile stress if strength (in terms of the energy necessary to produce failure) is to be maintained. However, the failure energy in polyethylene has a greater strain component than most other thermoplastics, so that the limits imposed on strain cannot be compensated for by a sufficient increase in stress. This results in a reduction of the impact strength and represents a significant difference compared with the behaviour of materials like polystyrene which are less dependent on tensile

Polymer Rheology

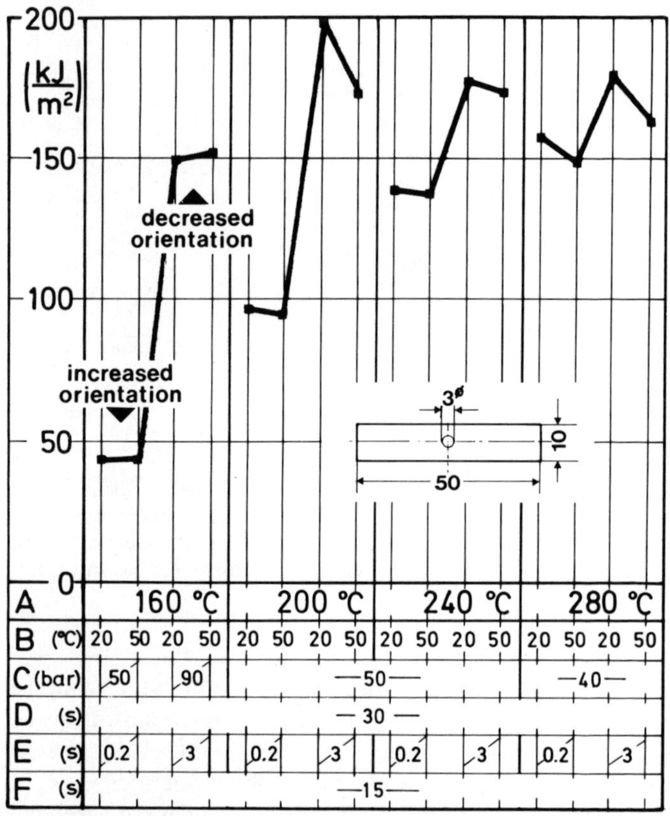

FIG. 18.4.   Tensile impact test on high-density polyethylene (HDPE) specimens (4 × 10 × 50 mm) notched with a circular hole: Effect of variation in the processing conditions. A—Melt temperature; B—Mould temperature; C—Injection pressure; D—Cycle time; E—Injection (shot) time; F—'Packing' time. The particular shape of the kinked curves is a function of E, whilst the shift to higher strength values is determined by A. (The points are joined together only in order to make the changes in impact strength values stand out more clearly.) The standard deviation was too small to enable it to be included in the figure.

strain and in which the tensile strength increases with increasing orientation.

However, we are here concerned not so much with impact strength *per se* as with the indications relating to orientation which an impact strength analysis provides. We may therefore also use anisotropy for examining the phenomenology of tensile failure itself. The application of a stress in the direction of orientation is characterised by a crack formation limit which is higher than that observed when applying a stress in the cross direction, especially in the presence of an aggressive medium. This general rule also applies to polyethylene.

The direction of orientation is not necessarily identical with the injection direction (see also Chapter 21) since polymers tend to orient biaxially or even in a direction normal to the axis of injection under certain conditions. This is illustrated by means of electronically recorded stress–strain curves of drilled specimens in tensile impact tests (Fig. 18.5). Figure 18.5 shows that the tensile strength parallel to the direction of orientation (left-hand side) at the lower processing temperature (a) is greater and that the tensile strain is smaller than at the higher processing temperature (b). In the cross direction (right-hand side) the tensile strength is lower because of the presence of lines—or rather planes—of weak points. However, failure no longer occurs in brittle manner: we have 'shear break' or 'bructile' failure.

The lower tensile energy exhibited along the lines, or planes, of weakness correspond to a lower resistance to fracture in tension as the crack formation limit (which has been thus reduced) is reached. The second material (Fig. 18.6) shows a similar effect of failure in a brittle manner at lower temperatures (a) when it is strained parallel to the orientation direction (left-hand side). In the cross direction however (right-hand side) the higher temperature effects brittle failure (b) because of orientation (left-hand side, (a)). The specimens in (b) did indeed break crosswise on application of an aggressive medium in an environmental stress cracking test. This clearly indicates orientation in the cross direction.

In the case of polyethylene the use of conico-cylindrical steel pins is preferred to the use of steel balls (Fig. 18.7). The pins enable one to attain a higher deformation and this is particularly appropriate where the specimen thickness is insufficient to retain a sphere. Furthermore, loci of weakness tend to develop near the surface in partially crystalline polymers: this results in shear between the rapidly freezing layer which is in contact with the mould and the deeper layers which are still in the molten state and which flow along the frozen-off portion. Not least, there are topographic differences between the detailed crystal formation close to the mould wall and the crystal formation of the bulk of the moulding which only freezes off much later. It is therefore desirable to select a method of impression in which the initial deformation is applied along the entire specimen thickness. In such a situation a cylindrical pin is clearly superior to a sphere. In very thick-walled specimens the same applies for the opposite reason: the affected equatorial region may be too small relative to the total cross-sectional area. On the other hand, steel balls are commercially available in a

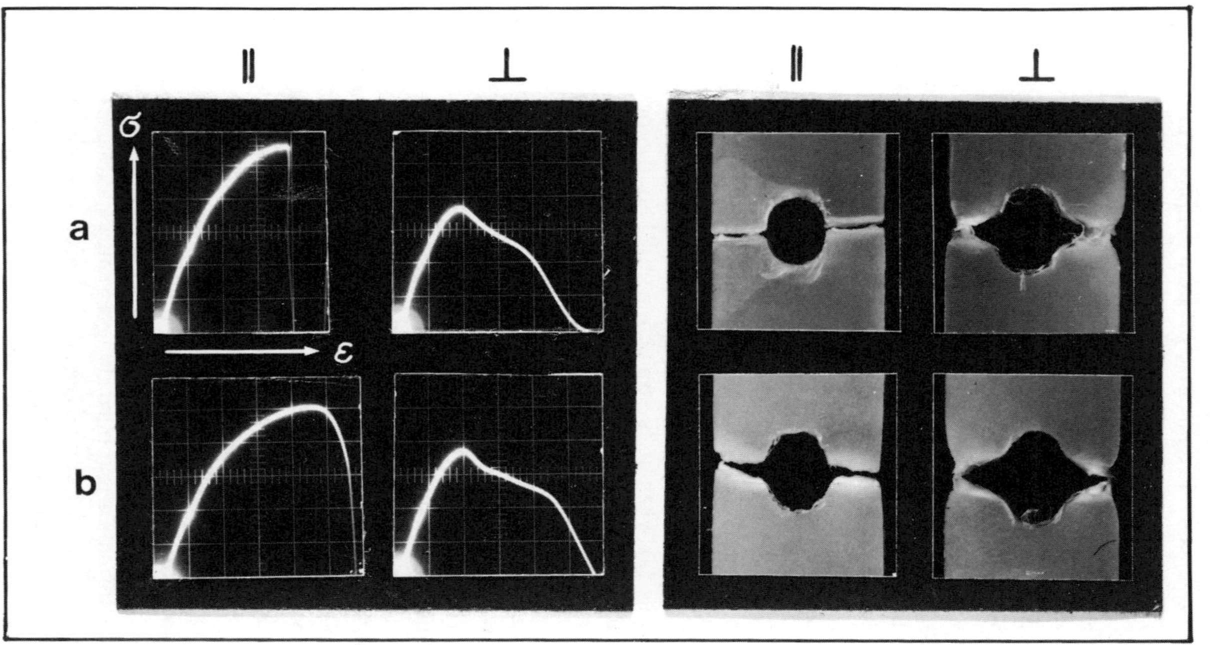

Fig. 18.5.   Mechanical behaviour of HDPE (MFI 0·22, density 0·95). Moulding conditions: a—Melt at 180 °C, injection pressure 50 bar; b—Melt at 270 °C, injection pressure 40 bar. *Left:* Electronically recorded stress–strain curves of high speed uniaxial tensile tests on specimens with hole notches. Straining rate about $3\,\mathrm{m\,s^{-1}}$; $2 \times 50 \times 80\,\mathrm{mm}$ specimens were cut from injection moulded sheet parallel to (∥) and normal to (⊥) the injection direction. *Right:* Appearance of similar specimens after rupture in a tensile impact test at a comparable straining rate.

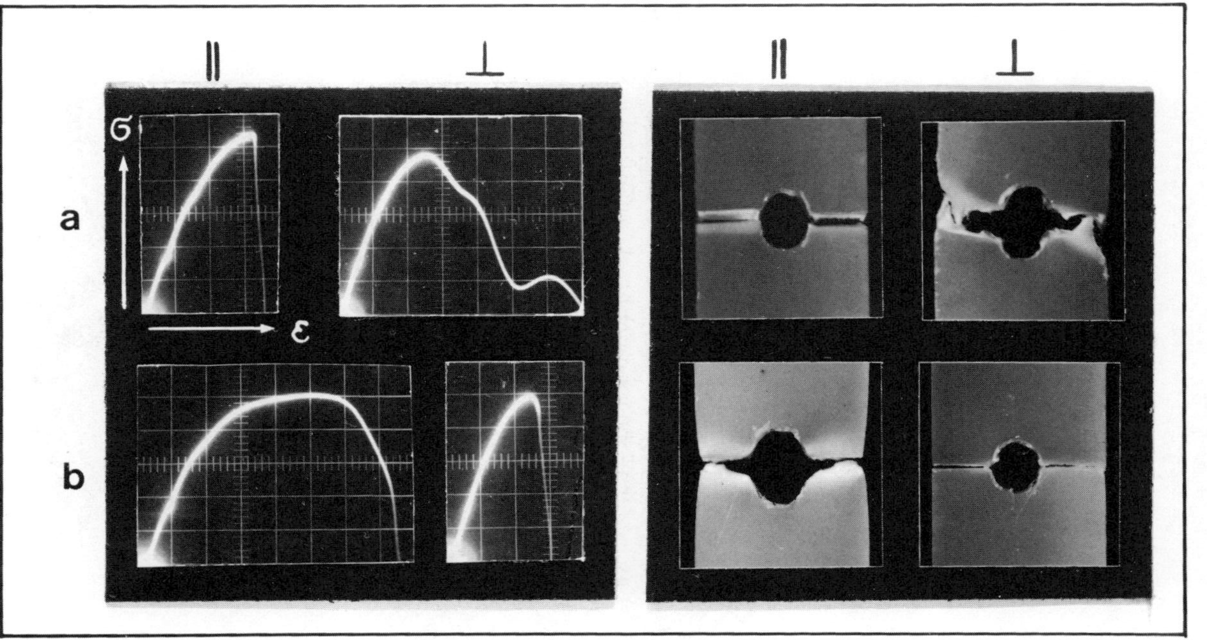

FIG. 18.6. Mechanical behaviour of HDPE (MFI 0·96, density 0·95). Moulding conditions: a—Melt at 180°C, injection pressure 50 bar; b—Melt at 270°C, injection pressure 40 bar. *Left:* Electronically recorded stress–strain curves of high speed uniaxial tensile tests on specimens with hole notches. Straining rate about 3 m sec^{-1}; 2 × 50 × 80 mm specimens were cut from injection moulded sheet parallel to (∥) and normal to (⊥) the injection direction. *Right:* Appearance of similar specimens after rupture in a tensile impact test at a comparable straining rate.

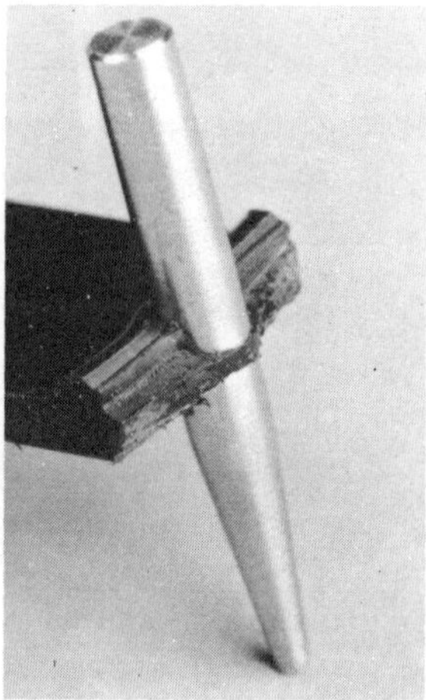

Fig. 18.7.   Pin impressed into a weld region.

close range of appropriate sizes with extremely narrow tolerances, while cylindrical pins have to be specially made. Figure 18.8 shows an example of pins being used for ascertaining the topography of weaknesses in a screw cap made from high-density polyethylene (HDPE).

On immersion in an aggressive medium break occurs in the direction of injection flow; this shows that failure has been due to orientation. The

Fig. 18.8.   Pin impressed into a moulded knob.

quantitative evaluation is based on the diameter of the steel pin which just causes fracture.

We obviously want mouldings to be free from intrinsic damage and to have as high a critical strain level as possible. This may be accomplished by the judicious selection of the polymer grade and by carefully adjusting the injection moulding conditions.

MOLECULAR WEIGHT

Drawn polystyrene

When considering the direction of weakness in an oriented structure we refer to the clearly visible extension of cracks parallel to the axis of orientation. The use of an aggressive medium produces this effect instantaneously; this is seen in Fig. 18.9 for drawn styrene homopolymer.

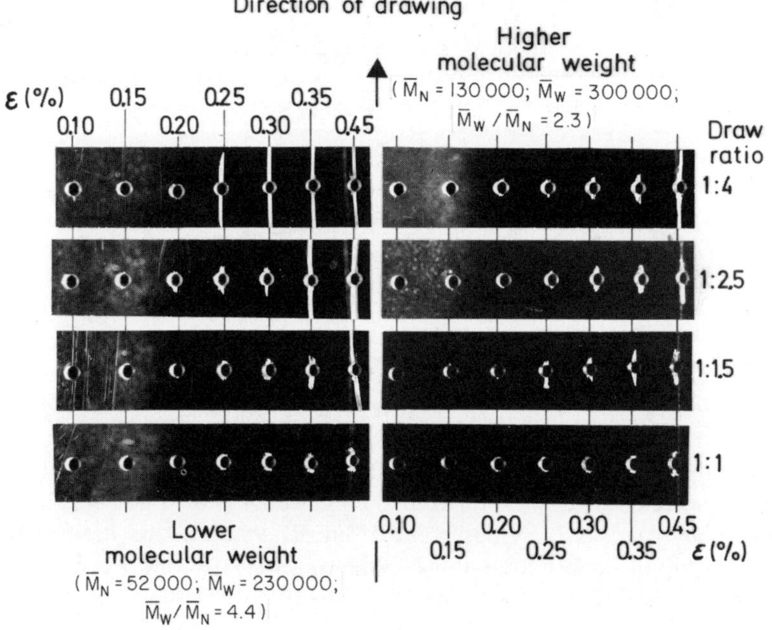

FIG. 18.9. Stress crack formation in drawn GP polystyrene sheet of different molecular weight after the impression of oversize balls and immersion for 60 min in *n*-heptane/isopropanol (1:5) at room temperature. Draw ratios range from 1:1 to 1:4. The percent strain was empirically taken to be the tenfold ball oversize.

The samples were impressed with balls of stepwise increasing oversize which corresponded to strains of 0·10 to 0·35 %. It is seen that the crack intensity of polymers of lower and higher molecular weight differs considerably (compare left-hand side and right-hand side).

When drawn at a ratio of 1:1 (lower line) no visible differences are as yet apparent. At a draw ratio of 1:1·5 the crack at the highest deformation level is much longer (left); the same occurs at the next lower deformation level after drawing to 1:2·5 and continues in the same way at a draw ratio of 1:4 (top left). Only at that very high draw ratio does the high molecular weight polymer begin to show a crack of similar length, and then only at the highest deformation level (top right). The cracks are clearly much shorter in the higher molecular weight material.

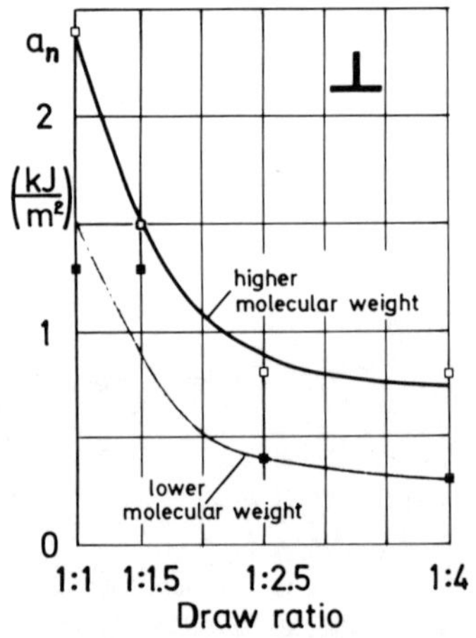

FIG. 18.10.   Flexural impact test (according to DIN 53 453); $4 \times 6 \times 50$ mm specimens cut from drawn GP polystyrene sheet normal to the draw direction ($\perp$).

Molecular weight is obviously relevant to crack behaviour. After all, a lower molecular weight implies the presence of more 'notches', on a molecular scale, as Lenk has pointed out. This also shows itself in the relationship of impact strength when the test is carried out in a direction normal to that of the orientation (Fig. 18.10).

Polycarbonate After Repeated Recycling of Injection Moulded Material

The repeated reprocessing of polymers leads to a reduction in molecular weight due to thermal and/or oxidative degradation which is greatly

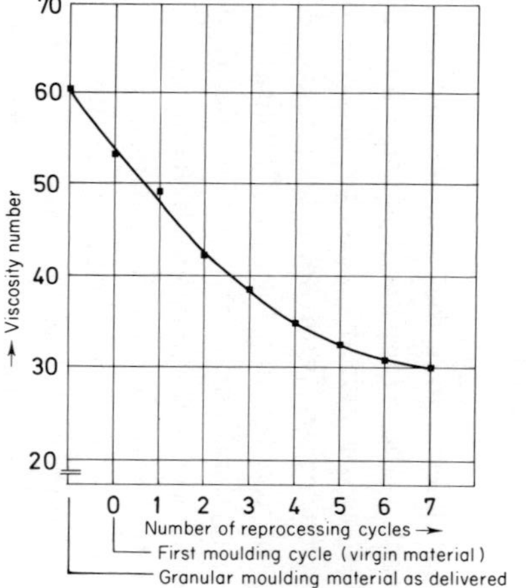

FIG. 18.11. Changes in the viscosity number (DIN 51 562) as a consequence of repeated reprocessing of the same material (glass-fibre-reinforced polycarbonate).

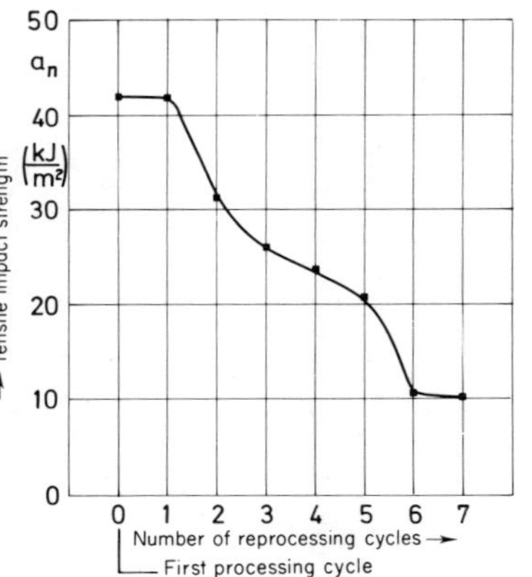

FIG. 18.12. Changes in the Charpy impact strength (DIN 53 453) as a consequence of repeated reprocessing of the same material (glass-fibre-reinforced polycarbonate).

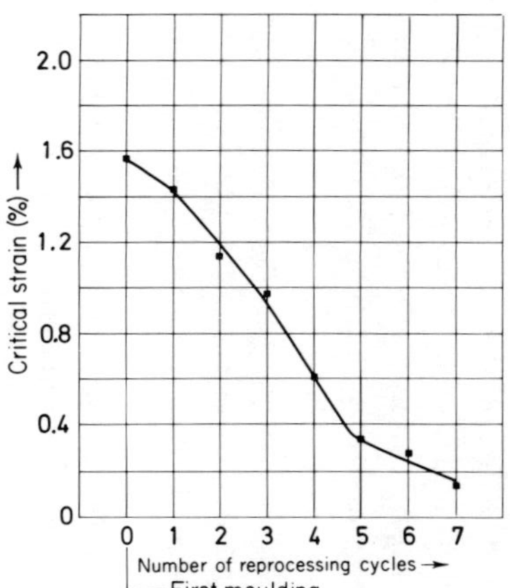

FIG. 18.13. Changes in the crack formation limit (critical strain) in air (DIN 53 449) as a consequence of repeated reprocessing of the same material (glass-fibre-reinforced poly-carbonate).

accelerated under high-shear conditions. This obviously affects the mechanical properties; in industrial production only small amounts of reground material are therefore added to the feedstock if the best product quality is to be achieved. Changes in molecular weight may be monitored by means of solution viscometry and this can then be correlated with the results of mechanical tests.

In order to show how such a correlation may be accomplished we remoulded a glass-fibre-filled polycarbonate in a number of successive cycles. Recycling 100 % of the material at each step served to demonstrate, as clearly as possible, the connection between the rheologically determined bulk condition and the constitution, as well as the influence that these have upon the mechanical properties of the material. Figure 18.11 shows the reduction in viscosity number with each successive cycle, while Fig. 18.12 gives the results of the Charpy impact test (DIN 53 453).

The crack formation limit, or critical strain, also diminishes steadily in air (Fig. 18.13). In this test $4 \times 6 \times 50$ mm bars (DIN 53 449) were ball-impressed and stored in air at 23 °C and 50 % relative humidity for 7 days. A flexural test according to DIN 53 452 followed and the limit of 5 % loss of strength relative to the strength of the unimpressed control bars indicated the critical strain.

Other polymers show a similar common trend of viscosity number, impact strength and critical strain. In some polymers stress cracking has been demonstrated even more spectacularly by the use of appropriate aggressive media.

IMHOMOGENEITIES DUE TO THE FORMATION OF SPHERULITES

Submicroscopic structural defects may be assigned to the detailed morphological structure such as the arrangement of spherulitic clusters in PVC.[1] Whether these clusters are indeed defects or the very opposite, namely regions of improved order, is of special significance for the processing of PVC and for the optimisation of plastification.

It is extremely fortunate that the flexural impact test for the identification of (internal) notches (Stuart) has once again proved to be suitable. The only disadvantage is the possibility that the impact energy necessary for reaching the critical fracture limit may not in fact be attained. Although notching may serve to bring about fracture this usually also involves the destruction of the integrity of the surface fibre even before the test proper is started, so

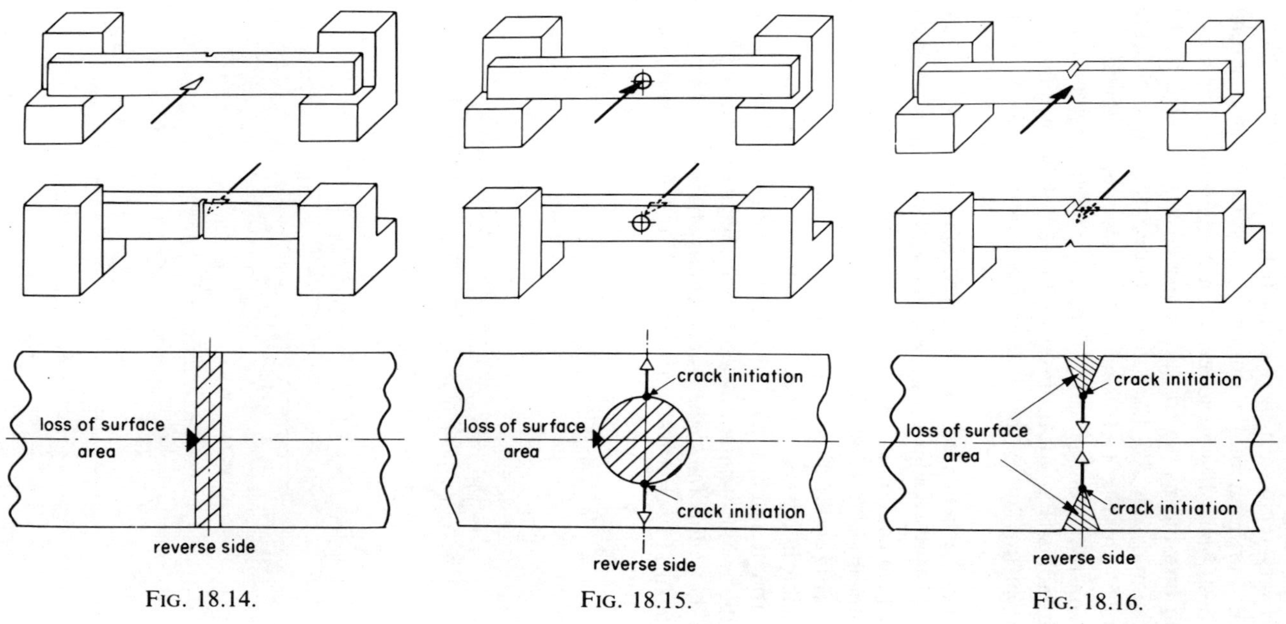

FIG. 18.14.

FIG. 18.15.

FIG. 18.16.

 Polymer Rheology

that the very region which is of particular interest in a flexural test is excluded. Furthermore, the standard notch radius is very small and this often causes break inside the relaxation time of the secondary maximum, which in turn means that brittle failure ensues and that further analysis becomes impossible. We have therefore discarded notches which cut into the surface fibre (Fig. 18.14) and use notches either in the form of a drilled hole of constant radius (Fig. 18.15) or in the form of a double V-notch with variable radius (Fig. 18.16) instead. These notches are tangential to the surface which is hit by the striker.[2] Fracture commences on the surface which faces away from the striker because that is the place where the stress is concentrated when notches are placed tangentially with respect to the surface. The polyaxiality greatly restricts any latent flow tendencies, so that the specimen cannot 'dodge' the stresses by shearing.

The hole is really a composite of the two laterally positioned half-round notches (Fig. 18.17). It represents the simplest and most reliably reproducible notch design. The double V-notch is a further development of the half-round notch. It has the advantage of allowing the notch radius to be modified while maintaining a constant effective cross-sectional area. The smaller the radius the higher the stress or strain rate.

The flexural impact test with double V-notches of variable radius has been electronically recorded in the form of a stress–strain diagram in flexure

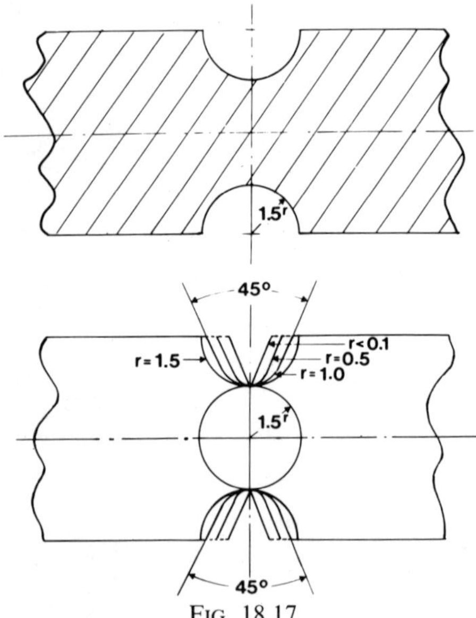

Fig. 18.17.

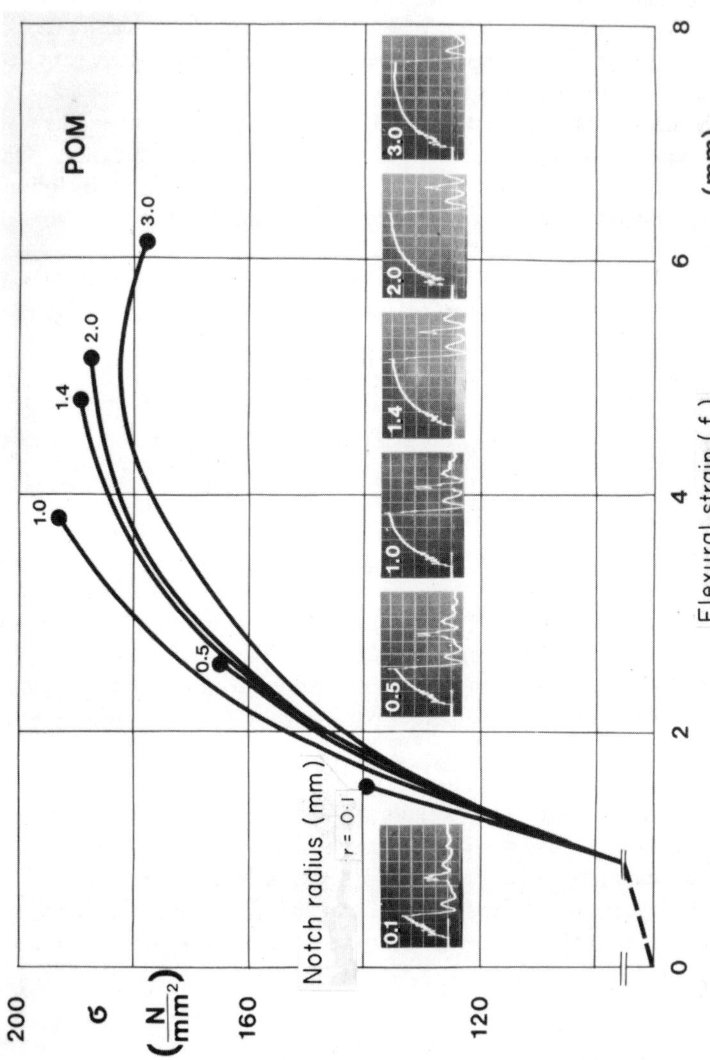

Fig. 18.18.   Extension of the electronic stress–strain record in tension (inset) by means of a plot of stress (σ) vs. flexural strain (f) in a flexural impact test on 4 × 6 × 50 mm acetal (POM) resin specimens with double-V notches of different notch radii.

in a manner which is comparable to the tensile impact test mentioned earlier on, but with a different clamp velocity (Fig. 18.18). In both cases the strain decreases with increasing stress until the tough/brittle transition is reached.

Changes caused by localised defects are much more significant in the region beyond the yield point where they manifest themselves by a reduction in strain than in the region below the yield point where they are indicated by a reduction in stress.

The flexural impact test with a double V-notch of 0·25 mm radius was used for the assessment of natural gas pipes of 2 mm wall thickness. The pipes had been extruded at four different levels of plastification, including some with much less plastification than is commonly used. The impact

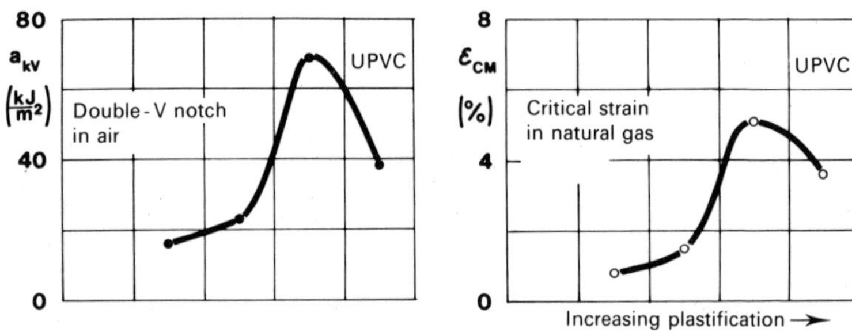

FIG. 18.19.

test results are shown in Fig. 18.19 (left). The critical strain was determined by a tensile test after ball impression and storage in an atmosphere of natural gas (Fig. 18.19, right). It agreed with the trend of the impact strength data, showing, once again, a parallel response to the presence of structural defects.

In a flexural impact test time to break is reduced by internal notches. On the other hand, the defects or weakness are under particularly concentrated attack when an aggressive medium is present. The actual damage done after subjecting the material to these 'static forces of disintegration' depends on the severity, e.g. the radii of the 'internal notches' which constitute the sum-total of all the internal weakness. The nature of these still awaits undisputed identification.

CRYSTALLINE INHOMOGENEITIES

The preceding example illustrated the relationship between the rheology of processing and the mechanical properties in the glassy state. PVC is an amorphous polymer and critical strain could be fairly readily correlated with weakness arising from the spherulitic microstructure.

In partially crystalline polymers, on the other hand, it is rather more difficult to design and adjust the processing conditions in a way which ensures that the component attains the level of strength which the nature of the polymer might entitle one to expect. Classical short-term testing methods are incapable of detecting the long-term effect of flaws and weaknesses which at the time of testing are as yet submicroscopic.

This is particularly evident when moulded articles are subjected to elevated temperatures. A similar problem is encountered when considering the weld strength of pipe made from HDPE.

We started by exploiting the fact that aggressive media can be used to indicate morphological changes which will cause a loss of strength in the long term. We have seen that these indications may be obtained in terms of changes in the crack formation limit, and within quite a short period of time. Thus, oversize pins inserted into a welded specimen will cause rupture in the welding zone, whilst the zone well away from the weld line shows no

FIG. 18.20.   Environmental stress cracking of HDPE pipe in the weld region after impression of oversize pins and 7 days immersion in a wetting agent at 70 °C.

sign of rupture when an identical pin is inserted there (Fig. 18.20). Since the crack tends towards the weld margin, it would appear in this case that a transition region exists which is especially susceptible to rupture; this has also been observed in the case of metals. It remains to be shown whether this is caused by a direct change in the crystalline order, or by the stresses which result from volume contractions due to the density changes which occur

FIG. 18.21.   Environmental stress cracking of HDPE sheet in the weld region after impression of oversize pins and 7 days immersion in a wetting agent at 70°C.

during crystal formation. In any case, however, one may start by using oversize pins to identify the regions of weakness (whatever their cause) and to determine their strength—or lack of strength. The information is obtained in terms of crack formation limits, or critical strains. Furthermore, we may also detect cracks in the weld itself when a cold-welding process has been used (Fig. 18.21).

Lastly, the built-up material in the weld may itself have a lower resistance to deformation (Fig. 18.22) so that the crack may start on one surface (Fig. 18.22(a)), penetrate the specimen, and emerge on the opposite side (Fig. 18.22(b)).

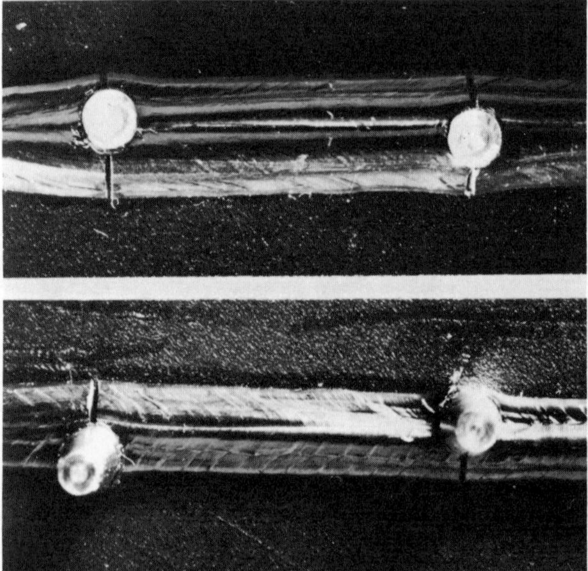

FIG. 18.22.   Environmental stress cracking of HDPE sheet in the weld region after impression of oversize pins and 7 days immersion in a wetting agent at 70°C.

MACROSCOPIC DEFECTS

Styrene Homopolymer

One of the most clearly visible weaknesses is associated with the welds which are formed when streams of molten polymer converge and coalesce. Such welds may be produced by injection moulding 4 × 6 × 50 mm bars in a mould containing a cylindrical insert of 3 mm diameter. The insert produces a corresponding hole with pronounced orientation in the material on its upstream side and with a well defined weld line on its downstream side. The impression of oversize balls, followed by immersion in an aggressive medium, causes the formation of crack patterns which are typical for orientational and weld inhomogeneities respectively (Fig. 18.23).

The cracks which are due to orientation range over an angle of 45° on either side of the longitudinal axis and have zero extension in the cross direction, whilst the crack in the weld line runs in a single plane and tends to fork at the tip, its length depending directly on the ball oversize. On

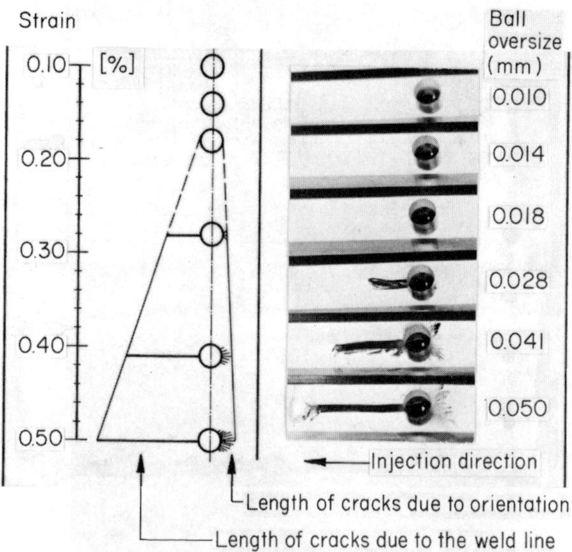

FIG. 18.23. Stress cracking in GP polystyrene induced by *n*-heptane/isopropanol (1:2) at room temperature. The specimens (4 × 6 × 50 mm) had been injection moulded and annealed for 1 h at 80 °C. The crack length as measured in the photographs on the right were then plotted against percent strain, the latter being taken as the tenfold ball oversize in mm (left).

converting that oversize into the corresponding percentage strain and plotting this against the crack length one obtains a straight line which gives the crack length limit for any given applied strain within the appropriate limits. In this particular case it was seen upon extrapolation to zero crack length that a strain of about 0·18 % can just be tolerated without causing a crack to appear in the weld. The function for the maximum length of orientation-induced cracks was a similar straight line which approached the same zero value of 0·18 % strain on extrapolation. In other words, the crack length is a function of the energy expended during deformation. The difference in the crack lengths in the weld line and in the oriented material along the axis of orientation shows that the flaws or defects caused by weld formation and by orientation respectively are very different in magnitude, and that they result in a corresponding difference in the loss of breaking strength which they cause.

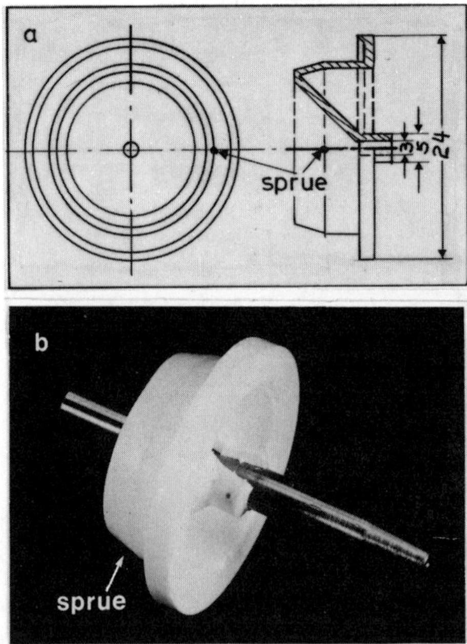

FIG. 18.24.   Crack formation in the weld line of a bottle cap moulded from HDPE after immersion in a solution of surface active agent at 70 °C. (a) Design sketch including a central 3 mm diameter hole in the moulding; (b) crack in the weld line on the opposite side (with respect to the sprue) of the hole.

Polyethylene

The use of a specific aggressive medium enables one to investigate the weld strength of moulded drip spout inserts for the necks of bottles. The moulding is made from HDPE and is seen in Fig. 18.24(a); the crack caused by forcing in an oversize pin is shown in Fig. 18.24(b).

It is possible to optimise strength by changing the processing conditions until the crack formation limit (or the critical strain) has reached the highest value which can be obtained.

SUMMARY

We began with a constructive evaluation of the stress limits by using the concept of critical strain. It soon became obvious that critical strain was largely determined by the rheological stress history. This, in turn, led to an inversion of the procedure of assessing changes in the inherent condition of components on the basis of critical strain.

In view of this inseparable duality it is easy to appreciate that any testing method based upon stress cracking must be open-ended both ways: only in this way does it become possible to quantify critical strain in standard test specimens in a manner which enables it to serve as a design parameter and to use any change in the latter to describe and understand the condition of a component as it leaves the production line.

Having first ascertained the critical strain under arbitrary standard conditions (and having accorded it the status of a design parameter) it is still necessary to test the end product in order to avoid any risk that the values may have changed due to changes resulting from varying the processing conditions.

Uniaxial tensile tests, although well defined, are not suitable for converting figures from standard test specimens to data which can be claimed to be acceptable parameters for assessing the quality of moulded articles. They appear even more irrelevant once one compares the results with those obtained by ball and pin impression methods. Neither can the technically simple flexural test according to ASTM 1693 be considered in this context because its time-dependent crack formation limits cannot be related to either an applied initial stress or to an applied initial strain. Both initial stress and initial strain, however, are of the utmost importance for the evaluation of crack formation processes. This will become even more evident following the investigations described in Chapter 19.

REFERENCES

1. G. MENGES and N. BERNDTSEN, *Kunststoffe*, **66**(11), 735 (1976).
2. J. POHRT, *Kunststoffe*, **66**(6), 348 (1976).

19

Critical Strain: An Engineer's View of the Energy Balance During Deformation

J. POHRT

Materials Testing Department (Polymers),
BASF, D-6700 Ludwigshafen, Federal Republic of Germany

INTRODUCTION

Stress cracking has been presented as the result of some critical strain being exceeded. Not only does this constitute a terminological dichotomy, but it reflects altogether the uncertainty as to whether the stress or the strain is ultimately responsible for crack initiation.

CONSTANT STRESS OR CONSTANT STRAIN?

Let us first consider the stress–strain curves for two different materials (Fig. 19.1) and see what responses arise when the materials are subjected to (i) a constant stress and (ii) a constant strain.

Figure 19.2 shows that at constant stress, material B absorbs more energy than material A because of the greater strain. The higher energy input increases the probability that material B will fail at an earlier stage, unless it is correspondingly stronger. On the other hand, material A suffers the higher energy input at constant strain because it is exposed to a higher stress (Fig. 19.3).

Strictly speaking, we should also include energy losses due to retardation and the generation of heat, but we shall ignore these losses here and confine ourselves to a consideration of fundamental principles.

Assuming that the areas under the stress–strain curves up to failure are equal (Fig. 19.4); material A then reaches the higher critical stress while material B attains the higher critical strain.

Let us assume that a further comparison of the materials shows that their critical strain is identical. Figure 19.3 may be used to illustrate this: the

     Polymer Rheology

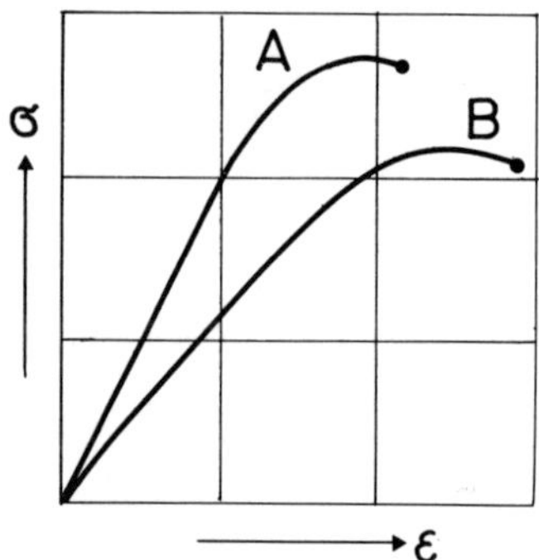

FIG. 19.1.   Tensile tests on plastics with different modulus; stress–strain curves.

constant strain $\varepsilon_{A,B}$ equals the critical strain ε_c. This time A is the material with the higher energy absorption (the higher 'critical energy'), because it has the higher critical stress.

The superiority of material A becomes obvious when we identify the constant stress $\sigma_{A,B}$ with the critical stress σ_c in Fig. 19.2. Material B is clearly subjected to heavy demands and is forced to absorb an excessive amount of energy right up to the failure limit. This limit is shown by the area under the curve at $\varepsilon_{A,B}$ $(= \varepsilon_c)$ in Fig. 19.3.

The practical significance of this is that a comparison of material quality becomes possible only when critical strain is considered side by side with the corresponding critical stress. Thus, glass fibre reinforcement may cause a thermoplastic to show a higher (critical) stress together with a lower

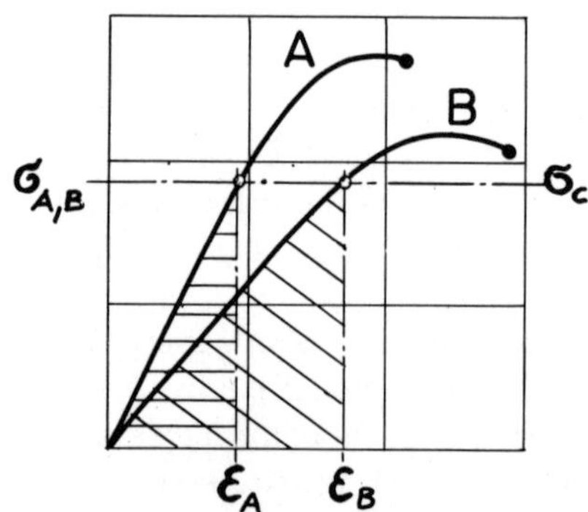

FIG. 19.2.   Tensile tests on plastics; the diagram illustrates the difference in strain at some arbitrarily chosen stress for the two materials, as well as the difference in the energy absorbed.

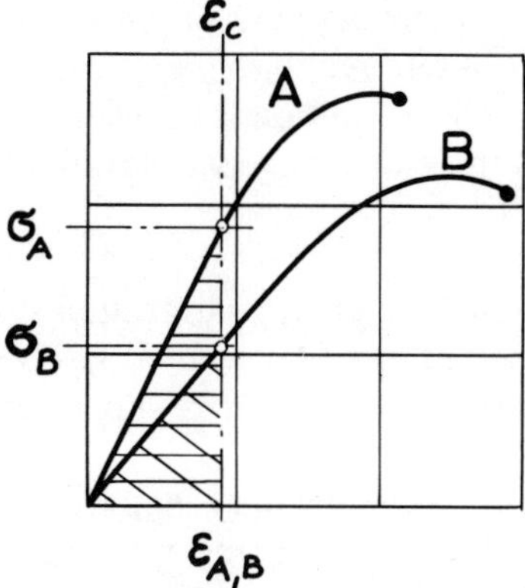

Fig. 19.3. Tensile tests on plastics; the diagram illustrates the difference in stress at some arbitrarily chosen strain for the two materials, as well as the difference in the energy absorbed.

(critical) strain. This may in fact shift the capacity to withstand an applied mechanical energy to a higher failure energy level.†

The area under the curve represents the energy which is instantaneously absorbed when a given stress or strain is applied. The applied energy makes demands upon the material which will fatigue under static or cyclic loading and suffer eventual structural breakdown as a result of the stepwise growth

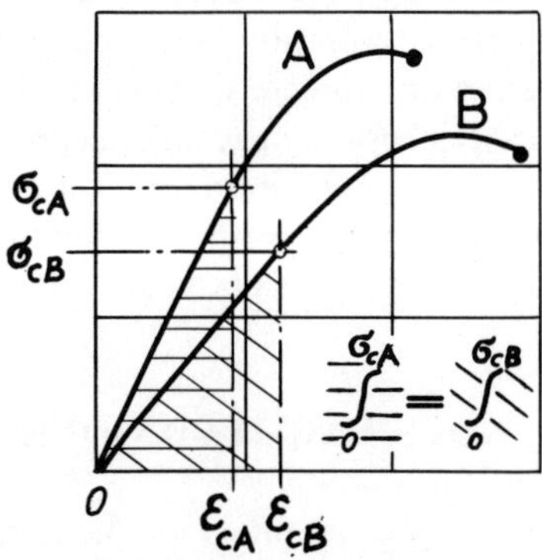

Fig. 19.4. Tensile tests on plastics; the diagram illustrates the differences in stress and in strain at some arbitrarily chosen energy absorption which is the same for both materials.

† Strain is an excellent design parameter. Its quantitative assessment in terms of a critical value is not affected by the restriction in its qualitative significance which has been referred to above.

of cracks which originate at weak points or flaws. The energy absorbed by the material in unit time is referred to as the 'energy efficiency'. Depending on whether one works under conditions of constant stress (creep) or under conditions of constant strain (stress relaxation) the input differs considerably.

In creep the energy absorbed increases as the strain increases in the course of the experiment. The energy absorption is equal to the loss of potential energy as indicated by a change in the level of the weight (if this is the stress-producing load). If the load is supported on reaching some arbitrary strain the initial energy reduces steadily as the stress relaxes.

If the same stress is simultaneously applied to the two materials considered before, then the demand made upon the mechanical resistance under constant load is greater than that made under conditions of constant strain. If the *deformation stress* just attains the crack initiation limit, then the continuing *loading stress* will cause rupture eventually, because of the increasing energy which is brought to bear upon the material.

The fact that break constitutes a convenient failure criterion explains the popularity of the conventional creep experiments carried out in air. At constant deformation it is virtually impossible to attain more than some level of damage without rupture, and this cannot be assessed at all easily; at constant strain that level must be greatly exceeded right at the commencement of the test in order to supply the necessary crack propagation energy over and above the energy needed for crack initiation. On the other hand, crack initiation tests, combined with the use of aggressive media, enables one to determine rupture levels under constant deformation. Using a special notching technique this can even be done with polyethylene which is the tough thermoplastic *par excellence*. The sections which follow describe this technique; the implications of establishing the crack initiation energy as a design parameter will then be considered.

NOTCHING TECHNIQUE

There are a number of suitable methods for the testing of thermoplastics which have a tendency to form cracks at low strains. Apart from the time-honoured tensile test there are the various flexural tests described in the literature. Here one tries in the main to identify some applied stress or strain which is supposed to indicate the crack initiation limits in absolute terms. Numerous publications have appeared on this subject and transparent polymers such as poly(methylmethacrylate) and polystyrene figure prominently. In practice, the applicability of such methods is restricted since the determination of absolute values often requires a considerable capital outlay for equipment. The Bell Telephone Test (ASTM 1693), on the other hand, is a method of stark simplicity; it makes no claim to produce absolute values but affords a comparative assessment of stress cracking resistance when this (and no more) is required. It should be noted that the Bell Telephone Test uses a notching technique in order to accelerate crack

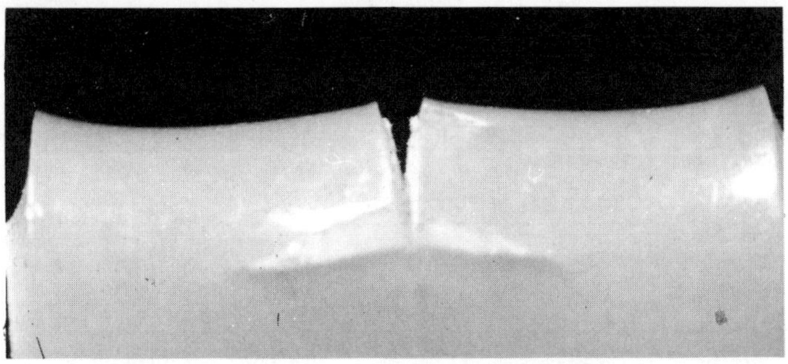

FIG. 19.5.   Crack formation in low-density polyethylene (LDPE) during a test according to ASTM 1693, carried out in a solution of a surface active agent at 50 °C.

initiation. The notch which is produced by the surface incision results in a cross-directional contraction in subsequent flexure, so that it assumes the shape of a V-profile (Fig. 19.5) and high multiaxial stresses arise at the root of the notch.

Carey[1] and Hittmair and Ullman[2] have carried out tensile tests on polyethylene samples which had been notched with a circular hole. They commented on the satisfactory results obtained for purposes of material ranking. Notching is a well known device for initiating cracks in flexural impact tests, but it has not been extensively used for experiments involving static loading. Basically, one is confronted by the same mechanism of increasing stress while the flow response is restricted as a result of the reduction in the ratio of shear stresses to normal stresses (see Chapter 15). A localised stress concentration is equivalent to exposing the material to conditions of increased severity. The same effect becomes immediately apparent when an aggressive medium acts on a polymer which has a closely defined crack initiation limit, such as styrene–acrylonitrile copolymer (SAN) on coming into contact with ethanol (Fig. 19.6).

Conventional dumb-bell specimens were cut from a compression moulded (and hence virtually isotropic) sheet of SAN. They were tested in uniaxial tension. The moving clamp was stopped when some arbitrary

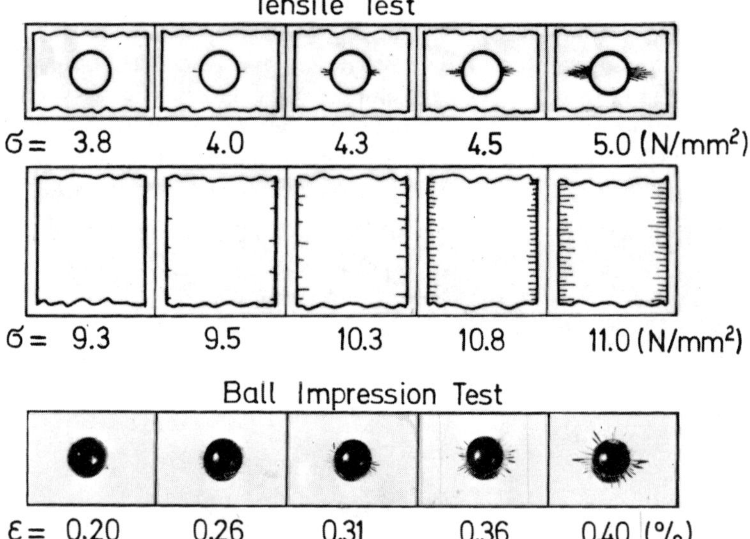

Fɪɢ. 19.6.   Behaviour in tension of SAN following wetting with ethanol. The illustrations show portions of dumb-bell specimens. *Top:* Tensile test with a hole notch of 3 mm diameter; *centre:* Tensile test without notch; *bottom:* Tensile test with impressed oversize balls. The first cracks appear at a stage intermediate between that corresponding to the first and second picture from the left. The modulus (E) was 3550 N mm^{-2}, so that the strain may be calculated from $\varepsilon = 100(\sigma/E)$ (%).

predetermined stress level was attained and the specimen was uniformly wetted with ethanol.

We compared the crack initiation in unnotched specimens (sections of which are sketched in the centre row of Fig. 19.6) with the crack initiation of specimens with a hole notch (top row). In the unnotched specimens the first cracks appeared at a (critical) stress of 9·5 N mm^{-2} whilst in the notched specimens they appeared at 4·0 N mm^{-2}, giving a ratio of approximately 2·4. This ratio may be taken to be equivalent to the notching factor which determines the ratio of the stress at the notch root to that in the unnotched region. For a 10 mm wide specimen with a centrally placed hole of 3 mm diameter the stress is thus increased by a factor of 2·4. It is therefore necessary to reduce the externally applied stress in order to adjust to the reduction in the crack initiation energy which was brought about by the action of the stress-corrosive medium. Incidentally, the factor of 2·4 in this quasi-Hookean stress–strain function is similar to that given for metals in Fig. 19.7 by Wellinger and Dietmann.[3]

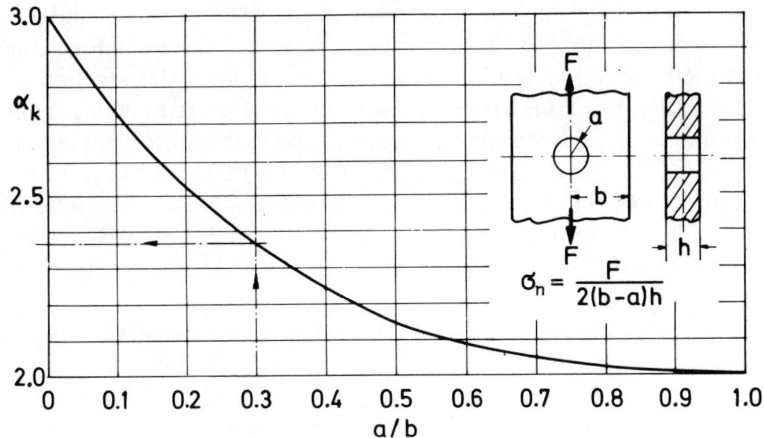

FIG. 19.7.   Notching factors for drilled sheet specimens under tension or pressure (after Wellinger and Dietmann[3]). Example: Determination of the notching factor α_k for specimens of 10 mm width, 4 mm thickness and 3 mm hole diameter. The top row of Fig. 19.6 suggests that in those specimens (a/b = 0·3) the notching factor

$$\alpha_k = 2\cdot37.$$

The bottom row of Fig. 19.6 illustrates the determination of the 'ethanol stress' by the ball impression method. The first cracks are seen at a ball oversize of 0·026 mm. Empirically, an oversize of 0·10 mm corresponds to a strain of 1 %, so that a critical oversize of 0·026 mm is seen to produce a critical strain of 0·26 %. As the tensile modulus is 3550 N mm^{-2} the critical stress works out at 9·2 N mm^{-2}; this is very close to the value of 9·5 N mm^{-2} for the crack initiation step which was identified as the critical stress in the uniaxial tensile experiment (centre row of Fig. 19.6).

We do not, at this stage, wish to do more than attempt a correlation of (i) uniaxial and multiaxial tests in tension (with or without a hole), and (ii) tests involving the stress generated by ball impression and followed by the application of a tensile stress. The task of producing a working hypothesis which applies to viscoelastic materials like plastics still remains to be undertaken.

THE FUNCTION OF THE HOLE NOTCH

Figure 19.6 illustrates the use of the hole notch. This notch provides a locus of deformation during ball impression and is also the place where the stress is concentrated in the course of the subsequent tensile test. Stress and strain considerations are thus brought to a common denominator. Crack initiation is triggered near the hole during the initial deformation at a (critical) stress which inflicts irreversible damage. The extent of this damage is revealed by determining the residual strength in the subsequent test. The top row of Fig. 19.6 shows the location of crack initiation at the edge of the hole during the tensile test. Any damage inflicted by the preceding ball impression reduces the residual strength. The sensitivity with which even submicroscopic changes are revealed makes this technique far superior to existing conventional testing methods. Furthermore, the deliberate initiation of cracks eliminates interference due to internal flaws which are randomly distributed over the entire specimen. Such flaws are caused by processing defects, covering a wide range of severity (in extreme cases large voids), or they may be edge discontinuities created during the machining of the test specimen.

Tensile Tests with Hole Notch Specimens
The hole notch is exceedingly useful in strength determinations. This had already become apparent when the flexural impact test was described (see Fig. 18.15) and supplemented by tests with double V-notches of small notch radius for 'very tough' materials (Fig. 18.16). In such materials we can also use a hole-notch specimen for tensile impact tests and this is particularly appropriate for polyethylene (Fig. 18.4); it takes advantage of the reduction in cross-sectional area and also exploits the presence of the notch *per se*. It is no longer necessary to clamp a dumb-bell specimen and parallel-sided specimen can be used instead.

The flow process is confined to the region of reduced cross-sectional area on either side of the hole. This is particularly useful in tensile experiments on materials which have a high rupture strain because it avoids large elongations and so saves much experimental time on the testing machine. This is quite obvious when looking at the results of straining tensile specimens of high-density polyethylene (HDPE) with and without hole notches (Fig. 19.8). Furthermore, hole-notched specimens without a shoulder also produce stress–strain curves which differentiate clearly between the properties of grades of the same polymer. Thus, we investigated two commercial samples of HDPE which were moulded into lids. Their mechanical properties were tested parallel and normal to the direction of injection flow. Figure 19.9 gives the results of uniaxial tensile tests with the strain axis parallel to the injection flow axis. It shows the result of orientation by registering a high tensile stress and a low strain at failure.

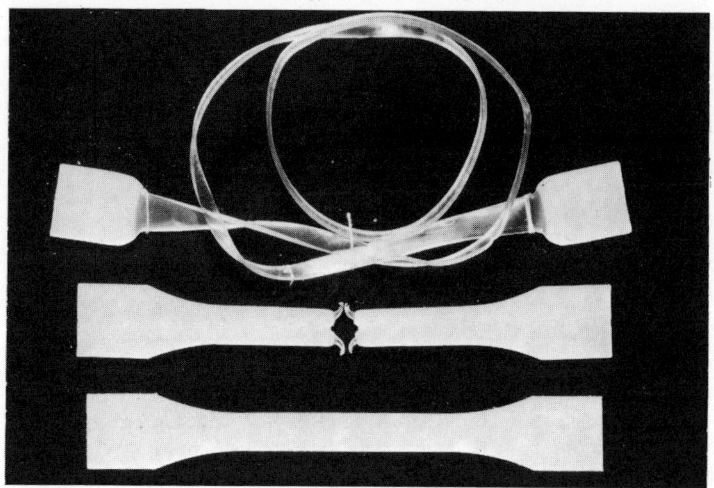

FIG. 19.8.

The 'unrestrained' direction has its strain axis normal to the flow direction and is clearly seen to give lower tensile stresses. The reduction of failure strain in the parallel direction has been eliminated in sample B when this was tested normal to the injection flow axis, despite the lines of weakness in the direction of orientation which were undoubtedly present and which largely influenced A.

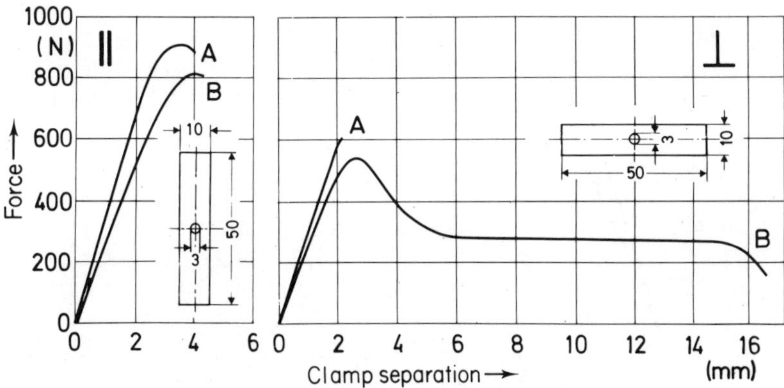

FIG. 19.9. Tensile tests on parallel-sided 10 mm wide specimens with 3 mm diameter hole notches, taken from moulded high-density polyethylene (HDPE) drum lids parallel to (||) and normal to (⊥) the injection flow direction. Polymers A and B have similar densities but differ slightly in their melt flow index. Strain rate: 10 mm min^{-1}.

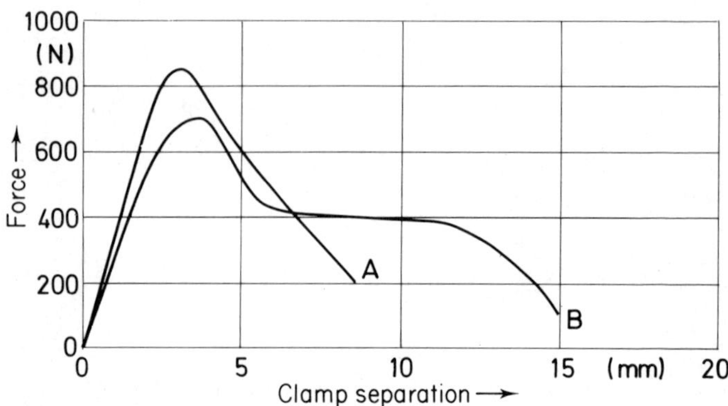

FIG. 19.10.   Tensile tests on parallel-sided 10 mm wide specimens with 3 mm diameter hole notches. The polymers were the same as A and B of Fig. 19.9 but the samples were taken from compression moulded sheet of 4 mm thickness after slow cooling. Strain rate: 10 mm min^{-1}.

It is possible to compare injection and compression moulded bar specimens of identical polymers and also to include the effect of the cooling rate after compression moulding by (i) allowing the compression moulded sheet to cool naturally after pressing at 180 °C (Fig. 19.10), and (ii) by quenching in cold tap water (Fig. 19.11). In addition to the further difference which arises due to different cooling rates the tensile stress and

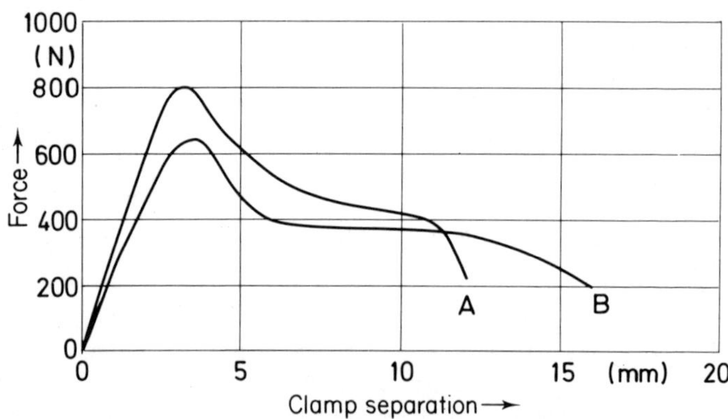

FIG. 19.11.   Tensile tests on parallel-sided 10 mm wide specimens with 3 mm diameter hole notches. The polymers were the same as A and B of Figs. 19.9 and 19.10 but the samples were taken from compression moulded sheet of 4 mm thickness after *fast* cooling. Strain rate: 10 mm min^{-1}.

the rupture strains are now quite different from those of Fig. 19.9. The initial linear portion of the curves provides a comparison of the moduli. Figure 19.12 shows the expected difference between the compression moulded polymer which was allowed to cool naturally and the one which was quenched in cold tap water, together with the results for specimens cut from injection moulded lids parallel and normal to the injection flow axis.

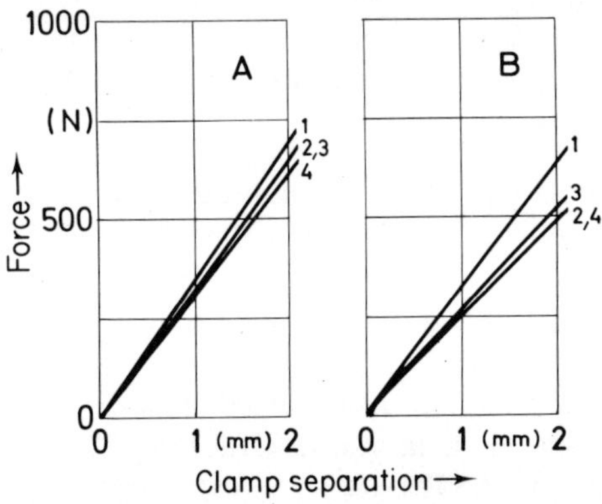

FIG. 19.12.   Comparison of the moduli of polymers A and B of Figs. 19.9, 10 and 11 (extrapolation of the initial linear portion). 1—Fig. 19.9 (‖); 2—Fig. 19.9 (⊥); 3—Fig. 19.10; 4—Fig. 19.11.

Comparing the two grades of HDPE in all four combinations of the two pairs of variables (method of moulding and direction of orientation) it was found that polymer B had the lowest modulus every time. This means that it cannot absorb the same stress as polymer A for any given strain up to the linear strain limit. The difference in the ratio of areas under the stress–strain curves further underlines the fact that polymer B was subjected to a smaller deformation energy. If this results in higher deformations for polymer B up to the point when stress cracking occurs, then one may well consider drawing up a balance sheet for the input and the dissipation of energy.

However, it is first necessary to determine the strain which causes crack initiation. This requires a stress cracking method which is sufficiently sensitive to differentiate between grades of polyethylene.

The Pin Impression Method
We have already mentioned conico-cylindrical pins as an alternative to impressing oversize balls when dealing with 'very tough' thermoplastics which require large oversize inserts (DIN 53 449), see Chapter 18. This has

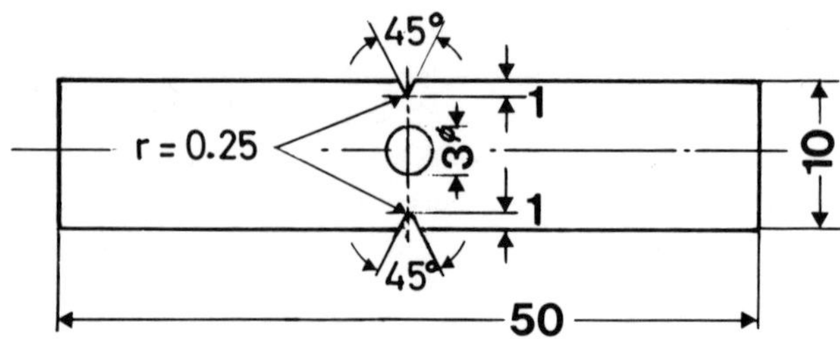

FIG. 19.13.   Dimensions of the specimens with double-V notches and a central
hole notch, as used for testing tough thermoplastics.

resulted in the development of a technique which is particularly suitable for
polyethylene and involves the additional positioning of V-shaped notches
on either side of the hole (Fig. 19.13); this minimises shear deformation and
so impedes flow while the stored stress is increased and the time to rupture is
reduced.

   We use the same principle which applied to the deliberate reduction in the
time to rupture by notching in a flexural test and in the stress cracking
method according to ASTM 1693.

   The deformation in a stress cracking experiment is generated by the
impression of oversize pins and particularly affects the area on either side of
the hole; this produces high multiaxial stresses at the root of the V-notches

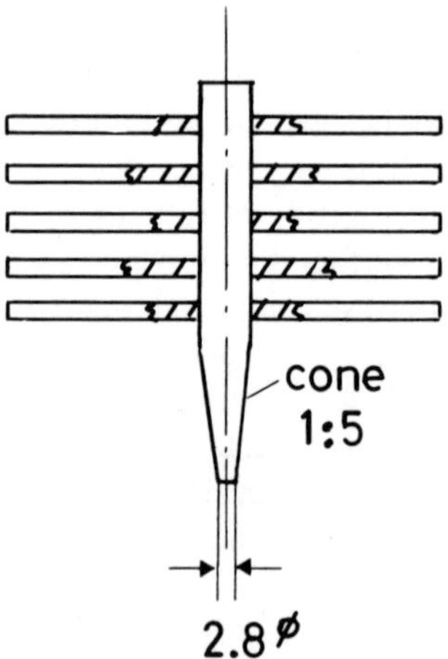

FIG. 19.14.   Dimensions of the standard specimen with double-V notches
(r = 0·25 mm) as used for routine testing (right). The diagram on the left shows a
group of five such specimens skewered onto one conico-cylindrical oversize pin.

which are dependent on the pin oversize. Several test specimens can be skewered onto one cylindrical pin (Fig. 19.14) in 'shish-kebab' fashion. This increases the statistical significance of the distribution of crack formation limits. We have to increase the number of specimens since polyethylene is a material which gives widely scattered results of crack formation limits, as is seen in the work of Kausch[4] who determined the time-to-rupture of polyethylene pipes under internal pressure (Fig. 19.15).

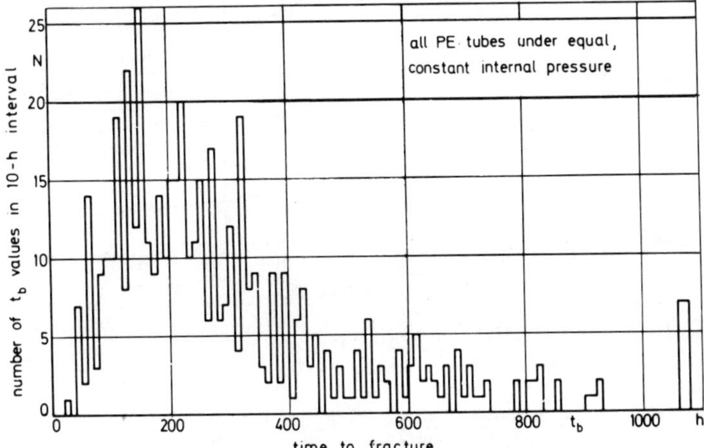

FIG. 19.15.   Scatter of times to fracture of polyethylene pipe under equal constant internal pressure $\sigma_\psi = 40\,\mathrm{kg\,cm}^{-2}$.

It is possible to express the average time which elapses before cracks appear following immersion in a surface active agent as a function of pin oversize. The precise moment of crack formation cannot really be determined in practice and one must make do with regular inspection at convenient and preferably short intervals, noting the time when cracks were first observed. This can then be used to generate an average time to crack from a suitably large number of observations.

The fact that notching constitutes an increase in the severity of the test and therefore causes a reduction in the time-to-rupture under static conditions (just as high rates of deformation do in tensile tests) can be proved by modifying the notch radii. This is illustrated by the result of tests with impressed oversize pins in an aggressive medium (Fig. 19.16). For purposes of routine testing one has to select a notch radius which produces data with a clearly identifiable time dependence; it should be neither too

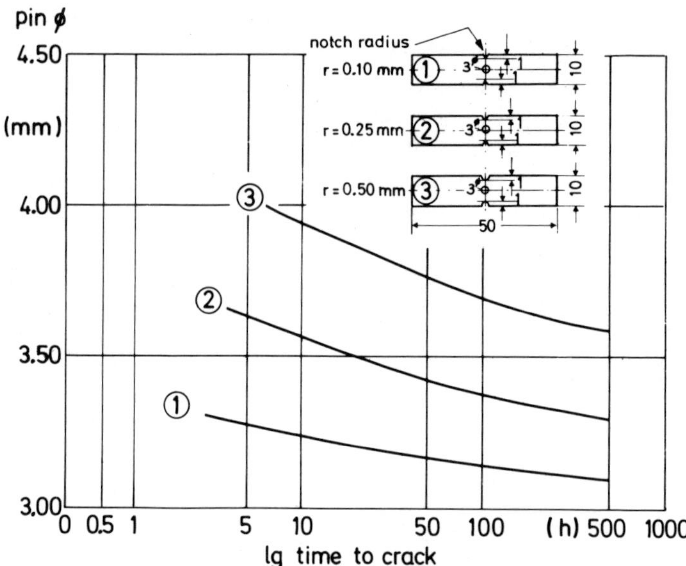

FIG. 19.16.   Creep test (HDPE in a solution of a wetting agent at 50 °C). The plot shows the mean times to crack formation as a function of the notch radius for specimens with otherwise identical dimensions under identical conditions of test. The zero point of the logarithmic time scale is obtained from the inverse hyperbolic function: $\log(t + \sqrt{t^2 + 1}) - 0\cdot301$.

large (which would take too long) nor too small (which would make the intervals too short to enable one to observe failure times with an acceptable degree of accuracy). Flexural impact tests with double V-notches had already revealed that the choice of the notch radius depends on the 'degree of toughness' of the product under test, i.e. by how much the time to rupture exceeds the relaxation time which characterises the tough/brittle transition. Generally we found that a notch radius of $0\cdot25$ mm was about right and this was used for the experiments, results of which were shown in Figs. 19.8–12. As expected, we found that polymer B had the greater strength in the long-term experiment, both for compression moulded specimens (Figs. 19.17 and 19.18) and in specimens taken from injection moulded lids (Figs. 19.19 and 19.20). Figures 19.17–19.21 show long-term experiments on HDPE at constant deformation (oversize pins) in a solution of a surface active agent at 50 °C. The specimens have double-V notches (notch radius r $= 0\cdot25$ mm). The results (mean time to crack) were obtained from deformation steps represented by pin diameters ranging from $3\cdot10$ to $4\cdot50$ mm (hole diameter:

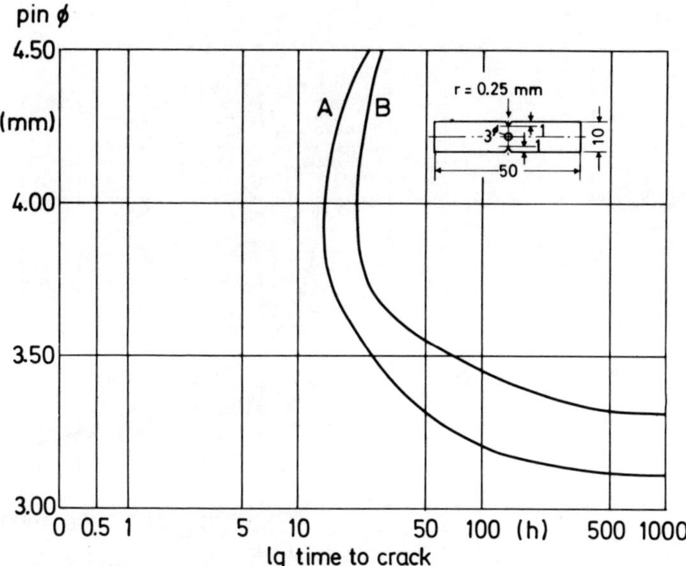

FIG. 19.17.   Specimens 4 × 10 × 50 mm from compression moulded sheet which was allowed to cool normally without chilling.

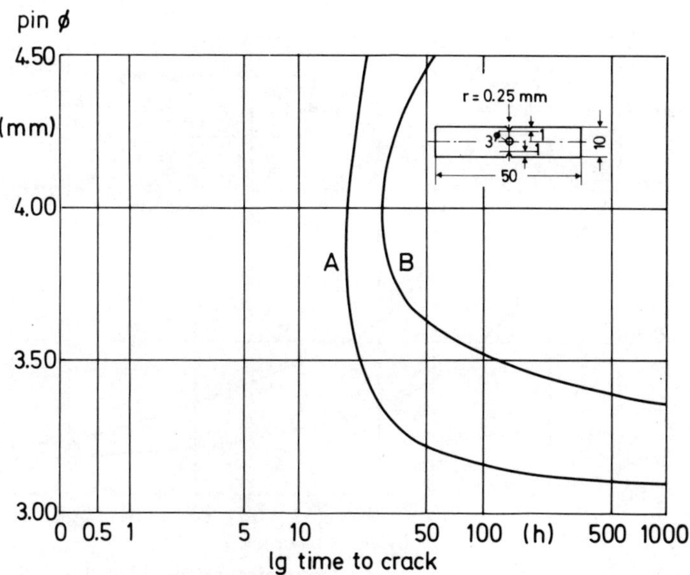

FIG. 19.18.   Specimens 4 × 10 × 50 mm from compression moulded sheet after crash cooling by quenching in cold tap water.

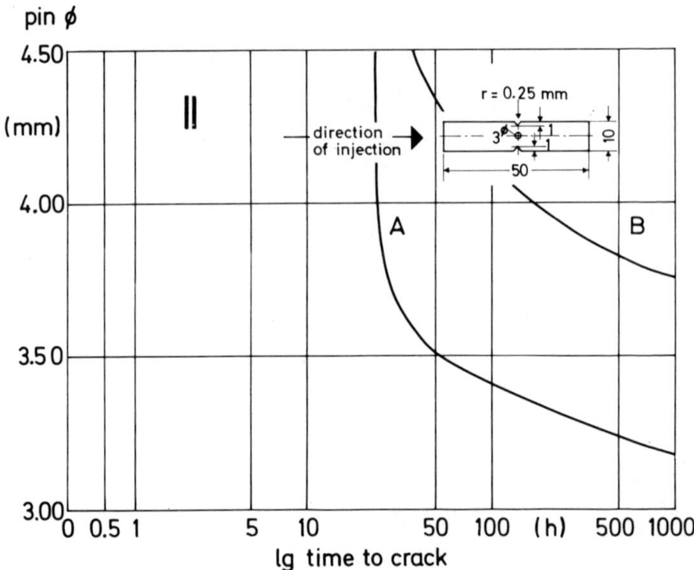

FIG. 19.19.   Specimens 4 × 10 × 50 mm from injection moulded sheet (plaques), after cutting them out with the long axis parallel to (‖) the injection direction (as shown).

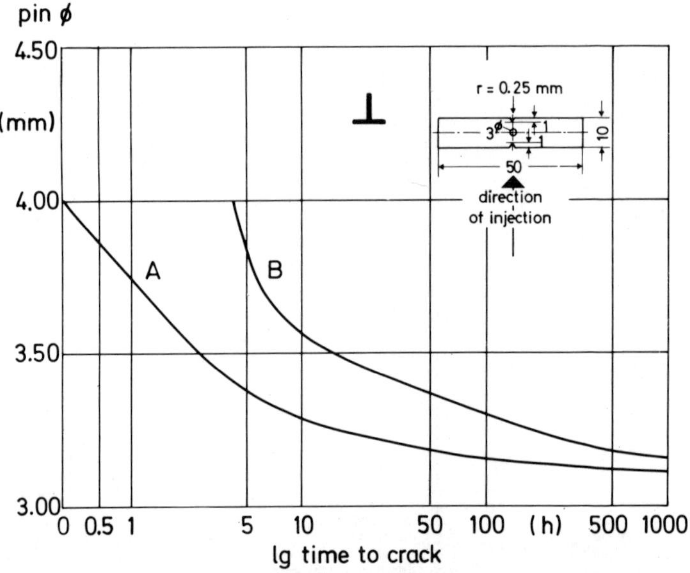

FIG. 19.20.   Specimens 4 × 10 × 50 mm from injection moulded sheet (plaques), after cutting them out with the long axis normal to (⊥) the injection direction (as shown).

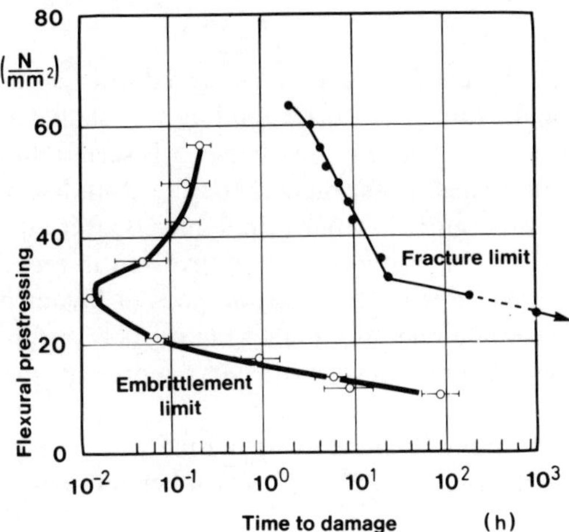

FIG. 19.21.   Damage limits of ABS in a 1:1 mixture of olive oil and oleic acid (after Haslett and Cohen[5]). The damage limit (embrittlement limit) is indicated by and taken to be the time required (as well as the magnitude of the load in flexure) which prevented the test specimens from surviving five loading cycles in a subsequent dynamic test at the appropriate stress amplitudes.

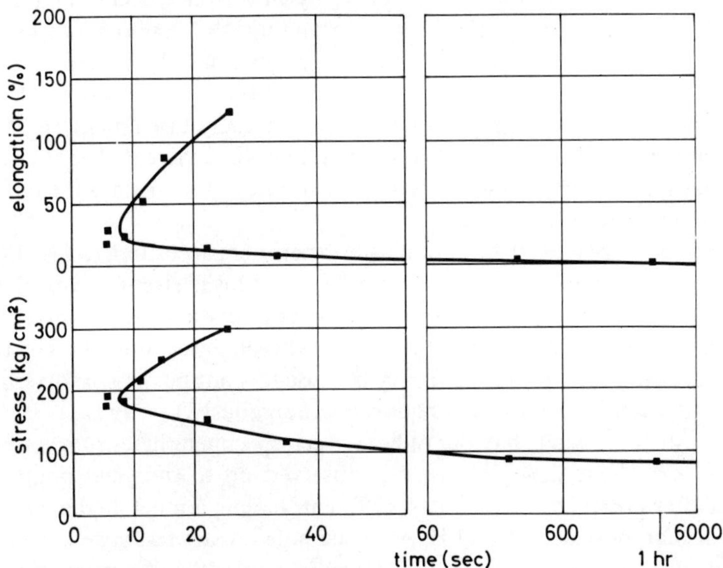

FIG. 19.22.   Delayed fracture of film grade ('P4') unsaturated natural rubber hydrochloride in 100 ppm ozone as a function of prestraining, a relaxation before exposure of 5 min being allowed. Each point represents a mean of 15 determinations. (Exposure to ozonised oxygen in cylinders according to G. Salomon and F. van Bloois[6]).

3·00 mm) five specimens having been skewered onto each pin. The times were then plotted against the corresponding pin diameter.

The lower long-term strength of polymer A is seen to be due to a higher stress absorption because of the higher modulus. But that by itself does not account for the superior behaviour of polymer B after injection moulding (Figs. 19.19 and 19.20). Without getting involved in any further detailed analysis it is quite clear that the decision goes in favour of polymer B.

The curvature in the region of higher pin oversizes leads one to suspect that the peak of a flexural stress–strain curve represents the point at which the deformation limit for the impression of oversize pins has been exceeded. Such a renewed strength increase with time has been reported in the literature for ABS (Fig. 19.21)[5] and for rubber film (Fig. 19.22).[6]

A FURTHER COMPARISON OF POLYETHYLENE GRADES

We continue with a further investigation which involved three grades of HDPE which were known to be polyethylenes with specially high stress crack resistance. We aimed at determining the limiting strain and establishing a limiting pin oversize for each polymer. An experiment with oversize pins in a surface active agent showed that polymers A and C had better long-term strength properties. The shortest failure times were found at a pin diameter of about 3·7 mm. The failure time increased again with a further increase in the pin oversize. This point clearly defines the energy-elastic maximum (Fig. 19.23).

Figures 19.24 and 19.25 show tensile experiment according to Fig. 19.23, at a straining rate of 10 mm per minute. The tensile strength σ is plotted against strain, the latter being expressed in terms of the distance travelled by the clamps, with the notched region lying between the clamps. (This method of expressing the strain is necessary because it is not possible to determine the extension in the notched region unambiguously.)

The stress–strain behaviour of hole-notch specimens in tension is seen in Fig. 19.24. Very little difference is observed up to the yield point, but thereafter polymer C is distinctly different, having a much higher rupture strain than polymers A and B. A tensile hole impact test gave analogous results for the energy to break as shown in Table 19.1. An increase in the multiaxiality by positioning additional V-notches of 0·25 mm radius on either side of the hole caused an increase in the tensile stress and a simultaneous reduction in the rupture strain (Fig. 19.25). Compared with polymers A and C, polymer B is now seen to have a substantially lower

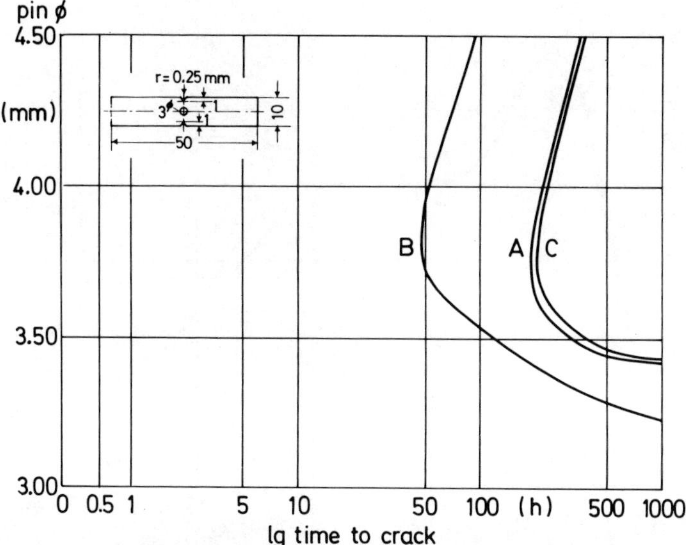

FIG. 19.23.   Long-term experiment on polyethylene of density 0·95. The three grades (A, B, C) have low melt flow indices which differ only slightly. Specimens 4 × 10 × 50 mm were taken from compression moulded sheet and had double-V notches of notch radius r = 0·25 mm. Seven steps of deformation by oversize pins ranging in diameter from 3·10 to 4·50 mm (hole diameter: 3·0 mm) were selected and 5 specimens were skewered on each pin. The time to break was taken as an average and plotted against the respective diameter of the oversize pin. The experiments were carried out in a solution of a surface active agent at 50 °C.

energy to rupture; this had already been foreshadowed in the course of tensile impact tests on specimens with a hole notch. The lower long-term strength of polymer B as seen in the oversize pin test (Fig. 19.23) confirms the view that the capacity of the polymer to absorb energy (and hence its energy to rupture) is smaller than that of polymers A and C.

Since the stress–strain curves for polymers A, B and C are virtually identical up to the yield point (Figs. 19.24 and 19.25) it is clear that the behaviour of polymer B is not due to a higher modulus. In order to decide

TABLE 19.1

Polymer	A	B	C
Tensile hole impact energy to break (N mm^{-2})	198	177	237
Standard deviation (N mm^{-2})	±5	±8	±10

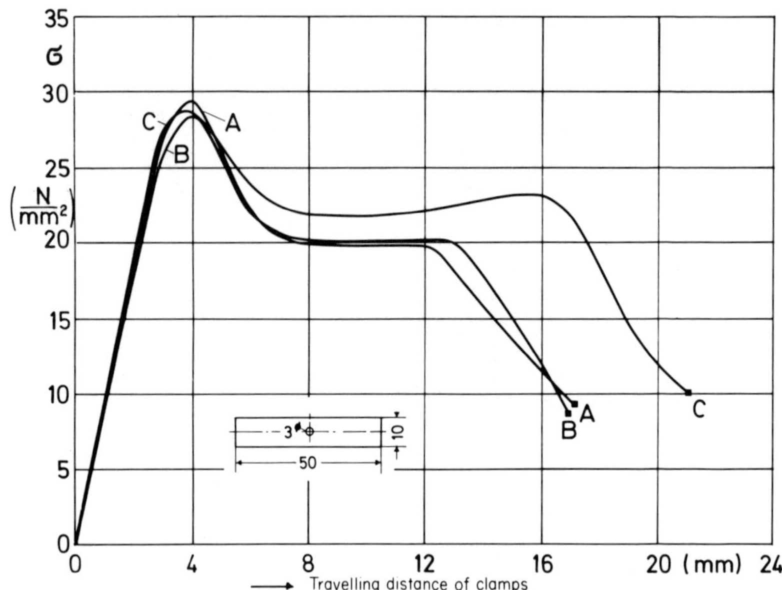

FIG. 19.24.   Tensile experiments on specimens with a hole notch.

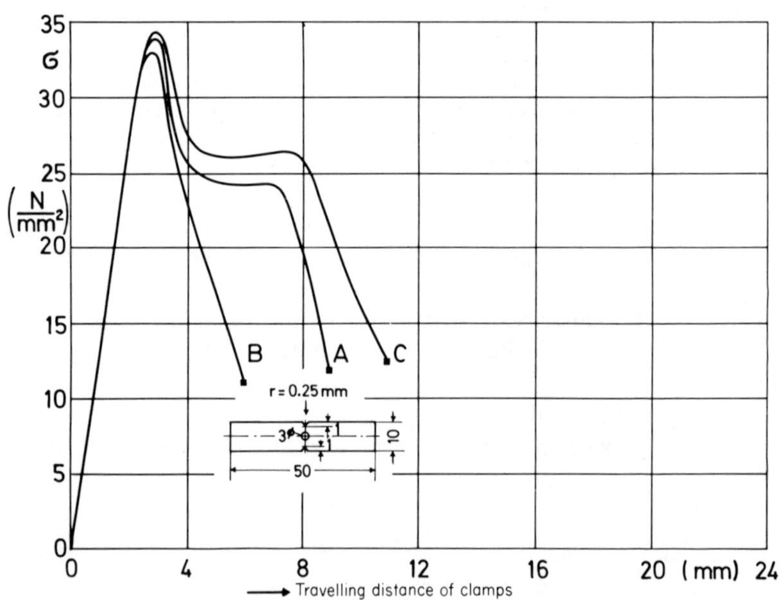

FIG. 19.25.   Tensile experiments on specimens with a double-V notch as well as a hole notch.

why polymer B does in fact behave in a peculiar manner we used the result of a method for characterising the *energy absorption* after impressing an oversize pin into a polyethylene specimen. The pin, however, was rather special: it was a conical spike. This method has since been developed for routine testing.

Spike Impression

The spike was a conical pin of ratio 1:5 (Fig. 19.26); this is similar to that of the conical portion of the conico-cylindrical oversize pins which had been used for impression in stress cracking experiments.

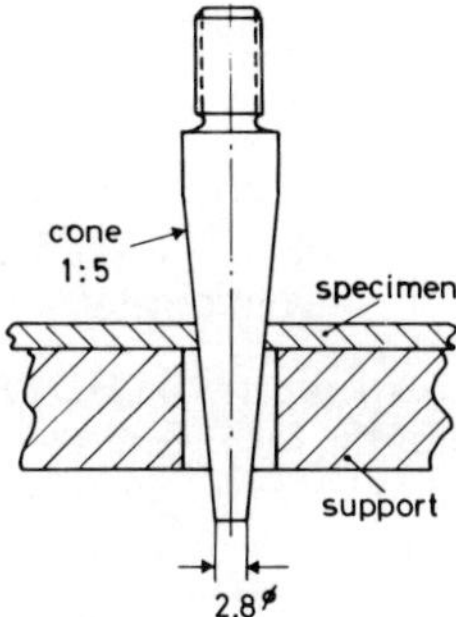

FIG. 19.26.

The spike is mounted on a tensile testing machine operating in the compression mode and is then forced into the specimen. A load–penetration curve is obtained in which the stress can be calculated making due allowance for the known changes in the cross-sectional area which are defined by the spike geometry. The curve commences as soon as the spike has made contact with the hole. (The initial slight curvature is due to taking up 'slack' until full contact is established around the circumference of the hole.) The curve then rises linearly up to the flow limit and thereafter up to the strain limit at the yield-point (Fig. 19.27). The experiment need not be continued any further because it is much more convenient to observe the subsequent behaviour in a notched tensile test (Figs. 19.24 and 19.25).

The results for polymers A, B and C show that the initial slopes, i.e. the initial moduli are identical. The depth of penetration of the spike, that is to say, its diameter at any stage, and hence the energy absorbed by the three specimens is therefore also the same up to the limit of linearity. Since the

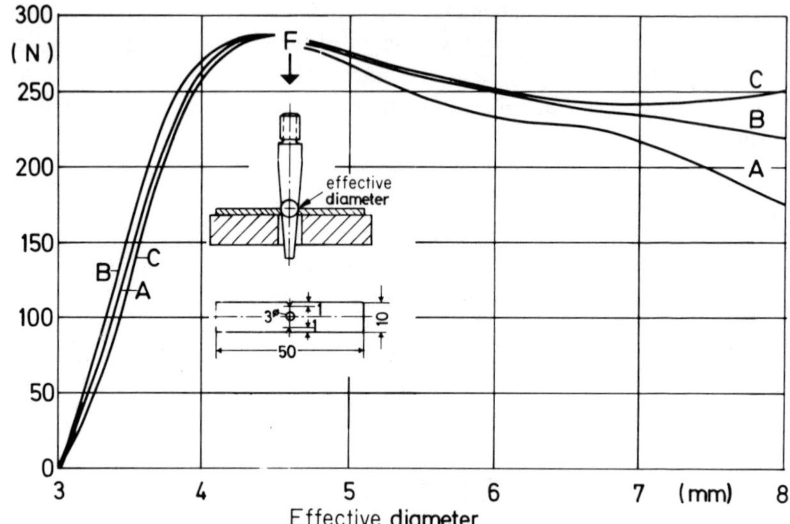

Fig. 19.27.   Spike impression experiments on HDPE (grade A, B and C as used in Figs. 19.23–25). This is a plot of the force generated as a function of the effective diameter of the conical pin (the spike) as the latter is pressed into the specimen at a rate of $10\,\mathrm{mm\,min^{-1}}$.

energy absorption in the stress cracking experiment is likewise the same, the difference in the strength test with oversize pins (Fig. 19.23) must be due to a difference in the crack initiation energy. In order to obtain the same crack initiation times for polymer B as for the other two polymers one would have to use a pin of smaller oversize, that is to say, one would have to apply less work.

The spike penetration experiment established the diameter which produces the yield strain. Figure 19·27 showed this to be about 4·5 mm, with a linearity limit of about 3·7 mm. This also identified the diameter oversize in the pin impression experiment of Fig. 19.23 which marks the shortest time to crack initiation. If the oversize is increased the viscoelastic region is exceeded, the increasing plastic deformation has a smaller energy-elastic portion and a stretching orientation is generated. Both these processes increase the stress cracking resistance until the instantaneous rupture limit is reached.

The relationship between the spike impression experiment and the stress cracking experiment with the conico-cylindrical pin can be established by means of the spike impression curve. Let us take the spike at a diameter of

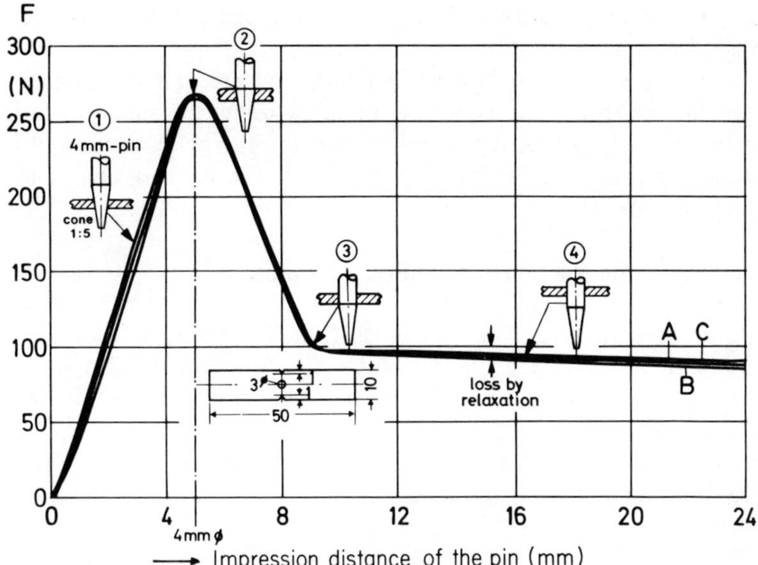

FIG. 19.28.   The load during the impression of a conico-cylindrical oversize pin of 4 mm diameter into HDPE (see Fig. 19.27). Penetration rate: 10 mm min^{-1}. Stage 1: The load increases as the conical portion forces its way in. Stage 2: The load maximum is reached as the cone reaches its maximum diameter (identical with that of the cylindrical portion) as it enters the top side of the test specimen. Stage 3: The cylindrical portion has reached the bottom side of the test specimen. The deformation process is now at an end. Stage 4: The work recorded from now on represents that involved in friction as the pin travels through the specimen. The stress drops slightly as a consequence of relaxation. *Note:* 5 mm of forward movement of the pin from the point of initial contact with the hole perimeter produces the maximum deformation to increase the hole diameter to 4 mm, given a 1:5 cone.

4 mm (Fig. 19.27). This is the diameter which just makes contact with the specimen. It produced a force of 270 N. We could equally use a conico-cylindrical pin of the same cylinder diameter (which is preceded by the usual conical section, which serves to facilitate insertion.) As soon as the cylindrical portion of 4 mm diameter enters the test specimen the stress falls off (Fig. 19.28). From that point onward the only energy expended is that which is necessary to force the conical portion through the remaining thickness of the specimen. As soon as the conical portion emerges on the other side of the specimen the stress–strain curve becomes almost horizontal and the stress merely represents the force necessary to overcome the

constant frictional resistance, the slight downward incline being due to stress relaxation.

The energy which has to be applied in order to increase the hole diameter in the surface from 3 to 4 mm, i.e. the area below the curve up to the maximum of Fig. 19.28, represents the specific energy absorption of the material and the conditions to which it has been subjected during processing.

Experiments with Specimens of Different Widths

The resistance against deformation by an oversize pin depends on the area of the cross-section on either side of the hole. It is therefore necessary to maintain the cross-section at a constant width. An oversize pin impression experiment with specimens of different widths indicates the difference in the stress which has been absorbed by the cross-section. This is expressed in terms of the time which elapses before cracks appear (Fig. 19.29). The energy at that point is not exactly equal to the work required for forcing in

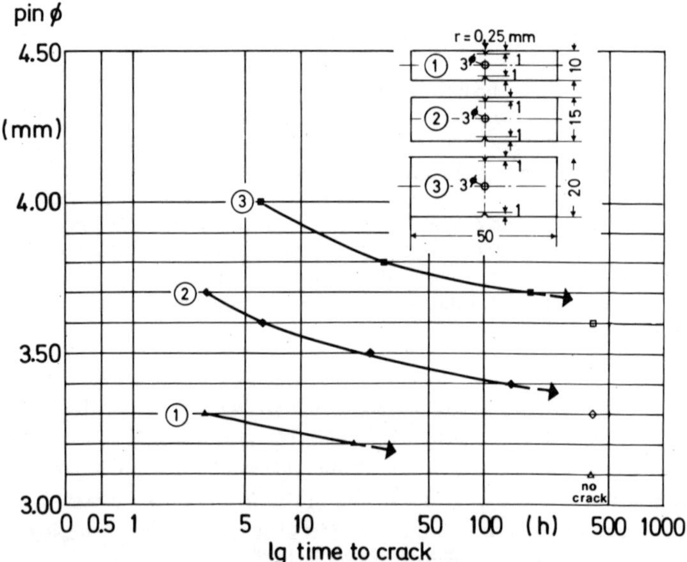

FIG. 19.29. The strain-time dependence of HDPE specimens immersed in a solution of a surface/active agent at 50 °C. The three curves illustrate the effect of the stress in the vicinity of the double-V notch. With increasing specimen width (i.e. with a reduction in the stress concentration) pins of correspondingly greater oversize are required to produce cracks at any given time.

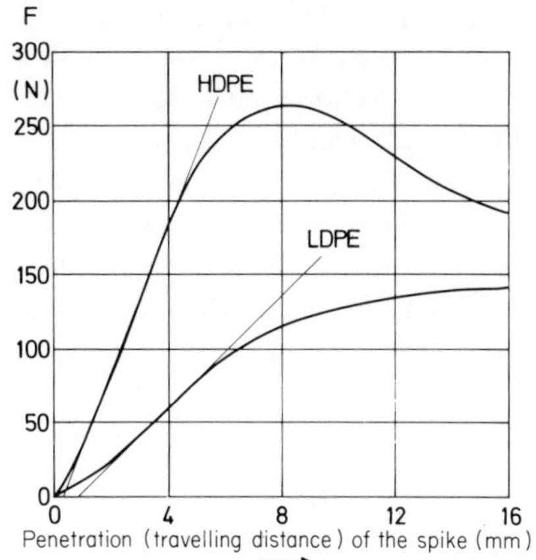

Fig. 19.30. Load/penetration distance curves of HDPE and LDPE in spike impression experiments at a penetration rate of $10\,\text{mm}\,\text{min}^{-1}$. A penetration of 5 mm is equivalent to an expansion of the hole diameter at the initial perimeter of contact between the spike and the hole of 1 mm. A slight curvature is observed initially until the spike has taken up the slack and becomes fully effective in its straining action, whereupon a linear portion follows the slope of which is a measure of the modulus of elasticity. In this experiment it is essential that comparisons be made only on specimens of absolutely identical geometry in order to prevent the energy absorption being affected by changes in the cross-sectional area and the consequent production of stresses of different magnitudes.

the oversize pins since the work needed to overcome the frictional resistance must be subtracted. These losses do not, however, represent variables which are sufficiently significant to distort the general conclusions; this is confirmed by the agreement of modulus and tensile stress for the three polymers A, B and C, which had already been observed in tensile tests on notched specimens earlier on (Figs. 19.24 and 19.25). If there are differences at all one might well get a clear difference in materials with different moduli. This is demonstrated by comparing several types of polyethylene (Fig. 19.30).

Utility of Spike Impression Test

The sensitivity of the method is such that one can even recognise changes in modulus and critical strain in one and the same material which arise from differences in the processing history. We do know that conventional methods for identifying and determining the magnitude of such changes are not always satisfactory. The spike impression method is not a substitute technique, but it does supplement standard procedures. The use of the spike impression technique was motivated by the need for a test which is suitable for establishing the suspected existence of flaws in closely defined regions. The introduction of the technique into stress cracking tests makes it possible to establish (i) which oversize is suitable for pin impression tests, and (ii) the energy which the specimen is forced to absorb (which is a function of the area under the stress–strain curve). It furthermore serves as a basis for checking the argument that cracks form because a specific limit in the energy absorbed by the test specimen has been exceeded, since neither the critical stress nor the critical strain are *per se* sufficient for the evaluation of the process by which the damage is inflicted.

SUMMARY

Stress cracking is considered as a process which involves the expenditure of energy. It can be triggered by applying either a stress or a strain. It must be borne in mind that the initially absorbed energy increases with time under constant load so long as the strain increases, and that it decreases at constant strain due to stress relaxation. If the performance of a material is defined by certain strain limits, then the instantaneous application of the given constant strain allows a higher initial stress and thereby a higher critical stress than that which is produced at constant load.

Assuming that the limiting damage is reached when a certain (critical) energy has been applied, the exceeding of some constant (critical) strain will cause permanent damage 'instantaneously' although an induction period must be allowed for. At constant load, however, permanent damage is inflicted only when a certain time interval has elapsed. This is why a test at constant strain is a much faster method for determining the critical level of damage.

There exists no physically determinable limit for the definition of 'critical damage'. It therefore becomes necessary to agree on some other criterion such as critical stess or critical strain.

The ball (or pin) impression technique establishes critical strain as the point at which a 5 % reduction in strength is observed compared to that for a specimen in which nothing was inserted into the hole. Alternatively, it may be defined by the unequivocal appearance of cracks, provided that tests do not reveal any loss of mechanical strength at even lower deformation levels at which no damage is as yet visually apparent. Such tests require measurements at a number of deformation steps, involving varying amounts of initial energy which the specimens are forced to absorb. This is done by impressing balls (or pins) of increasing oversize. The technique enables one to make an exceedingly accurate assessment of the mechanical properties of polymers, which is certainly not the case when the measurements are confined to a single-point determination in creep tests or to tests at constant strain.

The method of imposing an initial deformation with subsequent determination of the residual strength provides an alternative to the conventional creep experiment. The creep experiment does not enable one to identify a 'critical damage' and ends with rupture, which is not a very useful design parameter. Compared with rupture—which can certainly be identified without the shadow of a doubt—the determination of some characteristic limit of acceptable damage is less clear cut. It is, moreover, all the more difficult to determine such a limit at constant strain the greater the number of cracks that fail to develop due to relaxation acting as an energy sink. The number of cracks which fail to develop can, of course, never be known. One can therefore only use a method which is sufficiently sensitive to react to a level of damage which may well be submicroscopic.

It has been seen that strength tests on specimens with a cross-sectional area that has been disturbed by the impression of oversize balls (or pins) into drilled standard size holes produce reliable data on the mechanical characteristics of materials. These tests also introduce a new and exceedingly sensitive design principle.

The use of aggressive media contributes additional energy. This has the effect of increasing the damage beyond the critical level and of causing partial rupture. Applied to polyethylene it thus became possible to use a (suitably accelerated) time-dependent test at constant strain in which the damage became visually apparent without the need for any further mechanical tests. Partial rupture is the instantaneous consequence of the attainment of a critical level of damage.

The scattering of the crack formation limits which is so characteristic for polyethylene is reduced in oversize impression techniques by using several specimens at each deformation level. This does not cause any additional expenditure for multiplicate testing equipment since no additional apparatus is required just because more specimens have been impressed.

The use of aggressive media extends the role of stress cracking tests from the determination of stress or strain limits in air—a very useful parameter— to the identification and determination of flaws and weaknesses caused by the rheological realities of the manufacturing process, i.e. by the processing history of the component. Conversely, aggressive media which are capable of initiating cracks may be used to compare the performance of a material with the performance of the same material in a neutral environment such as air, and enable one to make good predictions for the long-term mechanical properties in air, once a correlation between the results of tests in air and tests in the aggressive medium has been established.

Finally, the development of stress cracking methods must also involve some thought concerning the extent to which they may be suitable for a routine solution of the kind of problems with which the design engineer is concerned. It is certainly possible to obtain results which are of scientific interest under specially defined conditions which require special apparatus and a long purse; the technologist, however, has to find a compromise between procedural simplicity and high precision. He has to design his testing equipment in such a way as to enable him to obtain meaningful data in the context of the expected service conditions, and to obtain themcheaply and routinely. Oversize impression techniques, in combination with aggressive media and standard mechanical tests, have been found to be highly successful in meeting those requirements.

REFERENCES

1. R. H. CAREY, ASTM Bulletin, 56, July 1950.
2. P. HITTMAIR and R. ULLMANN, *J. Appl. Poly. Sci.*, **6**(19), 1 (1962).
3. K. WELLINGER and H. DIETMANN, *Festigkeits Berechnung*, A. Kröner Verlag, Stuttgart (1976).
4. H. H. KAUSCH, *Materialprüfung*, **6**, 246 (1964).
5. HASLETT and COHEN, *SPE Journal*, **20**, 246 (1964).
6. G. SALOMON and F. VAN BLOOIS, *J. Appl. Poly. Sci.*, **7**, 117 (1963).

20

Rheo-Optics

The photoelastic effect was discovered by D. Brewster in 1816. He found that clear stressed glass gave coloured patterns when viewed in polarised light. By 1900 an underlying theory had been developed. Today photoelastic stress analysis is a valuable design aid.

When a load is applied to a clear plastic, brilliantly coloured bands will be observed in white light, or alternating bright and dark bands in monochromatic light. The latter are known as interference fringes. The interpretation of the optical effects enables one to map the stress distribution. Many problems which defy analytical solution can be readily solved by the application of photoelasticity.[1]

Photoelastic measurements enable one to:

(1) view the shear stress distribution at a glance;
(2) pinpoint the loci of maximum shear stress;
(3) determine actual stresses at any point, even in irregular geometries;
(4) observe the effect of minor changes in component geometry on the stress distribution with the aim of optimising design.

On the other hand, it is necessary to work with accurate scale models and any residual stresses must be annealed out. The experimental technique is readily applied only to two-dimensional models. The patterns produced are complex and it is possible to identify two types—isochromatic and isoclinic fringes. Of these the reasons for the existence of isochromatic fringes is more easily understood. The isoclinics can be used to determine the stress patterns present and these will be more fully discussed in the next chapter.

The refractive index of a material is defined as the ratio of the velocities of light in air and in the material respectively. If the material is ideally transparent the amount of light emerging will be unaffected; however, in polarised light the amount of light emerging will depend on the position of the polariser and the analyser, ranging from total extinction (crossed) to a maximum (parallel).

If the material is stressed the polarised ray is split into two parts vibrating at right angles to one another, the planes of vibration being those of the principal (maximum and minimum) tension stresses which are designated P and Q respectively. Since the light travels at slightly different speeds in the two planes compared to the speed in the unstressed material the two components will become increasingly out of phase. The retardation of one component relative to the other component makes recombination to the original plane wave impossible, but it enables one to calculate the stress in the material.

The difference between the principal stresses equals twice the maximum shear stress. (It should be noted that the third principal stress affects both components equally and does not therefore influence their relative retardation.) The relative retardation Δ, however, depends also on the specimen thickness and on the sensitivity of the material to photoelasticity. It is virtually constant for all wavelengths and is given by:

$$\Delta = Cd(P - Q) \tag{1}$$

where d = the specimen thickness, and C = the stress-optical coefficient (in $cm^2\,g^{-1}$), a material constant.

Once Δ has been determined experimentally, $(P - Q)$ may be calculated. A monochromatic light source impinges on the specimen in the crossed polariscope and a small load is applied. The light will be split by the specimen and two components will therefore emerge from the analyser. By increasing the applied load, the relative retardation Δ will be increased until the components are exactly half a cycle out of phase and the light will then be extinct. A further increase of the load (and thus of Δ) will eventually make the two components a whole cycle out of phase; they will then reinforce one another to give maximum intensity. As the load is further increased, the relative retardation also increases and extinction occurs a second time at a phase lag of $1\frac{1}{2}$ cycles and every subsequent half-cycle, whilst maximum intensity is again reached at 2 cycles and every subsequent full cycle.

In a two-dimensional case one therefore observes a series of monochromatic light bands or 'fringes', where each fringe represents a contour for a constant value of $(P - Q)$.

It should be noted that Δ is given in unit length. The phase difference m is given in cycles and represents a succession of integers. Δ and m are simply related by

$$\Delta = m\lambda \tag{2}$$

where λ is the wavelength of the monochromatic light used.

From eqns. (1) and (2) it follows that

$$P - Q = \frac{m\lambda}{Cd}$$

and setting $\lambda/C = S$

$$S = \frac{(P - Q)d}{m} \tag{3}$$

where m = the 'fringe order', S = the 'fringe stress coefficient' (in g/cm fringe), a constant for a given wavelength. This may be regarded as the tensile stress necessary to produce a change of one fringe in a material which is 1 cm thick, and d = thickness (in cm).

Equation (3) is of the utmost importance since it directly relates the number of fringes m to the shear stress of the specimen. Since C is a constant, S is directly proportional to the wavelength of the light used.

The photoelastic effect can equally be ascribed to the *strain* in the material instead of the stress. If

$$S = (P - Q)\frac{d}{m} \tag{3}$$

then a *fringe strain* coefficient $1/D$ can be similarly defined by

$$\frac{1}{D} = (\varepsilon_P - \varepsilon_Q)\frac{d}{m} \tag{4}$$

where $(\varepsilon_P - \varepsilon_Q)$ is the maximum shear strain.

Now, in an elastic material:

$$\varepsilon_P = \frac{P}{E} - v\frac{Q}{E}$$

and

$$\varepsilon_Q = \frac{Q}{E} - v\frac{P}{E}$$

where E = Young's modulus and v = Poisson's ratio.

Hence

$$\frac{1}{D} = (P - Q)(1 + v)\frac{d}{m}E$$

or

$$\frac{1}{D} = \frac{S}{E}(1 + v) \tag{5}$$

                    Polymer Rheology

If the material is sufficiently highly stressed to cause extension by *plastic flow* δ_P and δ_Q respectively, then the elastic and plastic strain becomes, respectively:

$$\varepsilon_P = \frac{1}{E}(P - vQ) + \delta_P \tag{6A}$$

and

$$\varepsilon_Q = \frac{1}{E}(Q - vP) + \delta_Q \tag{6B}$$

so that

$$\varepsilon_P - \varepsilon_Q = \frac{1}{E}(P - Q)(1 + v) + (\delta_P - \delta_Q) \tag{7}$$

where the first term on the right-hand side of eqn. (7) represents the elastic, and the second term the plastic strains.

It can be shown that the plastic strains are proportional to $(P - Q)$, specifically

$$(\delta_P - \delta_Q) = 3A(P - Q) \tag{8}$$

where A depends on time, temperature and material.

In pure tension Q degenerates to zero and

$$(\delta_{P_0})_{Q=0} = \xi = 2AP_0$$

so that

$$A = \xi/2P_0 \tag{9}$$

Substituting for A in eqn. (8):

$$(\delta_P - \delta_Q) = \frac{3}{2P_0}(P - Q)\xi$$

and substituting for $(\delta_P - \delta_Q)$ in eqns. (4) and (7) we get

$$\frac{1}{D} = \left[\frac{1}{E}(P - Q)(1 + v) + \frac{3}{2P_0}(P - Q)\xi\right]\frac{d}{m} \tag{10}$$

where ξ represents the longitudinal plastic strain which results from the application of a pure tensile stress P_0, preferably so chosen that P_0 approximates P so as to achieve reasonable accuracy.

It is, of course, obvious that $S = \alpha(1/D)$ and that α has the dimensions of a modulus.

The Effect of Temperature on the Fringe Stress and Fringe Strain Coefficients
This has been investigated for phenolic materials[2] and for cast acrylics.[3]

In phenolics there appears to be a gradual decline in S with increasing temperature. (The material is more sensitive at higher temperatures.) $1/D$ is practically constant within the temperature range investigated because there is a corresponding reduction in Young's modulus when the material is annealed.

The temperature dependence of the photoelastic properties in PMMA are anomalous and do not seem to be related to the other mechanical properties of the material.

The Effect of Creep on S and $1/D$
A general analysis of creep effects has shown that the photoelastic effect depends neither on S nor on $1/D$ exclusively but lies somewhere between, a condition which is satisfied by

$$\frac{1}{D_{\text{eff}}} = \left[\frac{1}{E}(P - Q)(1 + v) + \frac{3a}{2P_0}(P - Q)\xi \right]\frac{d}{m} \tag{11}$$

Equation (11) differs from (10) in that a material constant a appears on the right-hand side. This is defined by

$$a = \tfrac{3}{2}(1 + v)\mathrm{f}(t)$$

The overall strain can be expressed as a function of stress: elastic strain $\varepsilon_P| = (1 + \phi(t))(P/E)$ where

$$\mathrm{f}(t) = \frac{\psi(t)}{\phi(t)}$$

$\psi(t)$ can be ascertained very simply by applying a load to a tensile specimen and observing the change in the fringes at intervals of time t, using the equation

$$S = (P - Q)(1 + \psi(t))\frac{t}{m}$$

for further analysis of $\psi(t)$.

If the material behaves in such a way that the longitudinal plastic strain $(\delta_P)_{Q=0}$ is proportional to the applied stress P, then

$$(\delta_P)_{Q=0} = \frac{P}{E}\phi(t)$$

We can therefore calculate $\phi(t)$ and $\psi(t)$, hence $f(t)$ and thence a, and we now have all the data for calculating $1/D_{\mathrm{eff}}$ from eqn. (11).

We can also express the number of fringes m as a function of time according to eqn. (12) below:

$$m = (1 + \psi(t))\frac{Pd}{S} \tag{12}$$

A stress analysis thus becomes possible, despite mechanical and optical creep in the material, with a discrepancy of calculated and experimental results of less than 1% (for *PS*) at the maximum stress after 10 min of loading. The creep of phenolics is small and can usually be neglected; over $80 \times 10^3 \, \mathrm{kN \, cm^{-2}}$ are required for a strain of $\approx 0.1\%$.

Both fringe stress and fringe strain coefficients vary from one material to another. They also vary according to the history of the material and the testing conditions. The coefficients do not seem to be related to any material property, but they do depend on molecular structure. Therefore for the greatest accuracy in the determination of stress or strain the coefficients should be of low value, so that many fringes can be counted per unit length.

If light of different wavelength λ is used (commonly one uses yellow Na light, $\lambda = 5893 \, \text{Å}$), the coefficients must be corrected in direct proportion to λ, since it is the retardation in unit length that remains constant in a material, whilst the number of fringes depends on λ. When using white light, the resulting red and green series of isochromatic bands give an equivalent fringe stress coefficient which corresponds roughly to an interference of the red colour ($\lambda = 6900 \, \text{Å}$); coefficients are thus $\approx 17\%$ greater than those obtained with yellow Na light.

Comparing the different materials from the point of view of distortion produced through loading—the amount of distortion is directly indicated by the fringe strain coefficient (eqn. (4)).

Strain birefringence measurements have also been used to determine molecular entanglements in polymer solutions.[4] The method consists of the sudden application of a strain and an observation of the appearance and decay of the resulting induced birefringence. An application of the statistical theory of random-coil molecular networks can then be made, based upon the assumption that the observed birefringence in moderately concentrated solutions is due to the distortion of the molecular network formed from entanglements at mobile junction points, and that the relaxation of the anisotropic state is principally governed by relaxation processes at the junction sites. The theory of rubber elasticity should therefore be applicable.

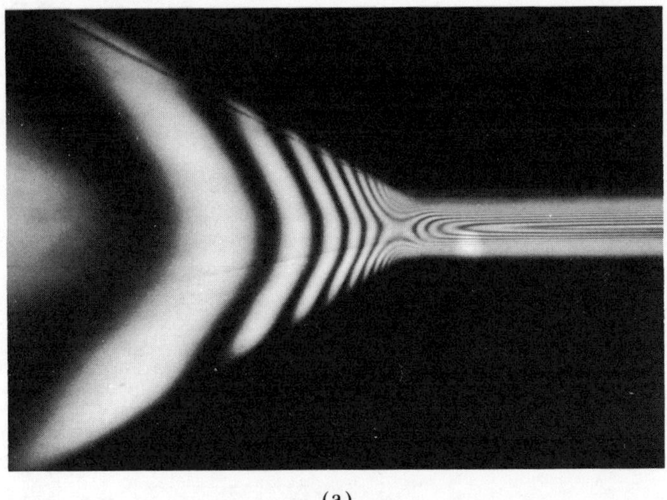

(a)

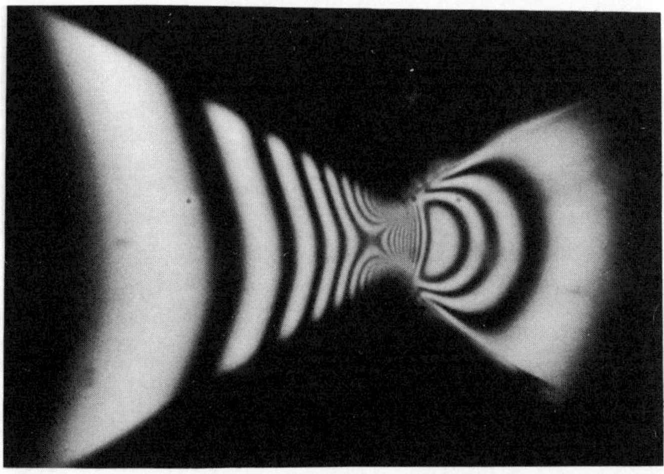

(b)

Fig. 20.1. Photograph of isochromatic fringe patterns of polystyrene melt at 200 °C flowing through a converging channel having a half-angle of 30 °(45): (a) die *with* a long slit section ($Q = 4\cdot33\,\mathrm{cm^3\,min^{-1}}$); (b) die *without* a slit section ($Q = 1\cdot5\cdot\mathrm{cm^3\,min^{-1}}$). Reproduced with permission from C. D. Han, *Rheology in Polymer Processing*, Academic Press (1976).

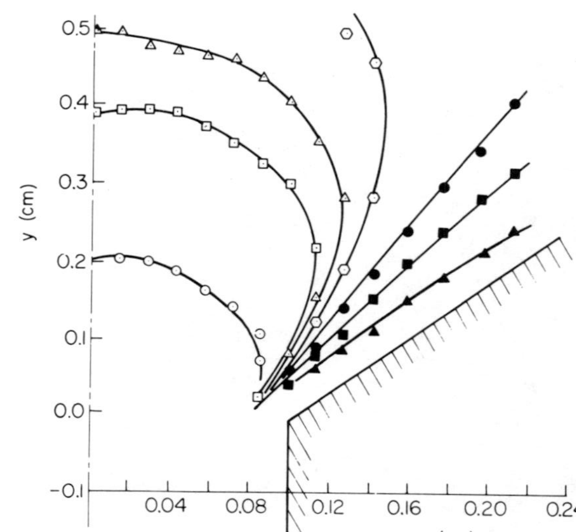

FIG. 20.2. Shear stress profiles of polystyrene melt at 200°C flowing through a converging channel having a half-angle of 30° ($Q = 5.36\,\text{cm}^3\,\text{min}^{-1}$) (30): (☉) $0.14 \times 10^5\,\text{dyn}\,\text{cm}^{-2}$; (⊡) 0.82×10^5; (△) 0.49×10^5; (⬡) 0.70×10^5; (●) 0.28×10^5; (■) 1.12×10^5; (▲) 1.40×10^5; (⬢) 1.54×10^5. Reproduced with permission from C. D. Han, *Rheology in Polymer Processing*, Academic Press (1976).

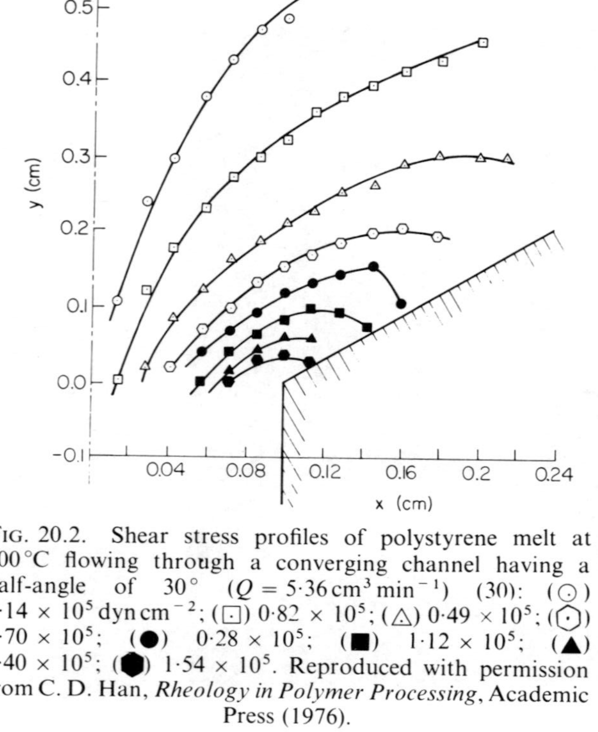

FIG. 20.3. Normal stress difference profiles of polystyrene melt at 200°C flowing through a converging channel having a half-angle of 30° ($Q = 5.36\,\text{cm}^3\,\text{min}^{-1}$) (30): (☉) $-1.12 \times 10^5\,\text{dyn}\,\text{cm}^{-2}$; (⊡) -0.56×10^5; (△) -0.42×10^5; (⊙) -0.28×10^5; (●) 0.0×10^5; (■) 0.28×10^5; (▲) 0.84×10^5. Reproduced with permission from C. D. Han, *Rheology in Polymer Processing*, Academic Press (1976).

Rheo-optics has proved to be a powerful tool for the analysis of the stress distribution in polymer melts, especially when melts are made to flow through geometrically complex channels. In this field, outstanding work has been done by Han and his school at the Polytechnic Institute of Brooklyn.[5]

Thus Han and Drexler[6] studied the stress distribution of polymer melts flowing through a rectangular slit die with a vertically tapered entrance. The isochromatic fringe patterns of polystyrene under stated conditions are shown in Figs. 20.1(a) and (b), whilst Figs. 20.2 and 20.3 give the shear stress and normal stress difference profiles for the melt flowing through the same die.

Han[7] has carried out an experimental study on melt flow through

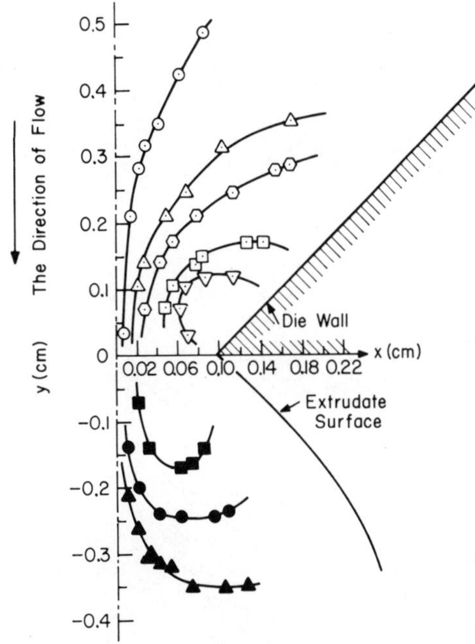

FIG. 20.4. Shear stress profiles of polystyrene melt at 200 °C flowing through a converging channel *without* a slit section ($Q = 1.5\,\text{cm}^3\,\text{min}^{-1}$) (45). Open symbols for stress in the die: (⊙) $0.68 \times 10^5\,\text{dyn cm}^{-2}$; (△) $1.76\ 10^5$; (⊙) 2.34×10^5; (□) 3.78×10^5; (▽) 4.78×10^5. Closed symbols for stresses in the extrudate: (■) $2.06 \times 10^5\,\text{dyn cm}^{-2}$; (●) 1.46×10^4; (▲) 0.912×10^5. Reproduced with permission from C. D. Han, *Rheology in Polymer Processing*, Academic Press (1976).

geometrically complex channels such as those sketched in Fig. 20.4 in which a rigorous theoretical analysis of the flow would be virtually impossible.

   Chapter 21 will deal qualitatively with the basis of rheo-optics and will give a number of examples which show the usefulness of the technique in engineering design.

REFERENCES

1. R. B. Heywood, *Designing for Photoelasticity*, Chapman & Hall, London (1952).
2. G. H. Lee and C. W. Armstrong, *J. Appl. Mech.*, **5,** A11–12 (March 1938).
3. H. A. Robinson, R. Ruggy and E. Slantz, *J. Appl. Phys.*, **15,** 343 (April 1944).
4. S. J. Gill and R. Toggenburger, *J. Poly. Sci.*, **60**(170), 569 (1962).
5. C. D. Han, *Rheology in Polymer Processing*, p. 151, Academic Press, London (1976).
6. C. D. Han and L. H. Drexler, *J. Appl. Poly. Sci.*, **17,** 2369 (1973).
7. C. D. Han, *J. Appl. Poly. Sci.*, **19,** 2403 (1975).

21

Rheo-Optical Techniques

F. Thamm

Professor of Polymer Physics and Engineering,
Technical University, Budapest

The apparatus required for the investigation of rheo-optical and photoelastic measurements is a polariscope as shown in Fig. 21.1. If white light is passed through, the image of the specimen projected on the screen will show coloured and dark fringes.

Double refraction depends on the wave nature of light which is described by the electro-magnetic theory. If a light beam is considered to be a harmonic vibration perpendicular to the direction of propagation, then it forms a sine wave along the axis of propagation. Its colour depends on the frequency, and the intensity on its amplitude A. Natural light such as sunlight consists of vibrations in every plane which contains the line of propagation. This also applies to light from an artificial source. In a polariser the beam becomes linearly polarised, depending on the plane of polarisation (vibration), the wavelength λ and the amplitude A. This is characterised by the light vector in Fig. 21.2.

If a second (analysing) polaroid is placed behind the polariser the amount of light transmitted depends on the angle between the planes of polarisation of the two filters. Since commercially available polaroids transmit a small amount of unpolarised light the field of crossed polaroids appears dark blue instead of black.

Whilst the frequency of any given colour remains constant, its velocity is retarded by passage through a medium. The ratio of this velocity relative to the velocity *in vacuo* is known as the refractive index.

Some transparent crystals show double refraction, with two mutually perpendicular planes (of double refraction) which have different refractive indices. If a beam of polarised light travels through such a crystal in a direction parallel to both planes, the beam is split into components vibrating in these planes and the components travel through the crystal with

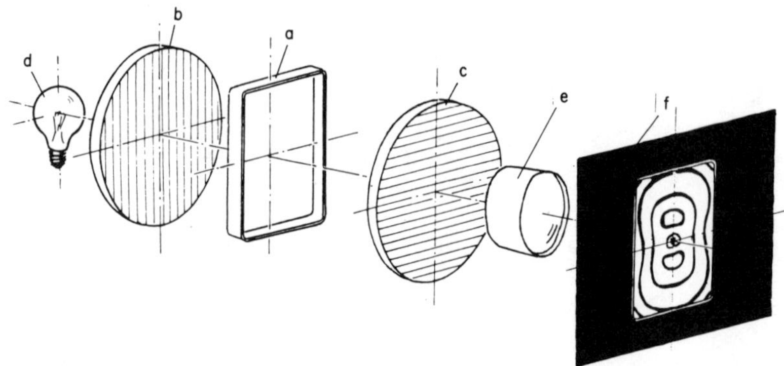

FIG. 21.1.   The polariscope—principle of design. a—specimen under observation;
b—polariser; c—analyser; d—light source; e—optical system; f—screen.

different velocities (Fig. 21.3). The faster component shows a relative shift Δ
against the slower one. The phase difference m is the ratio of that shift to the
wavelength of the beam measured *in vacuo* (or, in practice, in air). The
phase difference m can then be calculated from the two refractive indices,
the specimen thickness d and the wavelength λ, as shown in Chapter 20.

If two oscillations in perpendicular planes of the same wavelength with a

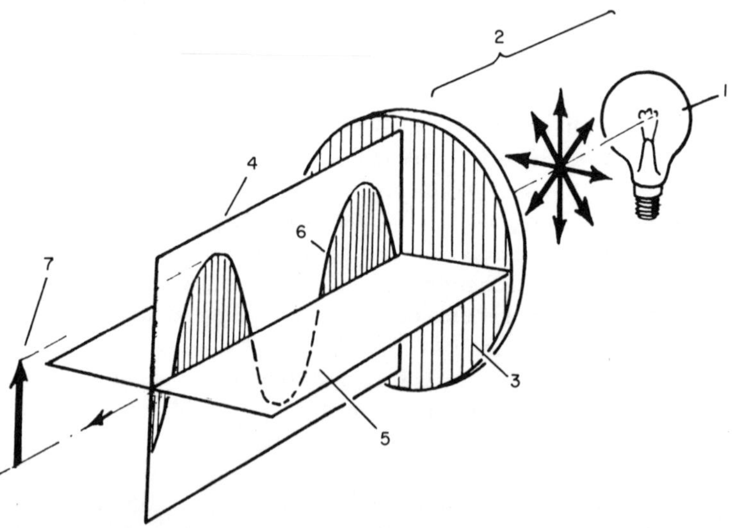

FIG. 21.2.   Light as a transverse wave function. 1—light source; 2—natural light;
3—polaroid; 4—plane of vibration; 5—plane of polarisation; 6—plane-polarised
light; 7—light vector.

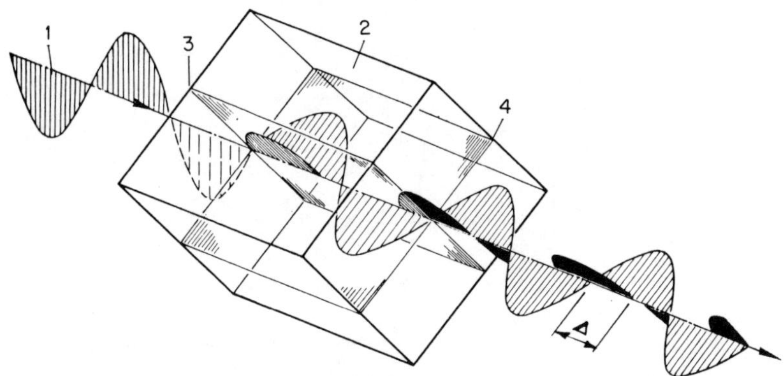

FIG. 21.3.   Passage of linearly polarised light through a double refracting crystal. 1—linearly polarised light beam; 2—double refracting crystal; 3 and 4—planes of double refraction.

shift Δ are vectorially added at any moment, a spiral curve results as shown in Fig. 21.4. If one considers one point on the path of the beam, the light vector is elliptical and the light is elliptically polarised. The shape of the ellipse depends on the angle ψ as defined in Fig. 21.4 and on the phase difference m (see Fig. 21.5).

If a beam of elliptically polarised light is passed through crossed polaroids, a linearly polarised component is transmitted. The light vector of this component is indicated in Fig. 21.5 (broad arrow) for the case of m having the value of 0·3. The intensity of light transmitted through the analyser may be evaluated assuming the polaroids to be ideal and the light losses negligible.[1]

So much for the double refraction associated with the molecular structure. A single crystal behaves uniformly throughout its volume. A double refracting crystal of uniform thickness therefore exhibits the same brightness and colour over the entire cross-section of the polariscope. Polymer chains are more or less optically anisotropic. If all the chains were oriented parallel to each other, a very high degree of double refraction would be observed. In a random distribution, however, the double refraction will cancel out and in fact be zero. Double refraction in plastics thus arises due to non-randomness of the chains and this may be due to the presence of an elastic strain (stress birefringence) or to orientation (orientational birefringence). A rigorous separation of these two types of double refraction is not always possible, especially in cases where much creep occurs. In both, the phase difference and the direction of the principal planes of polarisation may change from point to point in any given

Polymer Rheology

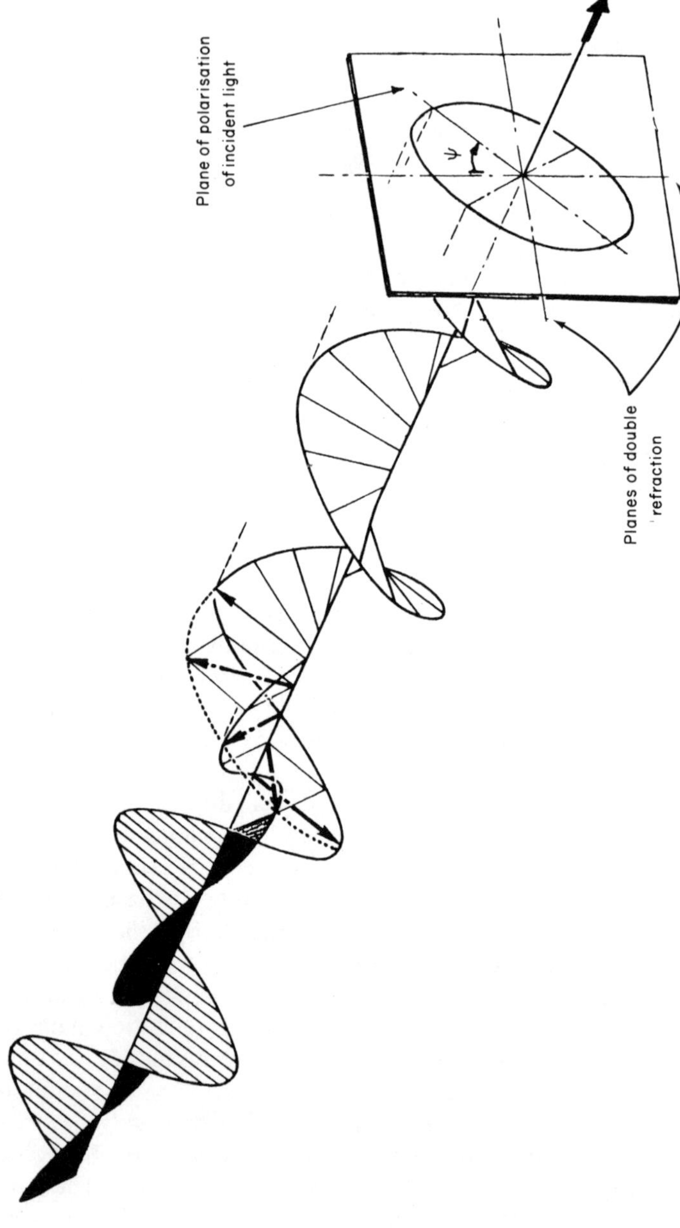

Fig. 21.4.   Addition of two linearly polarised light components with a phase difference to an elliptically polarised beam of light.

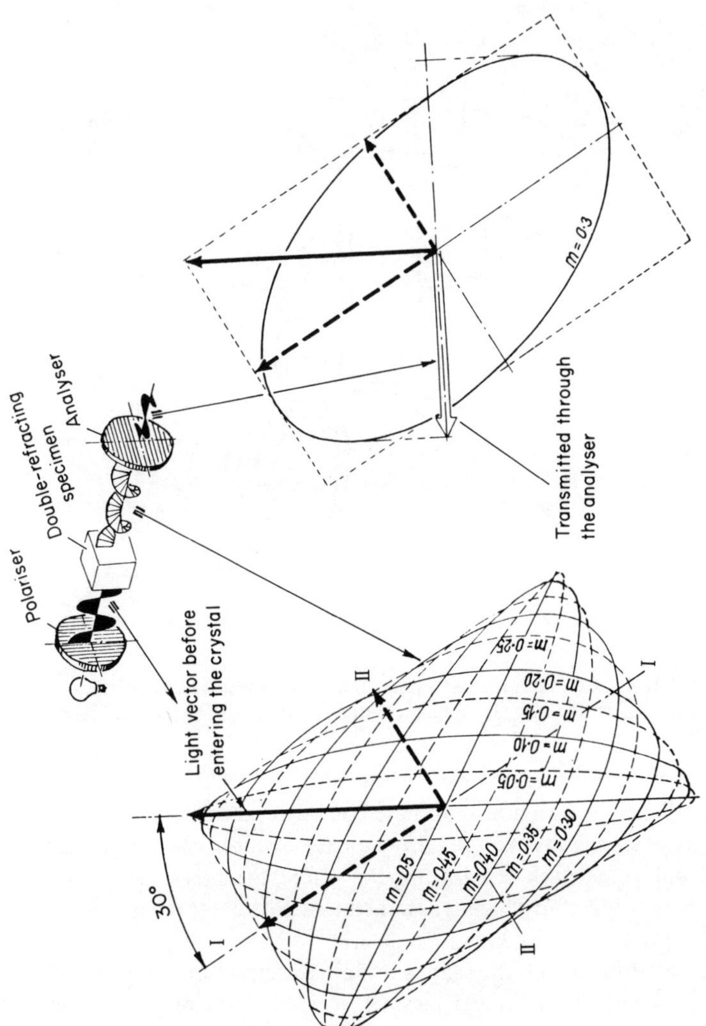

FIG. 21.5. The passage of light through a double refracting crystal between crossed polaroids.

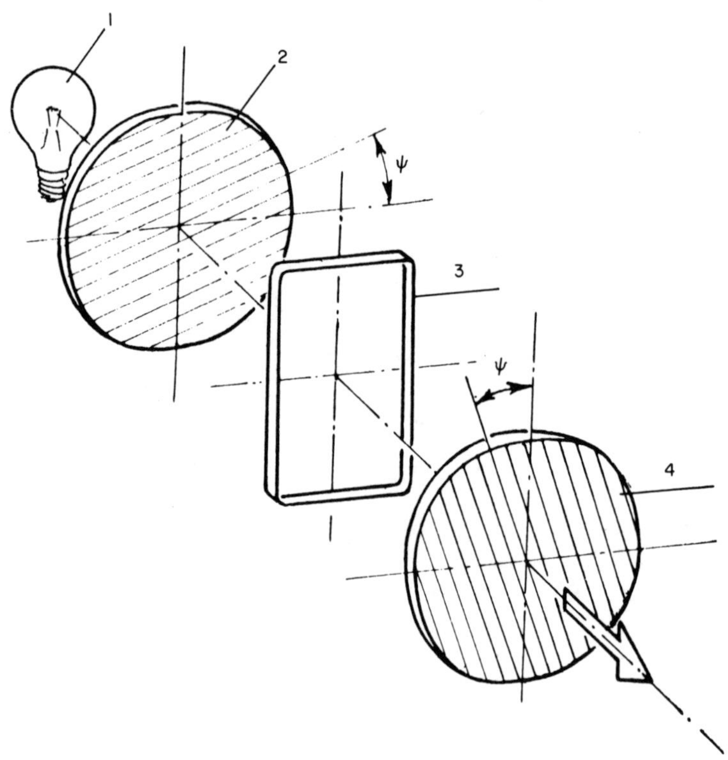

FIG. 21.6.   Rotation of polaroids to separate isochromatics and isoclinics. 1—light source; 2—polariser; 3—specimen under investigation; 4—analyser.

specimen. We can therefore observe lines along which the phase difference is constant. When using white light these lines are coloured; they are therefore known as isochromatic fringes. In the case of monochromatic light and crossed polaroids dark fringes appear where the phase difference is an integer. The value of m which is assigned to a particular fringe is the '*fringe order*'.

The lines along which the planes of double refraction are parallel to the planes of polarisation appear as dark fringes. They may be assigned to specific angles and are therefore known as isoclinics.

Both fringe types appear in the same pattern. They can be separated by rotating the crossed polaroids together round the optical axis of the system as shown in Fig. 21.6. During rotation the isochromatics remain constant while the isoclinics travel over an angle of rotation of 90°.

FIG. 21.7. A square flat plate under compression in the loading frame of a polariscope.

The appearance of isoclinics in a square flat plate under compression is shown in Fig. 21.7 and the photoelastic pattern for two different polaroid positions is seen in Fig. 21.8. Taking photographs at every 5 or 10° of rotation we obtain a set of isoclinics such as those which have been put together in Fig. 21.9.

When using white light the coloured isochromatics appear grey on black-and-white film, but in monochromatic light the isochromatics (like the isoclinics in either case) appear as dark fringes when m is an integer representing the fringe order.

The superimposed isoclinics always hide some of the isochromatic fringe pattern. To obtain a complete image of the isochromatics it is necessary to set up two extra filters known as quarter wave plates. These are filters with uniform double refraction with a phase difference of $m = \frac{1}{4}$ over the whole field. If two of them are introduced into a polariscope (Fig. 21.10) the image will contain no isoclinics. Such a polariscope is known as a circular polariscope. The light intensity may be calculated from well known equations and is seen to be independent of the angle ψ. If polariser and analyser are parallel then the intensity of transmitted light will become zero,

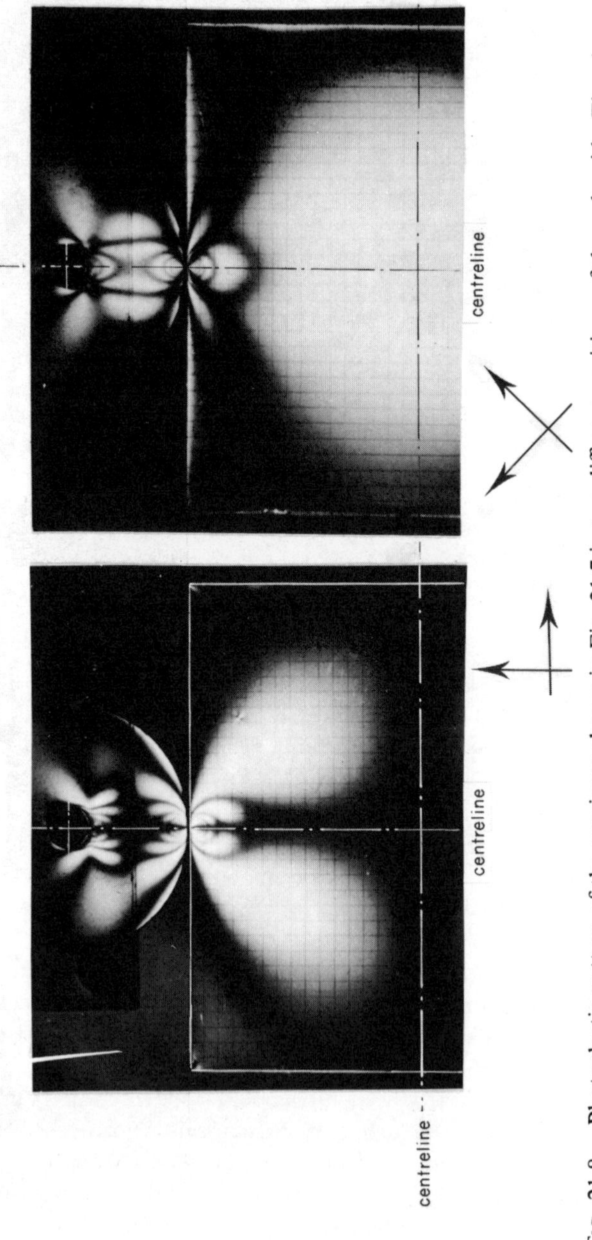

Fig. 21.8.   Photoelastic pattern of the specimen shown in Fig. 21.7 in two different positions of the polaroids. The planes of polarisation are shown by crossed arrows.

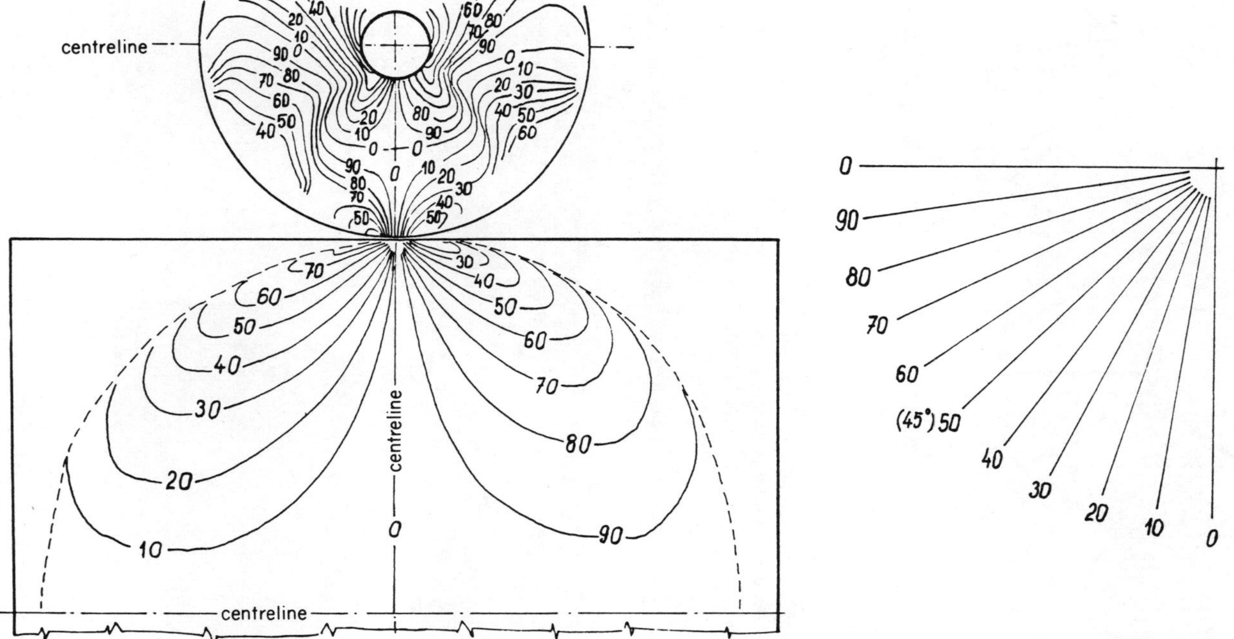

FIG. 21.9.   Isoclinic pattern constructed from photographs such as those shown in Fig. 21.8. The numbers on the isoclinics indicate the directions as given on the right-hand side, in decimal degrees.

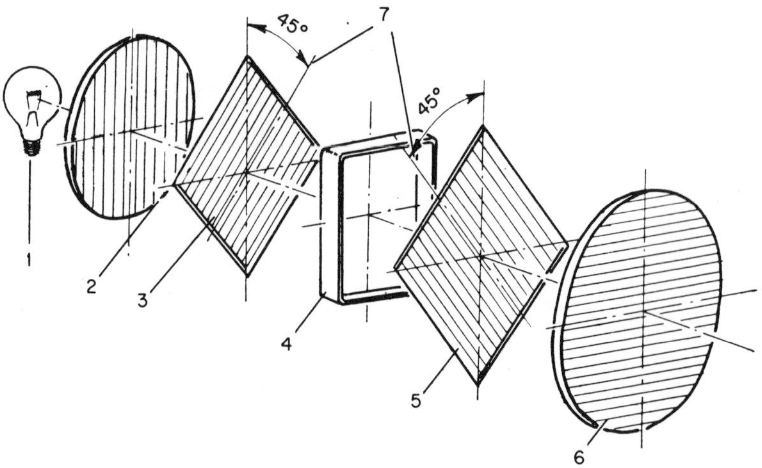

FIG. 21.10.   Circular polariscope. This is used to obtain isochromatic fringe patterns free from isoclinics. 1—light source; 2—polariser; 3—first quarter wave plate; 4—specimen under investigation; 5—second quarter wave plate; 6—analyser; 7—planes of polarisation of the quarter wave plates.

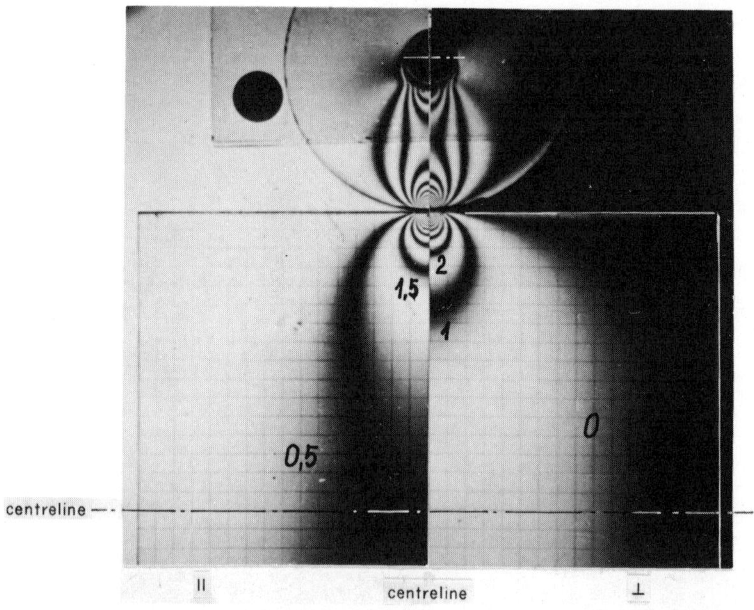

FIG. 21.11.   Isochromatic fringe pattern obtained in a circular polariscope: left-hand side—polaroids parallel, right-hand side—polaroids crossed. The numbers indicate the fringe order.

i.e. dark isochromatic fringes will occur when $m = \frac{1}{2}, 1\frac{1}{2}, 2\frac{1}{2}. \ldots$ Combining the isochromatic fringe patterns from crossed and parallel polaroids the number of fringes available for further evaluation is doubled. Figure 21.11 shows the isochromatic pattern of the same flat plate as that shown in Fig. 21.7 (circular polariscope, monochromatic sodium light).

Stress birefringence in the *glassy state* is induced by changing the relative position of neighbouring molecules in such a way that an increased energy level is attained ('energy-elastic deformation'). In the *rubbery state of cross-linked polymers* reversible orientation is caused by stretching the randomly coiled chains ('entropy-elastic deformation'). *In viscous and viscoelastic flow* the chains are oriented by the velocity gradient of the flow.

As the phase difference is a mean value of the birefringence along the path of a particular beam it can be simply related to the strain field only when the latter is uniform along the path of the light. This effectively restricts the investigations to plane states of stress, strain and strain rate. In the solid state the investigation of the three-dimensional states of stress is made possible by the method of stress freezing.[1] At moderate stresses or strain rates a linear velocity distribution was found to exist. The planes of double refraction have been found to be parallel to those of the principal stress or strain: along a particular isoclinic the principal stresses are parallel. On drawing these principal directions over the isoclinics a set of new curves— the *stress trajectories*—can be constructed, as shown in Fig. 21.12. In this figure the isoclinics and the principal planes are seen on the left-hand side;

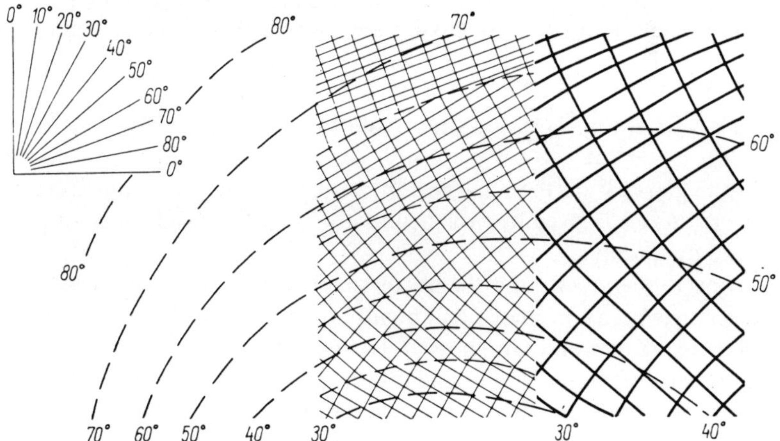

Fig. 21.12. The construction of stress trajectories from isoclinics.

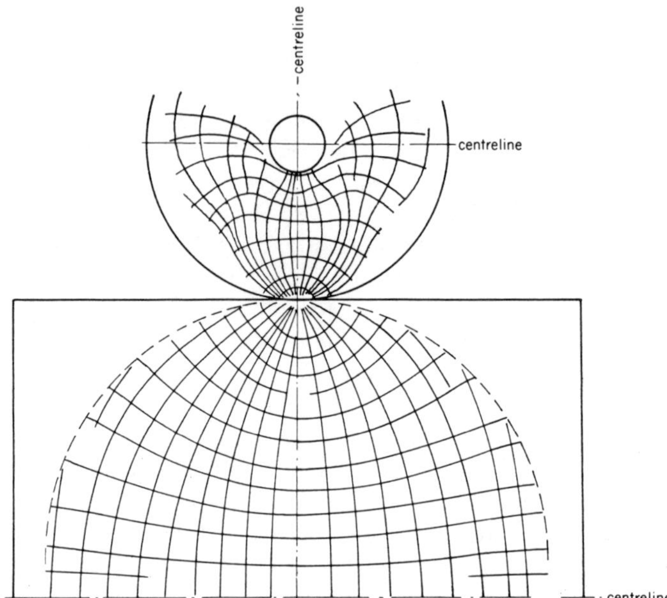

FIG. 21.13.   Stress trajectories, constructed from the isoclinics of Fig. 21.9.

in the centre the directions of the principal planes are drawn over the isoclinics, extending between the lines exactly bisecting the space which separates neighbouring isoclinics; and on the right-hand side, stress trajectories are drawn as smooth curves through the points of intersection of the isoclinics and the principal planes where the principal planes are tangential. The trajectories which have been obtained in this way give a qualitative indication of the stress flux. Using this method and the set of isoclinics from Fig. 21.9, the stress trajectories as seen in Fig. 21.13 were obtained.

In the glassy state the relationship between fringe order and stress (in terms of strain) is given by Wertheim's law[2] which is valid for both plane stresses and strains.

In the rubbery state the theory of small strains cannot be used. The problem has been considered and quantitative solutions have been offered by Fekete,[3] Kuhn and Gruen[4] and Treloar[5] for strains up to 50%.

In the case of viscous flow, Phillippoff[6,7] has found the fringe order to be proportional to the principal stress difference for laminar flow at moderate flow rates. The mathematical treatment also applied to viscoelastic fluids of moderate viscosity under a steady stress.

The stress-optical constant S (which is wavelength-dependent) and the strain-optical constant D (which depends on Young's modulus and Poisson's ratio) can be evaluated by means of a load/deformation test: for a glassy polymer a straight beam in pure flexure is convenient. The principal stress perpendicular to the axis of the beam vanishes and the stress calculated by the Navier equation is identical with the principal stress difference. The linear stress distribution along the y-axis (Fig. 21.14)

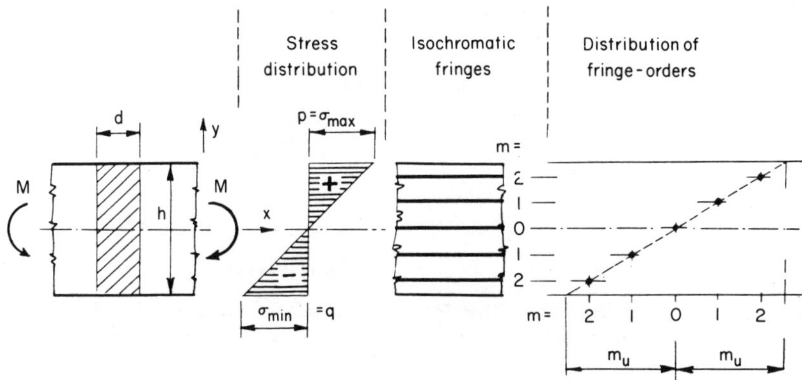

FIG. 21.14.   Evaluation of the stress-optical constant S by means of a straight beam in pure flexure.

produces an equidistant series of straight and parallel isochromatics with $m = 0$ at the neutral axis and increasing fringe orders on either side. The isochromatic pattern does not distinguish between positive and negative values of m. The linear distribution of m along the y-axis makes it possible to extrapolate the phase difference m_u accurately (see Fig. 21.14). If the applied bending moment M, the height h of the beam and m_u are known, the stress-optical coefficient can be calculated. The calculation is independent of the thickness of the beam, so that minor variations in the thickness of models can be tolerated.

In the rubbery state it is more convenient to use a rod shaped specimen which is subjected to a constant tensile strain. Because of the uniform stress distribution the phase difference will also be uniform throughout the field. The tensile strain at which complete extinction of the fringe order is achieved is determined and the two magnitudes together yield the strain-optical coefficient D, on the assumption that Poisson's ratio is 0·5.

The determination of the stress-optical constants of flow birefringence is

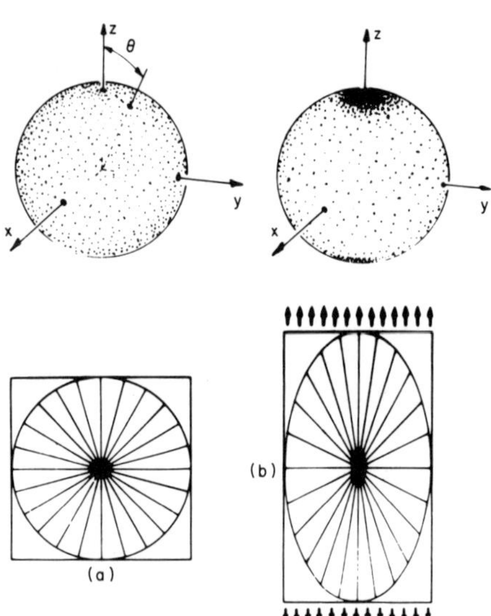

FIG. 21.15.   Orientation and the 'Polanyi sphere'. (a) Random distribution of polymer chains; (b) oriented structure. Top: representation of orientation at the Polanyi sphere.

described in Refs. 8–10. The flow birefringence of polymer melts in extrusion dies and in moulds has been analysed in depth by Han.[11]

Birefringence is often associated with the orientation of polymer molecules. This can be demonstrated by a 'Polanyi sphere' in which the molecular chains are represented by straight lines radiating from the origin of a Cartesian coordinate system (Fig. 21.15), where a random distribution represents the unstressed state. The refractive index of the polymer will be different for light travelling along the chains and for light travelling in a direction perpendicular to the chains. When there is radial symmetry in the plane normal to the stress-axis z, the shift in refractive index for a light ray travelling in this (the xy) plane can be calculated provided that the mean value of the orientation angle of the chains with respect to the z-direction is known. The latter also requires a knowledge of the density function of intersections of chains per unit surface of the Polanyi sphere; for uniformly distributed intersections the density function is equal to unity. To characterise the amount of orientation we use the orientation strain. This is the strain which would cause the observed birefringence in the *absence* of molecular orientation. The difference in the refractive indices $n_\perp$ and $n_\parallel$ of the material in the directions normal and parallel to the chains respectively is a function of the orientation strain, but this difference is rarely known. However, if the orientation is moderate and uniaxial, then it is practically proportional to the orientation strain.

This has been applied to thermosetting resins where the interconnected network prevents substantial molecular relaxation, and to thermoplastics and rubbers where a plastic stress with a completely different molecular configuration is superimposed upon an initial energy-elastic stress. In the case of melt flow it is more difficult to connect the orientational strain with any actual strain. On the other hand, flow orientation certainly affects the mechanical and thermal properties of plastics, as has already been discussed in considerable detail.

We shall now show how photoelastic techniques have been used in the optimisation of the engineering design of plastics products.

Example 1. The Stress Distribution in a Paper Based Laminate During Punching[12]

It was the purpose of this investigation to improve the structure of the laminate, and specifically, to prevent crack formation during the punching process.

FIG. 21.16. Stress distribution in a paper based laminate during punching. The model is in the loading frame of a polariscope.

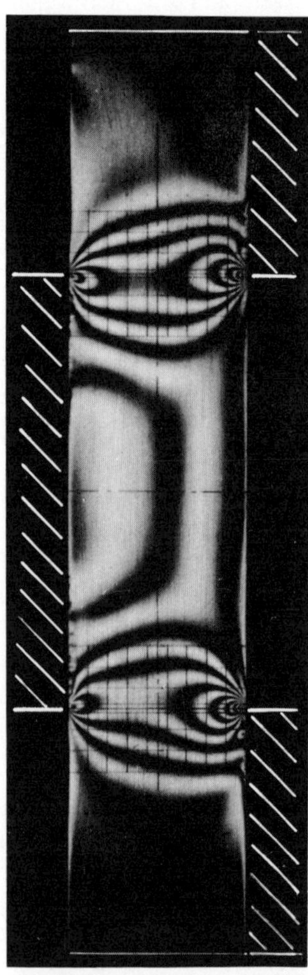

FIG. 21.17.   The isochromatic fringe pattern of the model shown in Fig. 21.16.

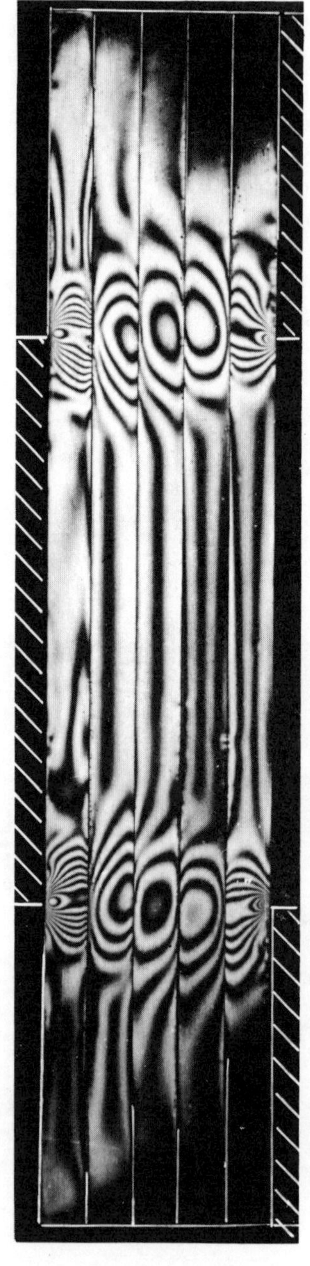

FIG. 21.18.   The isochromatic fringe pattern of a model consisting of separate sheets loaded as in Fig. 21.16.

A cross-section of the laminate was modelled in epoxy resin. In order to analyse the structure of the phenolic paper laminate which also had a copper backing four different models were used:

(i)   a homogeneous sample;
(ii)   a sample with a sheet of copper backing;
(iii) separate sheets;
(iv) sheets bonded with a resin of a lower Young's modulus.

Figure 21.16 shows the second model in the loading frame of a polariscope, while Fig. 21.17 shows the isochromatic fringe pattern of the same model between crossed polaroids, and Fig. 21.18 shows a similarly obtained pattern for model (iii).

The shear stress in the plane of punching can be calculated if the principal stress difference and the angle Φ between the plane of punching and stress p

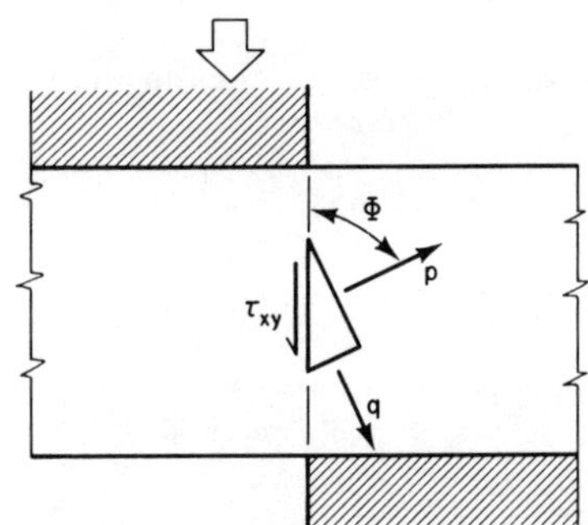

Fig. 21.19.   Diagrammatical illustration of the stress components in the shear plane.

are known (see Fig. 21.19). The first may be calculated from the isochromatic and the second from the isoclinic pattern.

The distribution of the shear stresses for the four models along the plane of punching is shown in Fig. 12.20. Maximum shear stress trajectories have also been constructed; these are inclined at a 45° angle to the trajectories of the principal stresses and can be obtained from the isoclinic pattern (see Fig. 21.21). One would expect the cracks during punching to occur approximately along these lines, as indicated by the thicker lines in Fig. 21.21; it is seen that the cracks tend to deviate from the plane of punching much more in the multilayered models than in the homogeneous models. This is borne out by practical experience.

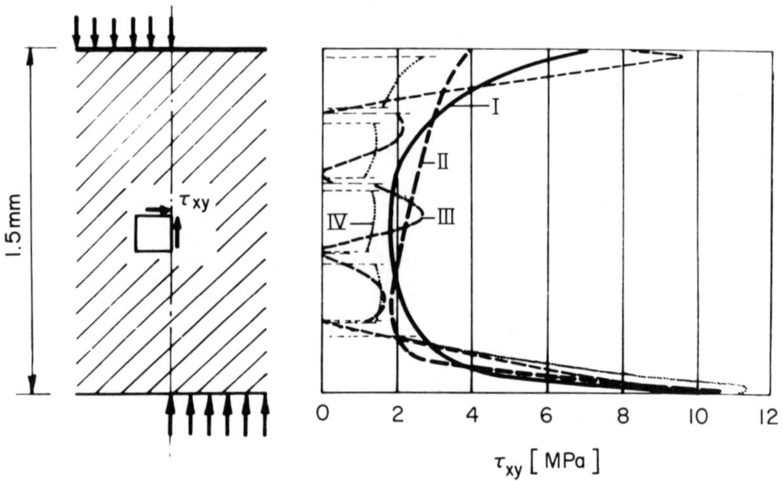

FIG. 21.20.   The distribution of shear stresses in the punching plane for four different models.

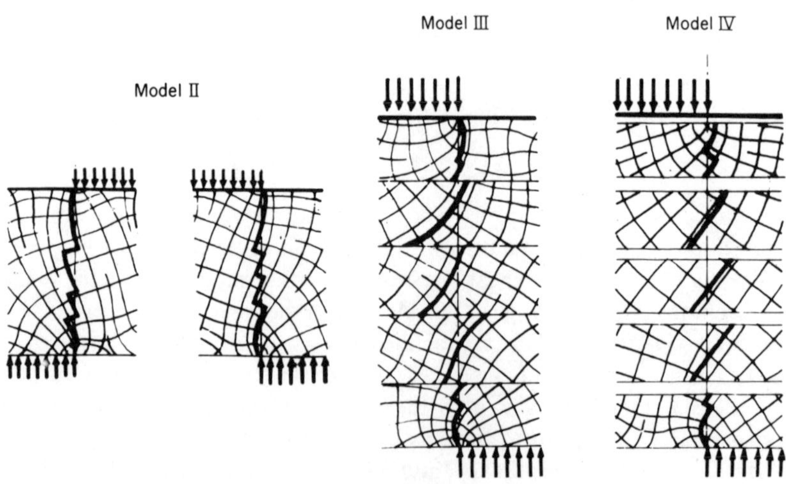

FIG. 21.21.   The trajectories of maximum shear for models II, III and IV. The lines of anticipated cracks during punching are marked with thick lines.

Example 2. Modelling the Stress Distribution in the Rubber Body of a Truck Tyre

The object of this investigation was to develop a shape for the rubber body of a tyre with a minimum tendency of the rubber tearing away from the plies. A model tyre cross-section was built up from linen cord and a gelatine/glycerine/water mixture (60/30/10). The model was placed in a

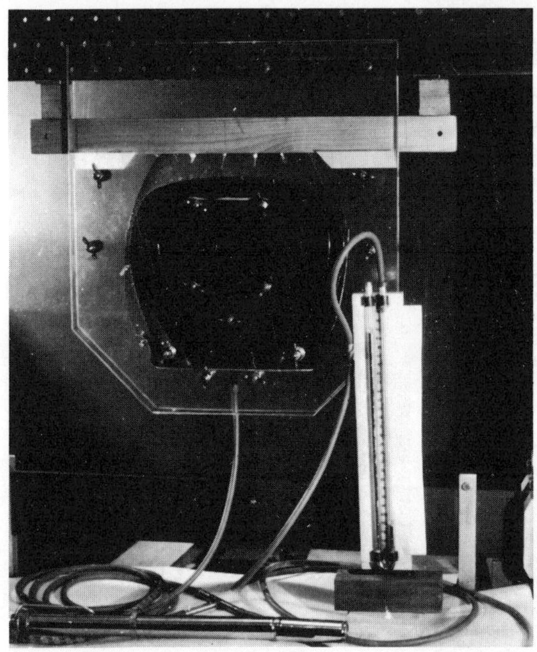

FIG. 21.22. Model of the cross-section of a truck tyre in the loading rig of a polariscope.

loading frame; internal pressure was applied as shown in Fig. 21.22 and an isochromatic fringe photograph was taken (Fig. 21.23). The distribution of shear stresses along the surface of the cord layer was calculated and is shown in Fig. 21.24. Using these photoelasticity results it was subsequently possible to design an optimum shape for the rubber body of the truck tyre.

Example 3. Stress Analysis of a Transport Chain Link of Complex Shape[13]

The chain link is shown in Fig. 21.25. It was to be made from a polyamide but doubts arose as to whether it would be able to withstand traction forces

Polymer Rheology

Fig. 21.23.   Isochromatic fringe pattern of the model shown in Fig. 21.22.

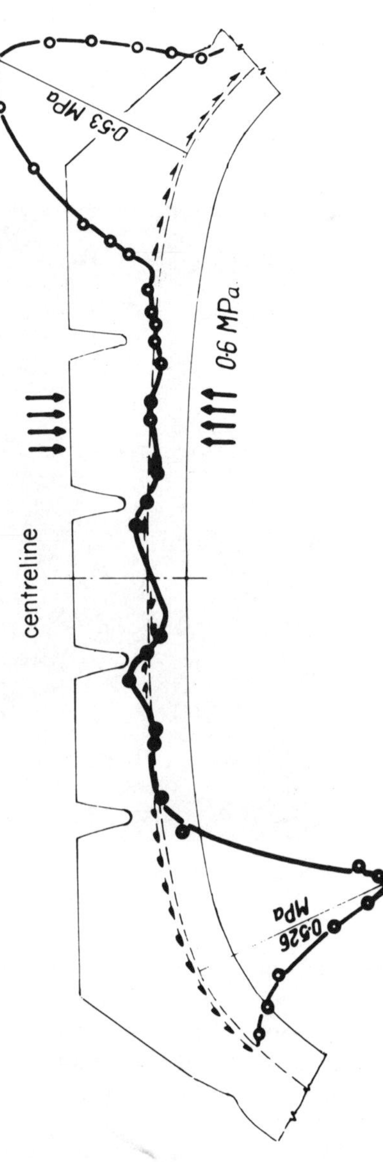

Fig. 21.24.   Distribution of shear stresses along the surface of the cord layer of the truck tyre as evaluated from photoelastic data.

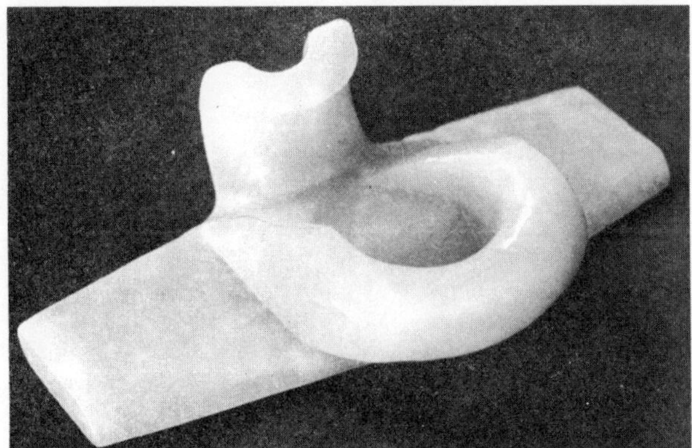

FIG. 21.25.   Model of a chain link which was the subject of a three-dimensional photoelastic investigation.

of up to 1 kN. A photoelastic stress analysis was therefore carried out before the mould was made. The complex shape of the component required the use of a stress freezing technique. A double scale model of the prototype was made and loaded as shown in Fig. 21.26. After heat treatment under load, the model was sliced as shown in Fig. 21.27, and isochromatic patterns of the slices were obtained (Figs. 21.28 and 21.29). The stresses at

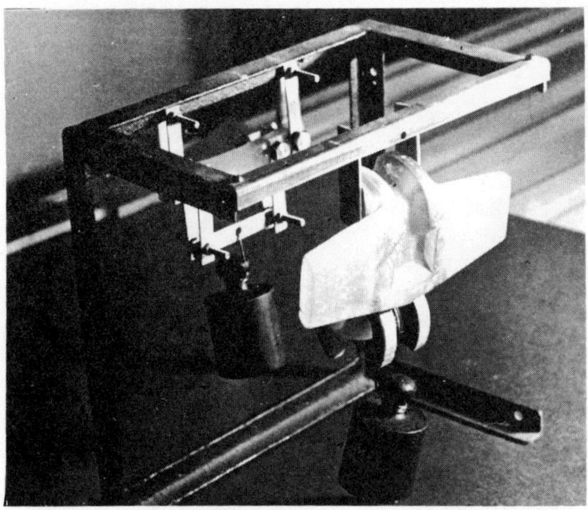

FIG. 21.26.   The chain link model (see Fig. 21.25) in the loading rig for frozen-in stress determination.

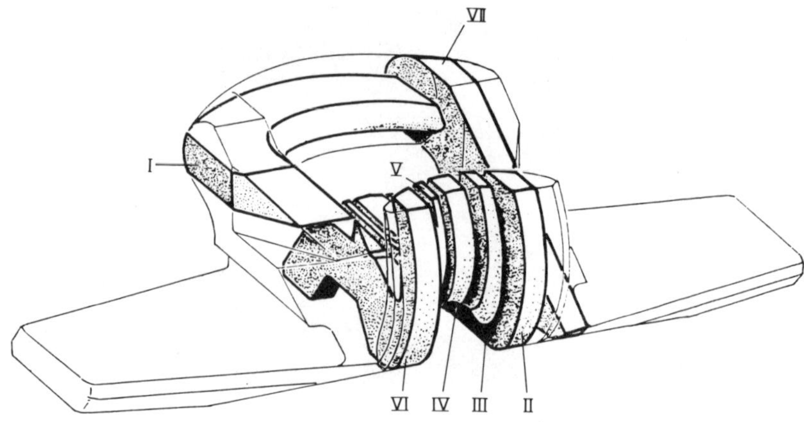

FIG. 21.27.   The position of the slices cut out of the chain link model as shown in
Fig. 21.25.

the surfaces are seen in Figs. 21.30 and 21.31. These results were converted
to the prototype using the laws of similarity[14,15] for a traction of 1 kN. The
peak stress values in the hook-like portion of the link in slice IV showed that
the design was faulty. The measured stress distribution indicated the
possibility of an approximate calculation which confirmed the results of the
experiment. The cost of this investigation was far less than the loss which

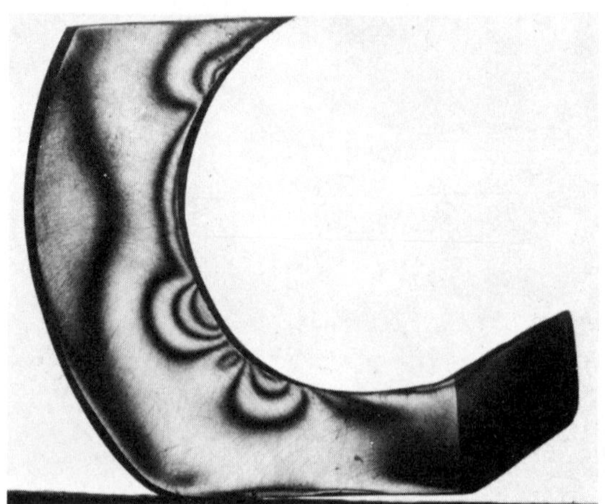

FIG. 21.28.   Isochromatic fringe pattern of slice I.

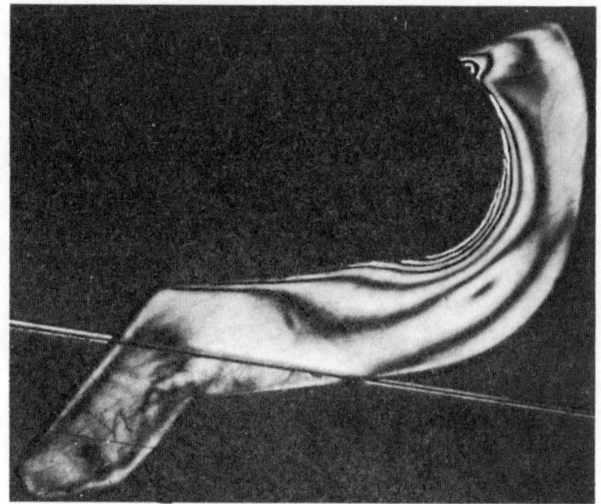

FIG. 21.29.   Isochromatic fringe pattern of slice IV.

would have been incurred if the component had been produced in its original shape.

Example 4. Investigation of the Frozen-in Stresses in the Insulation of High Voltage Cable

The low-density-polyethylene (LDPE) insulating layer of high voltage cables is extruded onto the conductor. The crystallinity of polyethylene makes it virtually impossible to obtain a macroscopic rheo-optical pattern. However, LDPE can be used for rheo-optical investigations because its lower crystallinity enables one to obtain suitable patterns, provided that the thickness of the section does not exceed 2 mm.

The aluminium conductor prevents thermal post-extrusion shrinkage. This causes residual stresses which are partially converted to orientation due to relaxation, thus leaving frozen stresses. These can be examined in the polariscope, using slices cut out of the cable insulation. The contrast between the dark and the bright fringes gives an indication of the degree of crystallinity (Figs. 21.32 and 21.33). The pattern obviously corresponds to the stress distribution in thick walled tube subjected to an internal pressure. The distribution is concentric, so that the isochromatic fringes should also consist of concentric circles. Any deviation from concentricity and/or a perfectly circular shape of the fringes indicates irregularities in the temperature distribution which in turn produces uneven shrinkage in the cooling melt. A cable with such an insulating layer has inferior electric strength.

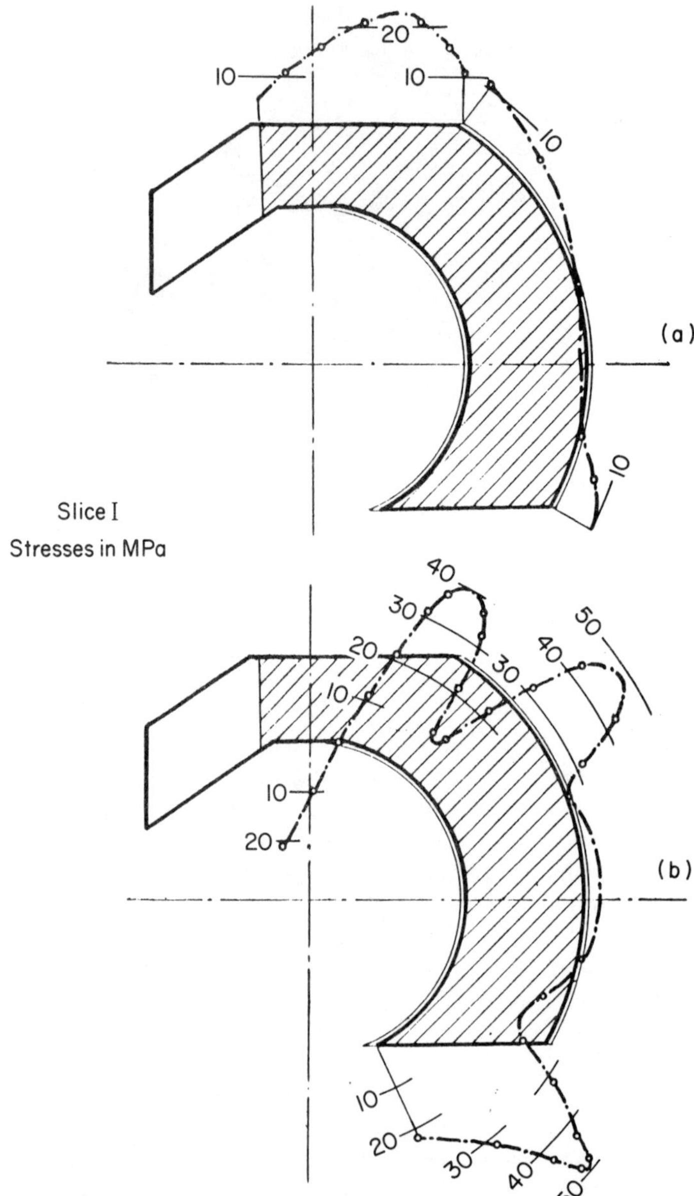

FIG. 21.30.   Stress distribution at the margin of slice I of Fig. 21.25. (a) Outer side; (b) inner side.

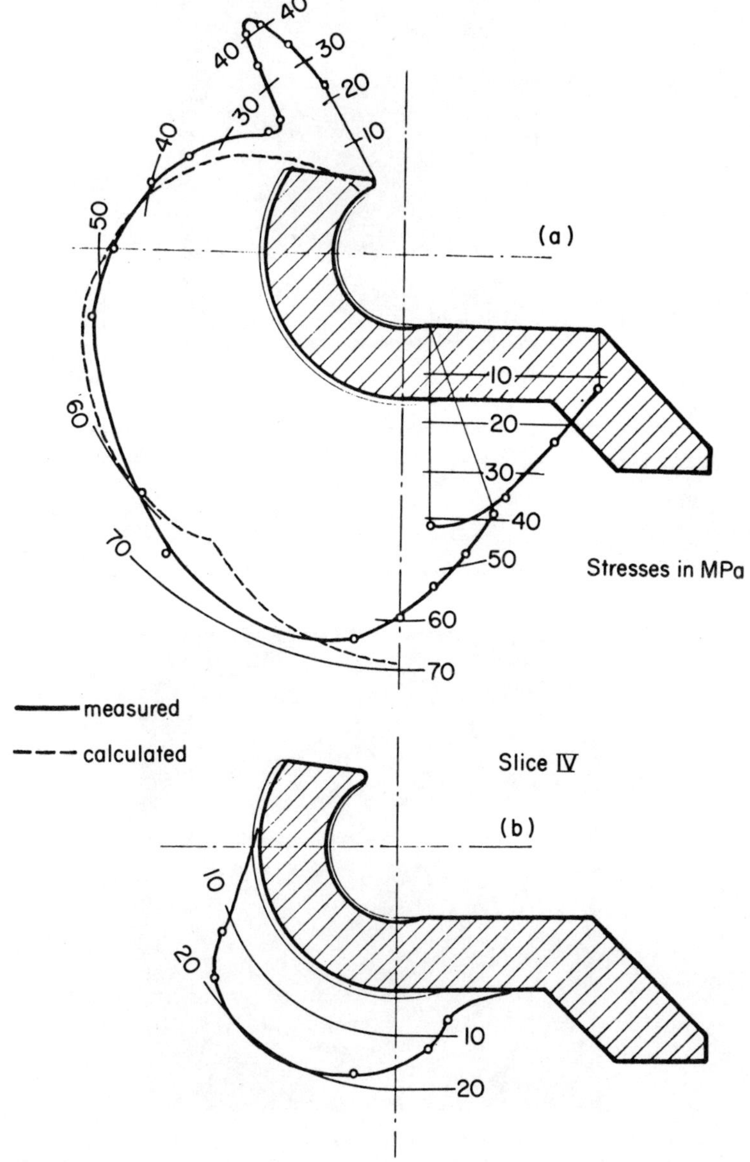

FIG. 21.31. Stress distribution at the margin of slice IV of Fig. 21.25. (a) Inner side; (b) outer side.

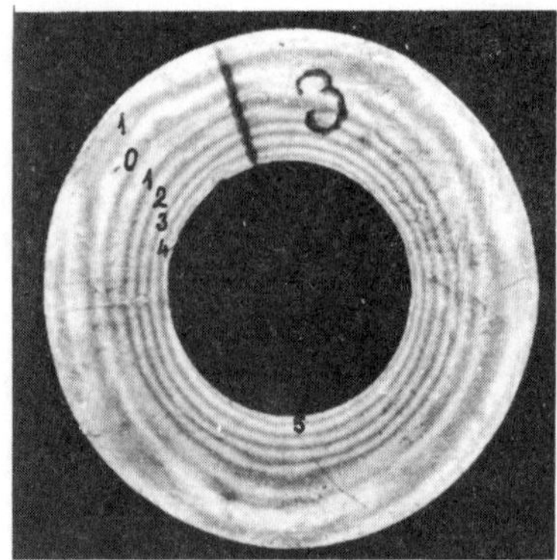

Fig. 21.32. Isochromatic fringe pattern in the cross-section of a slice from the insulating layer of a high voltage cable. The low contrast between bright and dark fringes indicates relatively high crystallinity. The written numbers show the fringe orders. (Courtesy of Hungarian Cable Works, Budapest.)

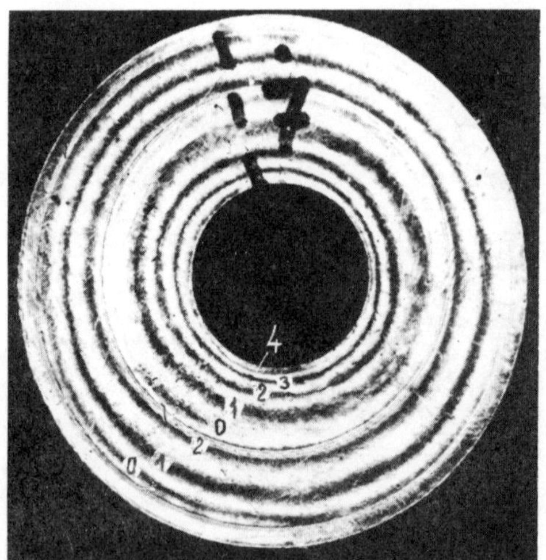

Fig. 21.33. Isochromatic fringe pattern as in Fig. 21.32. The high fringe contrast indicates low crystallinity (material: LDPE). Numbers indicate fringe orders. Extrusion of the insulating layer in two stages, made visible by the fringe pattern. (Courtesy of Hungarian Cable Works, Budapest.)

Rheo-optics has thus proved itself to be a valuable tool for quality control in cable manufacture.

Example 5. Molecular Orientation in Transparent Injection Moulded Parts

Most of the amorphous thermoplastics used for the mass production of injection moulded parts have far more orientational birefringence than birefringence due to residual stresses, so that the rheo-optical patterns are

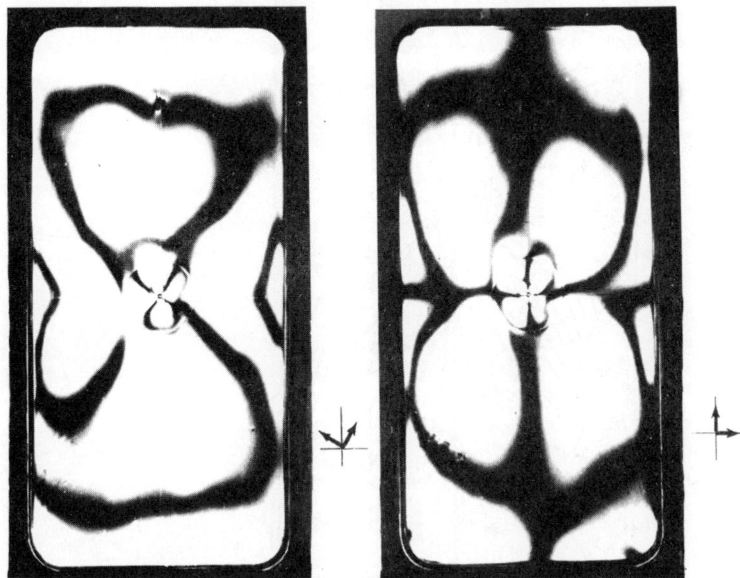

FIG. 21.34. Isoclinics of the pattern of orientation birefringence in an injection moulded polystyrene lid. The arrows indicate the planes of polarisation.

practically unaffected by the latter. Figure 21.34 shows the isoclinics of a polystyrene lid with a centre sprue, the polaroids being in two different positions. A fully developed construction of the isoclinics map is shown in Fig. 21.35 and the trajectories of molecular orientation in Fig. 21.36.

The extension of the flow front during injection causes a change in the direction of orientation from one *parallel* to the flow direction, to one which is *perpendicular* to the flow direction. This is illustrated in the furthest portion of the moulded lid (Fig. 21.36).

The isochromatic fringe patterns of two lids of somewhat different shape are shown in Fig. 21.37. Dynstat test specimens were cut from the lids, both parallel and perpendicular to the direction of orientation.

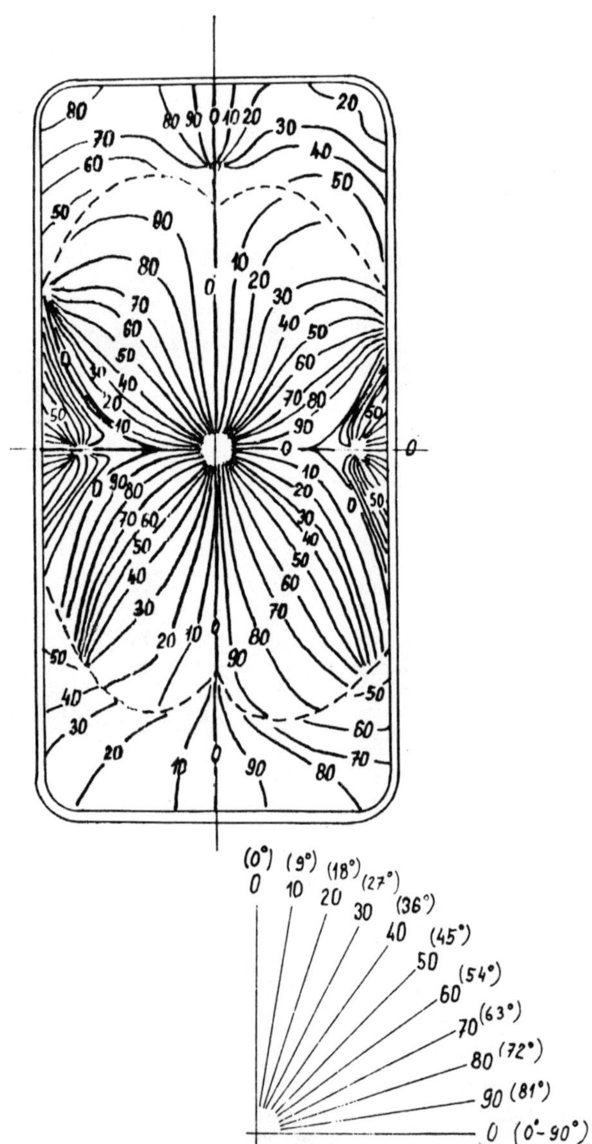

FIG. 21.35. Fully developed construction of the isoclinics of an injection moulded polystyrene lid.

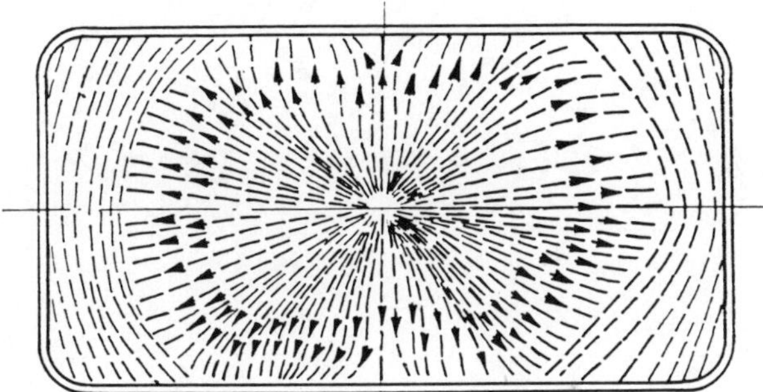

FIG. 21.36. Trajectories of molecular orientation constructed from Fig. 21.35.

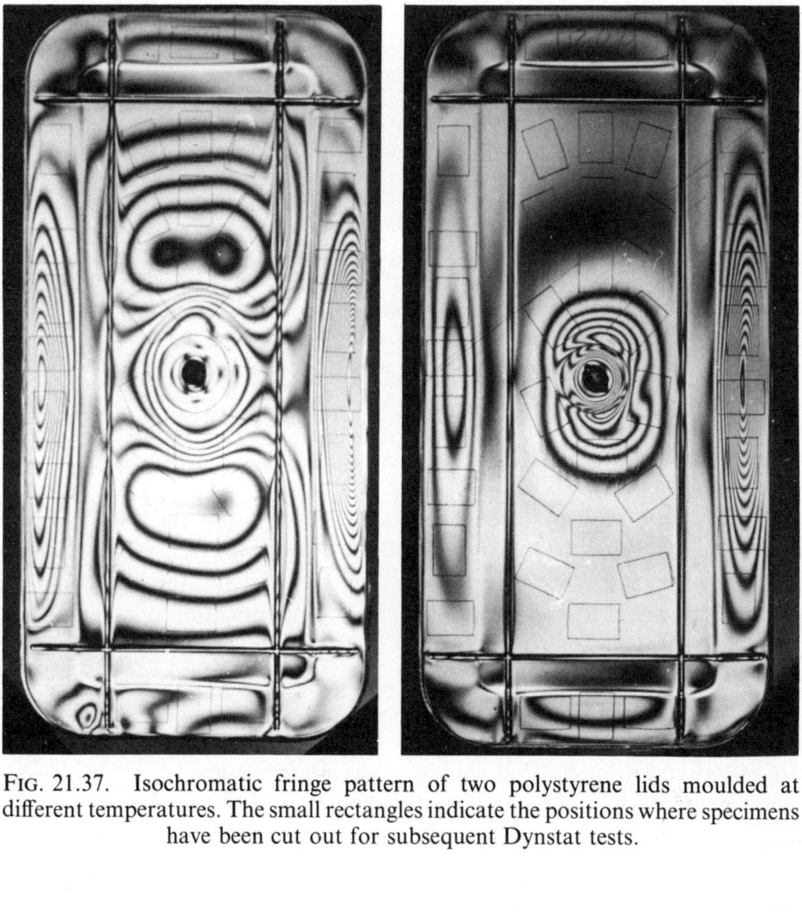

FIG. 21.37. Isochromatic fringe pattern of two polystyrene lids moulded at different temperatures. The small rectangles indicate the positions where specimens have been cut out for subsequent Dynstat tests.

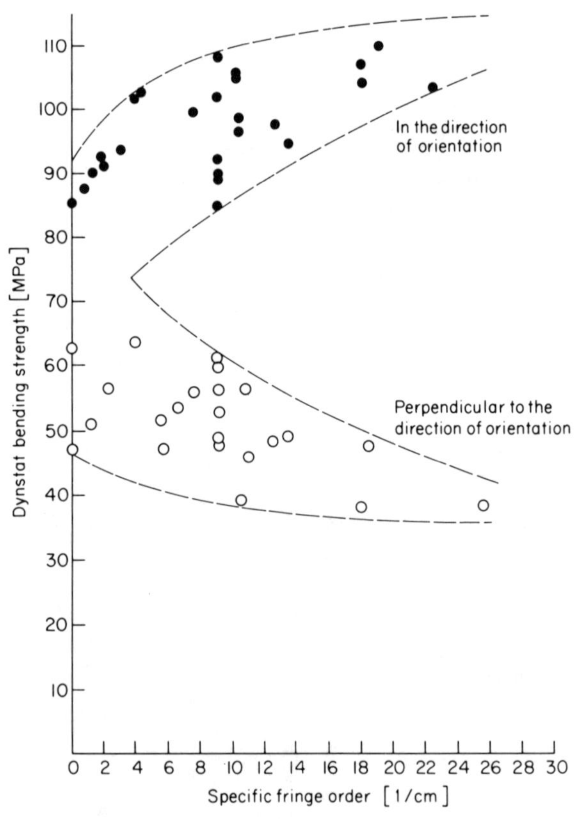

FIG. 21.38.   Dynstat flexural strength cut from the lids as marked out in Fig. 21.37
as a function of the specific fringe order.

The Dynstat flexural strength is shown as a function of the specific fringe order m/d, where d is the thickness of the specimen (Fig. 21.38). The strength is seen to increase with increasing fringe order in the direction of orientation, but it decreases *perpendicular* to the direction of orientation. On the basis of the concept of the weakest link, orientation is clearly undesirable in a moulding of this kind. The strength anisotropy, combined with the anisotropy in Young's modulus and the effect of thermal expansion is prone to produce cracks such as those shown in Fig. 21.39. Such cracks are often seen in mouldings from styrene homopolymer.

Since the processing temperature is known to affect orientation greatly, rheo-optical inspection constitutes a most useful non-destructive test for

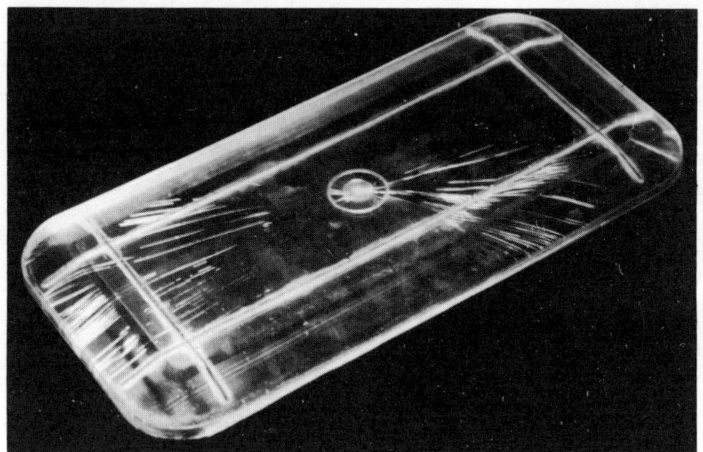

FIG. 21.39. Cracks in an injection moulded polystyrene lid due to anisotropy caused by molecular orientation.

assessing the quality of transparent mouldings and their anticipated subsequent performance in service.

APPENDIX—SOME EQUATIONS FOR RHEO-OPTICS

Phase difference

$$m = \frac{\Delta}{\lambda} = \frac{d}{\lambda}(n_1 - n_2)$$

Intensity of transmitted light
Ideal polaroid:

$$I = A^2 \sin^2(\pi m) \sin^2 2\varphi$$

Circular polariscope, crossed position:

$$I = A^2 \sin^2(\pi m)$$

Circular polariscope, parallel position:

$$I = A^2 \cos^2(\pi m)$$

Wertheim's Law
    Generally:

$$m = \frac{C}{\lambda}(P - Q) = \frac{C}{\lambda}\,d\,\frac{E}{1 + v}(E_P - E_Q)$$

Glassy state:

$$m = Dd(\varepsilon_P - \varepsilon_Q)$$

Rubbery state:

$$m = Dd(\varepsilon_P^* - \varepsilon_Q^*)$$

where

$$\varepsilon_P^* = \ln(1 + \varepsilon_P) \qquad \varepsilon_Q^* = \ln(1 + \varepsilon_Q)$$

(see Ref. 3), and

$$P - Q = \frac{2G}{1 + \varepsilon_R}\sinh(m/dD)$$

for strains up to 50% (Refs. 4, 5),

$$P - Q = 2G\sinh(m/dD)$$

for plane strains (Refs. 4, 5), and

$$G = E/2(1 + v)$$

Viscous flow:

$$m = \frac{d\eta}{S}(\dot{\varepsilon}_P - \dot{\varepsilon}_Q)$$

Material constants S and D
    Glassy state:

$$S = 6M/h^2 m_u$$

where M = applied bending moment, h = height of beam, and m_u = the known stress optical coefficient.

Rubbery state:

$$D = 2m^*/3(\varepsilon_P^*)_1$$

where $(\varepsilon_P^*)_1$ is the tensile strain at which complete extinction of the fringe order $m = m^*$ is achieved.

Flow: (cup–bob rotational viscometer)

$$S = \frac{2\eta r \omega d}{m_0(r_0 - r_i)}$$

Orientational birefringence:

$$n_1 - n_2 = (n_\parallel - n_\perp)(1 - \tfrac{3}{2}\sin^2 \theta_m)$$

where θ_m is the mean value of the orientation angle θ of the chains with respect to the x-direction and is a function of θ and a 'density function' which is zero when the Polanyi sphere is undistorted. $n_\parallel$ and $n_\perp$ are the refractive indices of light travelling along, and perpendicular, to the chains respectively. When there is radial symmetry in the plane normal to the stress-axis z, the shift in the refractive index for light travelling in this (the xy) plane is given by the above equation.[16] In practice $n_\parallel$ and $n_\perp$ are rarely known, but the difference is approximately equal to the orientational strain (if uniaxial), or to the orientational strain difference (if biaxial).

(All the symbols used that have not been explicitly defined above have already been defined and explained in the preceding text.)

REFERENCES

1. A. KUSKE and G. ROBERTSON, *Photoelastic Stress Analysis*, John Wiley & Sons, New York (1974).
2. E. G. COKER and L. N. G. FILON, *A Treatise on Photo-elasticity*, 2nd ed., Cambridge University Press (1957).
3. T. FEKETE, *Periodica Polytechnica*, **13**, 249–62 (1969).
4. W. KUHN and F. GRUEN, *Kolloid-Z.*, **101**, 248–71 (1942).
5. L. R. G. TRELOAR, *The Physics of Rubber Elasticity*, Clarendon Press, Oxford (1975).
6. W. PHILLIPPOFF, *J. Appl. Phys.*, **27**(9), 984–89 (1956).
7. W. PHILLIPPOFF, *Trans. Soc. Rheol.*, **5**, 163–91 (1961).
8. P. MUNK and P. KRATOCHVIL, *Collect. Czedpsl. Chem. Comm.*, **26**, 6 (1961); **27**, 9 (1962).
9. H. ABEN, *Izvestia AN SSSR*, **12**, 4, 370–5 (1963).
10. H. G. JERRARD, *Sci. Instr.*, **26**, 1007–110 (1966).
11. C. D. HAN, *Rheology in Polymer Processing*, Academic Press, New York (1976).
12. F. THAMM, *Proc. Conf. ICEM 72*, pp. 156–9, Prague (1972).

13. F. THAMM, *Beitraege zur Spannungs- und Dehnungsanalyse III*, pp. 101–12, Akademie-Verlag, Berlin (1966).
14. A. KUSKE and G. ROBERTSON, *Photoelastic Stress Analysis*, John Wiley & Sons, New York (1974).
15. J. T. PINDERA, *Beitraege zur Spannungs- und Dehnungsanalyse V*, pp. 103–30, Akademie-Verlag, Berlin (1968).
16. R. E. ROBERTSON and R. J. BUENKER, *J. Poly. Sci.*, **2**, 4889–901 (1964).

22

Rheology and Morphology

It is desirable to investigate the processes which occur in a solid polymer under stress or deformation. In order to do so it is necessary to state a number of well known generally accepted facts.

With some rare and untypical exceptions, polymer molecules are held together by primary valence bonds.

Ionic links do not normally occur in the main skeleton.

Intermolecular forces result from secondary attractions such as hydrogen bonding and London forces. These are weak compared with the primary valence forces.

The mechanical strength and stability of polymers is directly related to their high molecular weight.

Certain polymers have branched chains which may be the result of side reactions during polymerisation. However, the main chain is still recognisable as such. Regularly repeating side groups are not regarded as branched polymers in this sense.

Crosslinking and network structures occur during polymerisations in which at least some of the reactants have functionalities in excess of two. If these represent a substantial proportion of the reactants, then one obtains a dense network of primary valences, as in thermoset mouldings. Subsequent heat treatment can only result in degradation and never in a reversion to a material which can flow and be remoulded. The true melt state is restricted to thermoplastics which are capable of free translation in all directions as well as free rotation and vibration, within the limits set by the melt viscosity. In the solid state the freedom of both translation and rotation are largely lost.

The forces of attraction in solid polymers depend on the polymer type and differ greatly. In some cases the forces are so great as to produce the structural organisation of crystallinity. The principal forces of opposition are Brownian motion, the kinetic energy of which is temperature-dependent. If the intermolecular forces are weak or if the temperature is high the entropy of the system will be high. The entropy will also be increased by molecular dissymmetry—especially as regards the degree of regularity in chain substituent sequences.

There are two significant differences between crystallinity in low molecular weight substances and polymer crystallinity:

(1) In the case of polymers the formation of large ordered regions is rendered difficult because of the length and flexibility of the chains, as well as their unavoidable entanglements and environmental imperfections. Crystallinity can therefore never be complete and there is always a proportion of amorphous material present. By the same token, there also exists a certain probability of the formation zones of local order, so that even polymers which are regarded as typically amorphous possess a degree of crystallinity which may not be negligible.

(2) The average size of a crystallite is less than the length of a polymer chain. This means that one and the same polymer molecule may be a constituent of several crystalline and amorphous regions. The amorphous matrix contributes a high degree of cohesion in a partially crystalline polymer and acts like mortar in brickwork.

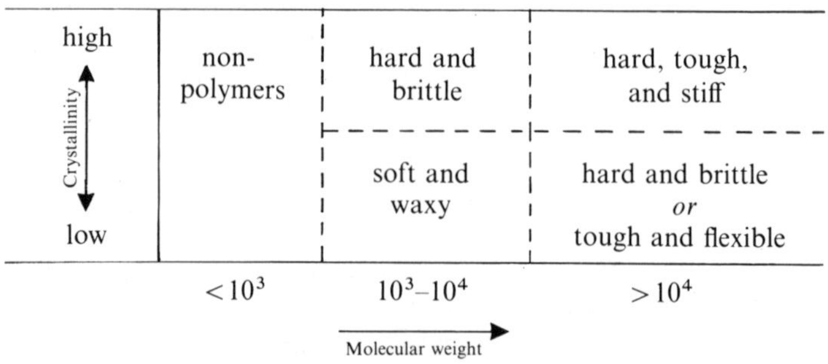

FIG. 22.1.   Nature of material as a function of crystallinity and molecular weight.

The influence of crystallinity and molecular weight on the mechanical properties which a polymer may show at any given temperature is roughly indicated in Fig. 22.1.

It should be remembered that crystals do not show any abrupt transition from ordered to disordered regions. On the contrary, marginal regions have degrees of semi-order. It should also be pointed out that the degree of crystallinity and the crystal size is often critically dependent on the processing history of the polymer, in particular on the cooling rate from the melt condition. In general, crystallinity will be high if the repeating units are short and highly symmetrical and when groups of great polar affinity are present. Under those circumstances the crystals which are about to be formed will find themselves in a favourable geometrical relationship. It is, of course, necessary that the time available for crystal formation and crystal growth be sufficient. Slow cooling—especially just below the crystalline melting point—is therefore a factor favouring the formation of large size crystals.

On the other hand, it is also possible to obtain normally crystalline polymers as metastable amorphous masses. This is achieved by rapid chilling through the region of maximum crystal growth just below the crystalline melting point. Crystallinity can be subsequently re-established by annealing just below T_m. Every crystalline polymer has a Gaussian distribution function which relates the rate of crystallisation to temperature. The temperature at which the last traces of crystallinity disappear represents the temperature at which the most perfectly formed crystals are in thermodynamic equilibrium with the amorphous phase. The temperature at which the rate of crystal formation is at a maximum is clearly one at which the mobility of the constituent units is high, but not so high that the frequency and amplitude of Brownian motion interferes with the regimentation of the units into ordered groups. Since the temperature difference between T_m and the temperature of maximum crystal formation is small, it is obvious that the crystals which form at the latter temperature will be small and frequently imperfect.

The disappearance of the crystalline phase causes distinct changes in properties. The material is more readily soluble in suitable solvents, it becomes a viscous liquid; density, refractive index and specific heat show abrupt changes when plotted vs. temperature. The birefringence due to the presence of two phases disappears. Any of these changes can be used for determining the crystalline melting point.

Apart from the inhomogeneities which arise from the coexistence of crystalline and amorphous phases, polymers almost invariably show some anisotropy which is rooted in the orientation of polymer chain elements within the amorphous regions and in the orientation of crystals. Orientation occurs along the principal axes and planes of strain and is always caused by stresses which act on the melt during processing or on the moulding or extrudate subsequently. Such orientation may or may not be desirable. In fibre and monofilament it is an essential part of the process which aims at achieving a high uniaxial tensile strength. It is also of advantage in the extrusion of profiles such as rod, tube and tape. When extruding film, biaxial orientation is usually aimed at. In mouldings, orientation is generally undesirable since it favours strength in one direction at the cost of weakness in another, but there are exceptions, such as the moulded polypropylene hinge.

Orientation increases both strength and stiffness (modulus) in the orientation direction, with corresponding decreases in the other directions. This is fully in accord with a fact already noted earlier, namely the considerable variations in viscosity in the liquid state which also depends upon, amongst other things, the direction of the plane of measurement relative to the direction of flow.

Orientation always exists when a moulding is kept at a suitably low temperature at which the free volume is small and both frequency and amplitude of the Brownian motion are negligible. Such orientation can, however, be annealed out at a temperature near the glass transition. If the material is transparent (e.g. polystyrene or polymethylmethacrylate) it is possible to demonstrate photoelastic strain fringes in polarised light.

If a partially crystalline material is oriented by cold-drawing, one may surmise that deformation affects only one of the two phases present, namely the one which represents the continuum. Crystallites are always the disperse phase, even when they contribute the major volume fraction. It is thus the amorphous phase which absorbs the entire deformational stress. At the same time, the crystallites will orient in the same direction as the amorphous chain segments. They may be viewed like raisins in a cake. No matter how many raisins are present, the mechanical properties of the cake will depend on the properties of the dough rather than those of the raisins, since the latter are merely fillers whose only contribution on the mechanical properties of the cake is the indirect one of determining the volume fraction of the dough. The reality of the raisin cake as a model for partially crystalline polymers which can be cold-drawn has been investigated by Lenk[1] who used various grades of low- and high-density polyethylenes and measured the instantaneous elastic recovery of extruded and subsequently cold-drawn polyethylene rod. The following assumptions were made:

(1)   Cold-drawing orients all polymer chains at least partially.
(2)   All fully extended chains after instantaneous retraction lie in a coplanar zigzag line with a tetrahedral bond angle ($109 \cdot 5°$).
(3)   At least *one* chain (more probably entangled chain elements belonging to different molecules) and, in any case, only a few chains

in a given drawn cross-section have had their bond angles distorted to 180°. (This is merely an imaginary model—see later.)

(4) The crystallites cannot suffer deformation in the presence of amorphous material.

Based on these assumptions it can be readily shown that the instantaneous retraction of a chain with a 109·5° zigzag shape after straightening (180° angles) should amount to 18·4%.

Crystallites account for a polyethylene volume fraction of between 50 and 90+ %. If they are not to suffer deformation, then the instantaneous retraction after draw must be a function of the volume fraction of amorphous material only, provided that retraction is measured *immediately* after drawing or in any case within a time interval which is less than that required for stress relaxation of the chains. This is easily done since the tensile viscosity of polyethylene is above 10^9 p and the retardation time is therefore of the order of days or hours, rather than minutes or seconds. It was found that the instantaneous recovery for low density polyethylene was 9%—about one-half the theoretical value. If the crystallinity of polyethylene is somewhat above 50 % (as is the case), then an instantaneous recovery of 9 % is precisely what would be expected.

The situation changes when the deformation is maintained over an extended period. In that case the stored elasticity of the distorted tetrahedral bond angles has enough time to dissipate energy by creeping. The mechanism for this is provided by slippage of entanglements at the entanglement points. The remaining elasticity can then only be due to retraction of residual distortions remaining in the bond angles. This makes it possible to calculate the angle of distortion α which was present before stress release by recording the magnitude of retraction, using the equation

$$\alpha = 2 \arcsin\left(\frac{100 - X}{100 - X - R}\sin\frac{109\cdot5}{2}\right) = 2\arcsin 0\cdot816\frac{100 - X}{100 - X - R}$$

where R = retraction (%) and X = crystallinity (volume–percent). The crystallinity must also be known. It also becomes possible to draw a curve of the bond angle as a function of time. That value is theoretically 180° at zero time and 109·5° at infinite time.

Setting $\alpha = 180°$ in the previous equation, then one may calculate the crystallinity X which a polyethylene should possess when an instantaneous recovery of R has been recorded:

$$X = 100 - 5\cdot44R$$

The maximum crystallinity at full extension thus follows a linear function of R. Experimental results on high-density polyethylenes confirm the arguments just presented and are shown in Table 22.1.

The arguments demanded that the C—C bond angles be distorted from $109.5°$ to $180°$. This, however, cannot be, since the energy required for such distortion would exceed the energy necessary to break the bond. If one thinks in terms of a single chain, then the model lacks all physical reality. Nevertheless, its quantitative application remains intact since it is quite easy

TABLE 22.1

Polyethylene (grade)	Stated crystallinity (%)	R (%)	Calculated crystallinity (%)
Vestolen A 15	—	2·9	84
Carlona 60-004 BB	90	1·7	89–92
Hostalen GF 5250	—	2·3	87

to modify the model in such a way as to render it physically acceptable: after all, it is not essential that one solitary chain should store the entire energy which is required for instantaneous retraction. It is perfectly acceptable that the distortion of bond angles and the corresponding stored energy be distributed over many—probably all—amorphous chain elements of the specimen. The chain which is originally expected to accept the entire strain can pass a part of it to the other chains and so distribute the stress in the form of a limited bond angle distortion all round. Provided that recovery is instantaneous, that is to say, provided it is made to take place before some of the stored energy has been dissipated by creeping slippage of entanglements, one would obtain a recovery value equal to that which should theoretically occur if a single chain had returned from a straight configuration ($180°$ bond angle) to the original tetrahedral bond angle zigzag of $109.5°$.

The original model is therefore applicable because it is energetically fully equivalent to the modified model. The distribution of angle distortions and the corresponding stored energy distribution may be compared to the statistical distribution of crystalline volume elements. We know that the crystallites are fairly evenly distributed over the total volume—and yet there would be no mathematical difference if the entire crystalline portion were concentrated into large local aggregates, perhaps even into one single hypothetical lump.

MODULUS AND CRYSTALLINITY

Krigbaum *et al.*[2] have developed a method for calculating the initial modulus of partially crystalline polymers on the assumption that the amorphous chains form an elastomeric network. The crystals can act as if they were cross-links, but the authors point out that the crystallites are bound to stress the amorphous chains surrounding them even in the absence of an externally applied stress. It can therefore be argued that the chains are randomly ordered so long as the crystallites are very small, but as they grow, so the stress on the amorphous phase will also grow until it becomes so large—and the corresponding entropy so low—that any further crystal formation ceases and the crystalline and amorphous phases are in equilibrium.

The theoretical treatment of this model resulted in an expression for the initial modulus G as a function of crystallinity X. The temperature dependence of G and X were also investigated and it was found that the initial modulus depends directly upon $1/(1 - X)$, provided that the temperature is well below T_m. This was confirmed experimentally. It should be pointed out that X is normalised between zero and unity. The relevant equations are as follows:

$$\frac{1}{1 - X} = \frac{2N_u \Delta H_u}{3R}\left(\frac{1}{T} - \frac{1}{T_m}\right)$$

where N_u is the number of repeating units in an amorphous chain segment between crystalline nuclei, ΔH_u is the heat of fusion per mol of the repeating unit, and R is the gas constant.

$$G_0 = \rho\frac{RT}{M_0 N}\frac{3}{5}\frac{1}{(1 - X)^3} + \frac{12}{5}\frac{1}{(1 - X)^2}$$

where M_0 is the molecular weight of a statistical segment, ρ is the density of the sample, and N is the number of statistical segments in an amorphous chain with its ends attached to the surface of growing crystallites. Combining these equations one obtains

$$G_0 = \frac{2\rho T \Delta H_u}{5M_0}\left(\frac{1}{T} - \frac{1}{T_m}\right)\left[4 + \left(\frac{2N_u \Delta H_u}{3R}\right)^{1/2}\left(\frac{1}{T} - \frac{1}{T_m}\right)^{1/2}\right]$$

It is noteworthy that in the last equation the parameter N has been eliminated.

INTER-CHAIN FORCES, CHAIN FLEXIBILITY AND T_m

Fusion involves separation of chains or chain segments which had been intimately associated in crystalline regions. The crystalline melting point T_m will therefore be high if the forces of inter-chain attraction are large. This is the case for polymers with secondary valence forces such as hydrogen bonds (polyvinyl alcohol, cellulose polymers, polyamides). Polar groups have a similar effect. If the number of polar groups is small, then T_m approaches the T_m for linear polyethylene, namely 135 °C. This is shown

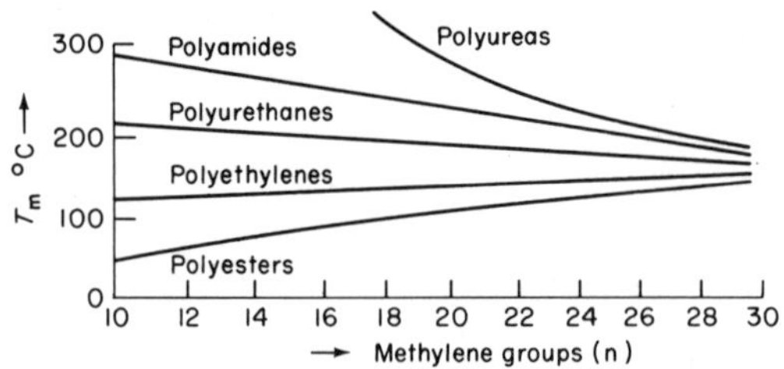

FIG. 22.2.   Melting point as a function of molecular weight for various aliphatic homologous series.

very clearly for homologous series of polyesters, polyurethanes, polyamides and polyureas (Fig. 22.2).

A comparison of T_m for linear polyethylene and PTFE gives values of 135 °C and 327 °C respectively, although the inter-chain forces are weak in both polymers. This is explained by the much greater stiffness of the PTFE chains. The slopes of the individual curves in Fig. 22.2 relative to the polyethylene curve are interesting. Polyureas have very high initial values because there are a large number of polar —NH— groups available for hydrogen bonding. In polyurethanes the effect is small despite a comparable number of polar groups, because the presence of oxygen atoms in the chains introduces an element of flexibility which is not present in homocarbon chains to anything like the same degree. Polyamides have, initially, high hydrogen bonding densities and tend to have high values of T_m. The aliphatic polyesters have low values of T_m in spite of the presence of polar ester groups because the presence of oxygen hetero-atoms in the chains produces flexibility which counteracts and dominates over the effect of the polar groups; moreover, the polarity of ester groups is far less than that of amide groups.

Aromatic ring systems have a pronounced chain stiffening effect. This is apparent when poly(ethyleneglycol adipate) is compared with poly(ethyleneglycol terephthalate):

$$\text{+O—CH}_2\text{—CH}_2\text{—O—CO—(CH}_2)_6\text{—CO+}_n \qquad T_m = 45\,°C$$

poly(ethyleneglycol adipate)

$$\left[\text{O—CH}_2\text{—CH}_2\text{—O—CO—}\!\!\bigcirc\!\!\text{—CO}\right]_n \qquad T_m = 265\,°C$$

poly(ethyleneglycol terephthalate)

Even greater differences are observed when the hexamethylenediamine polyamide of adipic acid (Nylon 66) is compared with the *p*-phenylene diamine polyamide of terephthalic acid. The former has a melting point of around 262 °C, the latter decomposes without melting at temperatures above 500 °C.

Flexibility (or stiffness) of chain molecules is governed by the degree of freedom of rotation round saturated bonds in the chain. The potential energy which opposes this rotation is between 1 and 5 kcal mol^{-1}. This is of the same order of magnitude as that of the molecular forces of cohesion. The replacement of six aliphatic methylene groups in polyethylene by a non-flexible group such as *p*-phenylene causes a substantial increase in T_m.[3]

$$\text{+CH}_2\text{—CH}_2\text{—CH}_2\text{—CH}_2\text{—CH}_2\text{—CH}_2\text{+}_n \qquad T_m = 135\,°C$$

linear polyethylene

$$\left[\bigcirc\!\!\text{—CH}_2\right]_n \qquad T_m = 380\,°C$$

poly(*p*-toluylene)

Suffice it to point out the importance of polar groups and chain stiffness and to emphasise that stiffness on a molecular scale must determine the mechanical properties of bulk polymers. Furthermore, the various factors which affect T_m also influence the *temperature* function of mechanical properties. For a treatment in depth of this subject the reader is referred to the standard works of polymer science[4] and papers by Hoffmann[5] and Natta.[6]

ENTANGLEMENT OF LINEAR CHAIN MOLECULES

Entanglement and disentanglement are reversible dynamic processes which take place above the glass transition temperature T_g. The process rate, naturally, is temperature-dependent. Any given temperature above T_g is characterised by an entanglement equilibrium which defines the most probable entanglement density per unit polymer volume.

Hoffmann[7] studied the relaxation phenomena of stretched entanglements and argued that the segments between entanglement knots relax at a faster rate at chain ends than they do in the centre of chains. This concept leads to a strain distribution along the chain after applying an external stress at zero time. The result is non-Newtonian behaviour both in the solid state below T_g and in the melt or solution, producing an abnormally fast rate of stress decay. This was supported by experimental evidence.

In another paper Hoffmann[8] derives a function of the time dependence of mechanical relaxation on the coupled diffusion of entangled segments between the entanglement knots. The relationship applies to the viscous flow rate which depends on the molecular weight of the chains and their free volume; it can be applied to solutions but also to relaxation generally. Deviation from Newtonian behaviour occurs when the molecular weight reaches a level at which entanglement becomes significant. Hoffmann also studied the effect of molecular weight distribution on the relaxation spectrum and the flow behaviour of polymers. He obtained good support for the molecular weight concept of entanglement and relaxation phenomena in strained polymers above T_g.

The concept opens up possibilities of molecular tailoring. In the absence of complicating factors such as crystallite formation this would enable the organic chemist to produce amorphous polymers with predetermined properties such as stiffness in the solid state above T_g, viscosity characteristics for melt processing (including deviation from Newtonian behaviour in terms of critical shear rate or critical shear stress), and viscosity in solution.

CRYSTALLINITY IN DRAWN POLYMERS

Drawing of a polymer orients the chains in the draw direction and should therefore lead to a configuration which is geometrically more favourable for crystallite formation. This presupposes that the repeating units of the polymer have sufficient symmetry to enable them to form crystalline aggregates. In a sense, this leads to strain-induced crystallisation which may be compared to shear-induced crystallisation in the melt, discussed earlier.

Strain-induced crystallisation in certain elastomers is indeed a firmly proven and distinctly observable process. Crystallisation of natural rubber is rapid at 400 to 800 % elongation and the T_m of the crystallites may be increased by almost 100 °C, because they have fewer imperfections than any crystallites which may be present in the undrawn condition. The melting of crystals is the result of an increasing tendency of the chains to react against the growing energy input, a reaction which involves rotation and entanglement. If the chains are under tensile constraint, then this molecular mechanism is rendered more difficult and the resulting stresses are transferred to the clamps which grip the sample. The initiation of crystallisation in a crystallisable elastomer above its normal T_m is marked by an increase in the slope of the stress–strain curve. This region of strain-hardening corresponds to the dilatant region of the generalised flow curve for the liquid state. Non-crystalline elastomers do not show strain-hardening and their tensile strength is low.

If a polymer which has been made crystalline by drawing is allowed to relax at a temperature slightly below its T_m, then the crystals which had been produced before 'melt' instantaneously. This is characteristic for a typical rubber at room temperature. On the other hand, if the polymer is at least 100 °C above its T_m, then the draw-induced crystals may be stable after relaxation and it is possible that the sample does not contract at all. This is the typical behaviour of crystalline fibre forming polymers if the fibre has been produced in the amorphous condition by rapid chilling from the melt. Subsequent drawing then produces crystallinity and the fibre will not shrink thereafter. This is the essence of the important cold-drawing operation. The polymer is now in a thermodynamically metastable condition and therefore not in true equilibrium, depending, as it does, on its thermal and mechanical history. A true equilibrium state can only be achieved by normalisation without any mechanical constraint whatever. However, this would then sacrifice the fibrous nature of the polymer.

It is a precondition for the cold-drawing of a fibre-forming polymer that a potentially crystalline monofilament should exist in its normalised amorphous elastic state just above its T_g. Whether cold-drawing is truly 'cold' is of no consequence, since 'hot' and 'cold' are relative terms. The lower temperature limit for cold-drawing is set by T_g. Thus 'Nylon 66' can be cold-drawn at room temperature. The cold-drawing temperature for 'Terylene' polyester is somewhat higher since its T_g is 67 °C.

The rate of crystallisation of stretched vulcanised natural rubber may be reduced by cooling. Thus, it is sufficiently slow at −20 °C to enable it to be readily measured. If crystallisation occurs at constant strain, then the initial stress decays gradually and the sample which was tautly stretched initially becomes limp and sags after the crystallites have formed. Thus, crystallisation *during* the draw process increases the stiffness; crystallisation *after* the drawing process reduces the stress because the cross-linkages, the amorphous chain segments and the crystallite bundles absorb the stress during the process of transformation.

VISCOSITY IN THE SOLID STATE—T_g

Clearly, the glass transition temperature T_g has an important influence on the mechanical properties of solid polymers. It is not intended to repeat earlier statements, but let it be said that the mobility of chain segments is greatly restricted below T_g whilst the free volume above T_g is such as to confer a substantially higher degree of kinetic freedom upon the chain segments. This brings about corresponding viscosity changes which are generally about three decades in magnitude.

We start with an Arrhenius type relationship

$$\ln \eta/\eta_0 = -E/RT$$

where η_0 is a constant viscosity at a reference temperature, and E is the activation energy for viscous flow. Williams, Landel and Ferry found that the following equation applied for high polymers in the temperature region from T_g to $(T_g + 100\,^\circ\text{C})$:

$$\log \frac{\eta}{\eta_{T_g}} = -\frac{17 \cdot 44(T - T_g)}{51 \cdot 6 + (T - T_g)}$$

and hence

$$-E = R\frac{\mathrm{d}\log\eta}{\mathrm{d}(1/T)} = \frac{4 \cdot 12 \times 1000\,T^2}{(51 \cdot 6 + T - T_g)^2}$$

These equations assume that the denominator of the exponential Arrhenius expression is $0 \cdot 025$ for all glassy polymers at T_g, and further that the coefficient of thermal expansion at T_g is $4 \cdot 8 \times 10^{-4}\,^\circ\text{C}^{-1}$. Experimental results confirm the correctness of these assumptions and they also received theoretical backing.

Values of T_g vary greatly among polymers. Thus we have values of $-123\,^\circ\text{C}$ for silicon polymers and $+208\,^\circ\text{C}$ for polyvinyl carbazole; the basic chemical structure is evidently decisive. This is adequately treated in most standard textbooks. For the present purpose it is sufficient to enumerate the factors which determine T_g:

—The mobility of the polymer chain
—Steric factors such as bulky side groups
—Molecular symmetry
—Polarity or cohesive energy
—Copolymerisation and plasticisation.

T_g is often found to relate empirically to T_m. Thus:

—In unsymmetrical polymers $T_g/T_m \cong 2/3$
—In symmetrical polymers $T_g/T_m \cong \frac{1}{2}$.

Other theoretical aspects involving free volume theories were briefly discussed in a previous treatise.[9]

So far we have excluded two aspects of viscoelasticity which are sources of fundamental difficulties for formulating a logical theory:

(1) the fact that a linear treatment can only be applied at small deformations;
(2) the necessity of relating the mechanical properties of materials under static conditions to those under dynamic conditions.

The first demands a precise quantification of 'small deformation'. What is the deformation at which a viscoelastic material deviates from linearity? This depends entirely upon the nature of the polymer in question. Thus deformations of up to 20% may be considered 'small' in rubbers, since linearity remains essentially intact up to about this limit. In the case of stiff thermoplastics the limit might be around 1% and in the case of thermosets it is substantially less than 1%. In order to take account of this, one would have to consider the molecular and morphological structure of each polymer on the basis of statistical mechanics. In any case, linear models of viscoelasticity are far from satisfactory. On the other hand, the problems involved in the formulation of a satisfactory and rigorous nonlinear viscoelastic theory are great and altogether outside the scope of this volume. However, it is worth noting that Edwards[10] believes that a solid which consists entirely of entangled long chain molecules possesses elastic properties which resemble those of a chemically cross-linked (vulcanised) material, provided that the elastic moduli are studied over a time interval which is short compared to the times necessary for creep to occur. This is an acceptable proposition which has been exploited by Lenk in his work on instantaneous elastic recovery of polyethylene rod after drawing (see earlier). Edwards derived a formula for the free energy of a model at low stress in which every chain is locally impeded by neighbouring chains. It is significant that Edwards also developed an expression for strains of *any* magnitude, that is to say even for strains which are outside the limits of proportionality.

As regards the correlation of static and dynamic parameters, a great deal of work has been done. Techniques of considerable mathematical complexity were employed which are altogether outside the scope of this book. The reader is referred to the detailed treatment of Ferry[11] and refs. 12–23.

Finally, attention is drawn to a statistical theory of the glass transition of polymers which has been worked out by Brunt.[24] This theory is quite rigorous and yet comparatively simple. It successfully avoids the limitation of 'small' deformations. Brunt defines the T_g ('freeze-point') as the total complex of changes in the mechanical properties of a polymer in the temperature region where they occur. The molecular interpretation of retarded rubber elasticity is considered first and this is followed by the temperature-dependence of relaxation phenomena. Brunt avoids rebound and damping and their inclusion in phenomenological or other theories, considering transition phenomena in the first place as special aspects of Brownian motion with a mode and amplitude of a Gaussian distribution.

An imposed deformation causes a change in the distribution of molecular motions. This increases the probability of certain molecular states at the expense of others. Relaxation is then considered as a gradual restoration of the Gaussian distribution. The mechanism of this process depends on the following assumptions:

(i)   The probability of a given molecular state is a function of the time during which it is found in that state.

(ii) The properties depend on free volume and average entanglement length. The latter may be strongly temperature-dependent if inter-chain attractions are so weak as to be liable to be effectively opposed by the Brownian motion.

Particularly interesting is Brunt's characterisation of the glassy and rubbery states.

It is claimed that the cohesive forces cannot be overcome below T_g. Molecular-sized motions are mostly vibrational. The spatial order of molecules is fixed and cannot therefore contribute to the entropy of the system. However, a few molecular clusters may oscillate between two energy levels. If the frequency of an externally imposed sinusoidal deformation comes into resonance with the eigenfrequency of these vibrations, then the relative distribution between the two energy levels is altered. Absorption maxima are created which show themselves when the dynamic–mechanical properties in the glassy state are investigated.

James and Guth[25] considered the rubbery state as a fluid mass containing a network of chains which opposes changes in shape. The small compressibility of rubbers is considered proof that the matrix of the network may be considered a liquid. The external shape of the rubber is fixed by the forces which act upon it and which operate through the agency of the entropy springs of the network, whilst in an ordinary liquid the shape is determined by the container and by surface tension. The 'liquid matrix' has an inner kinetic pressure which is in equilibrium with the secondary cohesive forces, and this is also true for conventional liquids. The matrix has its own viscosity and compressibility which is comparable to that of a low molecular weight liquid. There are discontinuities at T_g which are similar to those which occur at the freezing point of ordinary liquids. Furthermore, the units of the 'liquid matrix' can oscillate (or flow) within the limitations of the network because the free volume is greater than in the glassy state and because the secondary cohesive forces are reduced relative to the opposing forces generated by the thermal energy.

Brunt then considers molecular motions and analyses relaxation statistically. The detailed mathematical treatment leads to the following conclusions:

(1)   Elasticity is a consequence of entropy changes only. This is in accord with the theory of rubber elasticity.

(2)   The theory is not confined to 'small' deformations. The problem of non-linearity appears to have lost some of its serious difficulties.

(3)   Although a distinction is made between small and large deformations, this is independent of the linearity problem. The same deformation may be considered large at low temperatures (cold-draw) and small at high temperatures (creep deformation).

(4)   The glassy region is characterised not only by low temperatures or short experimental time scales, but also by the degree of deformation. At low temperatures the rate of change of successive configurations as well as the oscillatory amplitude is much reduced. The time/temperature superposition is therefore confined to the region for which the Williams–Landel–Ferry (WLF) equation is valid.

(5)   The position and shape of the creep curve along the logarithmic–time axis depends upon the magnitude of the flow

units and their homogeneity. If the material contains only *one* type of flow unit, then the creep curve covers about two decades. If the magnitude of the flow units is more widely distributed, then a proportionally wider log–time region is covered at any given temperature.

(6)   Creep and relaxation experiments may be represented by one and the same model.

(7)   The theory is claimed to be generally valid without being confined to any particular molecular structure.

Brunt's statistical theory has the following advantages:

(1)   It rests on a concept of continuous Brownian motion, so that frictional damping need not be considered and the theory is freed from the unreality of springs and dashpots. The restriction to small deformations (i.e. small with respect to the vibrational amplitudes) which confines the spectra to creep only and which prevents an analysis of elongational deformation, is removed.

(2)   All free volume theories remain fully valid.

(3)   Theoretically based on Brownian motion it removes the concept of a potential barrier between two 'holes'. This represents a fundamental difference from other theories which are based on activation energy, since there cannot be any exponentially time-dependent Arrhenius type relationships without an activation energy.

(4)   A distinction is made between reversible and irreversible changes in free energy, based upon the separation of 'true' and 'most probable' positions.

(5)   The probability distribution does not rest upon an equation of continuity which is—according to Brunt—of doubtful value.

Brunt's theory is original and should certainly be subjected to rigorous test. Its extensive qualitative presentation in concluding this chapter is justified by the stimulus which it provides for a quantitative study of the processes involved in structural rheology. This is bound to lead to a better understanding of the general properties of polymer materials.

REFERENCES

1. R. S. LENK, unpublished work.
2. W. R. KRIGBAUM, R. J. ROE and K. J. SMITH, JR, *Polymer*, **5**(10), 533 (Oct. 1964).

3. O. B. EDGAR and R. HILL, *J. Poly. Sci.*, **8**(1) (1952).
4. A. V. TOBOLSKY, *Properties and Structure of Polymers*, John Wiley & Sons (1960).
5. J. D. HOFFMANN, *IRE Trans.* (component parts), **4**(2) (June 1957).
6. G. NATTA, *SPE Trans.*, **3**(2), 99 (April 1963).
7. M. HOFFMANN, *Rheol. Acta*, **6**(1), 92 (1967).
8. M. HOFFMANN, *Rheol. Acta*, **6**(4), 377 (1967).
9. R. S. LENK, *Rheologie der Kunststoffe*, p. 220, Carl Hanser, Munich (1971).
10. S. I. EDWARDS, *Proc. Phys. Soc.*, **92**, 9 (1967).
11. J. D. FERRY, *Viscoelastic Properties of Polymers*, John Wiley & Sons, Inc., New York (1961).
12. R. D. ANDREWS, *Ind. Eng. Chem.*, **44**(4), 707 (April 1952).
13. J. D. FERRY, E. R. FITZGERALD *et al.*, *Ind. Eng. Chem.*, 703 (April 1952).
14. F. SCHWARZL and A. J. STAVERMAN, *Physica*, **18**(10), 791 (Oct. 1952).
15. F. SCHWARZL and A. STAVERMAN, *Appl. Sci. Res.*, **A4,** 127 (1953).
16. F. C. ROESLER and W. A. TWYMAN, *Proc. Phys. Soc.*, **68,** 2B 97 (1954).
17. H. LEADERMAN, *Trans. Soc. Rheol.*, **1,** 213 (1957).
18. H. FUJITA, *J. Appl. Phys.*, **29**(6) 943 (June 1958).
19. I. L. HOPKINS, *J. Appl. Poly. Sci.*, **7,** 971 (1963).
20. B. HLAVACEK and V. KOTRBA, *Rheol. Acta*, **6**(3), 288 (1967).
21. V. SINEVIC and B. HLAVACEK, paper presented at the annual meeting of German rheologists, Berlin-Dahlem, 1968.
22. G. A. AJROLDI, C. GARBUGLIO and G. PEZZIN, *Polymer Letters*, **6,** 119 (1968).
23. R. N. HAWARD and G. THACKRAY, *Proc. Roy. Soc.*, A 302 (1968).
24. N. A. BRUNT, *Kolloid-Z.*, **209**(1), 5 (1966).
25. H. M. JAMES and E. GUTH, *J. Chem. Phys.*, **11,** 455 (1943); **15,** 651 (1947); **15,** 669 (1947).

Index